BIBLIOGRAPHIC RECORD TARGET

Graduate Library
University of Michigan

Preservation Office

Storage Number: ______________

ACM0041
UL FMT B RT a BL m T/C DT 09/12/88 R/DT 09/12/88 CC STAT mm E/L 1
035/1: : |a (RLIN)MIUG86-B24227
035/2: : |a (CaOTULAS)160429619
040: : |a MiU |c MiU
100:1 : |a Byerly, William Elwood, |d b. 1849.
245:00: |a Elements of the integral calculus, |b with a key to the solution of differential equations, and A short table of integrals. |c By William Elwood Byerly.
250: : |a 2d ed., rev. and enl.
260: : |a Boston, |b Ginn and company, |c 1892.
300/1: : |a xvi, 339 p., 1 L., 65-76, 32 p. |b tab. diagrs. |c 22 cm.
500/1: : |a A sequel to the author's work on differential calculus, chapter v of which (Integration) is here reprinted as an appendix. cf. Pref.
500/2: : |a "A short table of integrals," compiled by B.O. Peirce, p. [1]-28 at end has separate t.-p.
650/1: 0: |a Calculus, Integral
650/2: 0: |a Functions
650/3: 0: |a Differential equations
700/1:1 : |a Peirce, Benjamin Osgood, |d 1854-1914.
998: : |c RSH |s 9124

Date work Began: ______________
Camera Operator: ______________

ELEMENTS

OF THE

INTEGRAL CALCULUS,

WITH A

KEY TO THE SOLUTION OF DIFFERENTIAL EQUATIONS, AND A SHORT TABLE OF INTEGRALS.

BY

WILLIAM ELWOOD BYERLY, Ph.D.,

PROFESSOR OF MATHEMATICS IN HARVARD UNIVERSITY.

SECOND EDITION, REVISED AND ENLARGED.

BOSTON, U.S.A.:

PUBLISHED BY GINN AND COMPANY.

1892.

Typography by J. S. Cushing & Co., Boston, U.S.A.

Presswork by Ginn & Co., Boston, U.S.A.

PREFACE.

THE following volume is a sequel to my treatise on the Differential Calculus, and, like that, is written as a text-book. The last chapter, however, a Key to the Solution of Differential Equations, may prove of service to working mathematicians.

I have used freely the works of Bertrand, Benjamin Peirce, Todhunter, and Boole; and I am much indebted to Professor J. M. Peirce for criticisms and suggestions.

I refer constantly to my work on the Differential Calculus as Volume I.; and for the sake of convenience I have added Chapter V. of that book, which treats of Integration, as an appendix to the present volume.

W. E. BYERLY.

CAMBRIDGE, 1881.

PREFACE TO SECOND EDITION.

In enlarging my Integral Calculus I have used freely Schlömilch's "Compendium der Höheren Analysis," Cayley's "Elliptic Functions," Meyer's "Bestimmte Integrale," Forsyth's "Differential Equations," and Williamson's "Integral Calculus."

The chapter on Theory of Functions was sketched out and in part written by Professor B. O. Peirce, to whom I am greatly indebted for numerous valuable suggestions touching other portions of the book, and who has kindly allowed me to have his Short Table of Integrals bound in with this volume.

W. E. BYERLY.

Cambridge, 1888.

ANALYTICAL TABLE OF CONTENTS.

CHAPTER I.

SYMBOLS OF OPERATION.

CHAPTER II.

IMAGINARIES.

CHAPTER III.

GENERAL METHODS OF INTEGRATING.

CHAPTER IV.

RATIONAL FRACTIONS.

CHAPTER V.

REDUCTION FORMULAS.

CHAPTER VI.

IRRATIONAL FORMS.

CHAPTER VII.

TRANSCENDENTAL FUNCTIONS.

CHAPTER VIII.

DEFINITE INTEGRALS.

CHAPTER IX.

LENGTHS OF CURVES.

CHAPTER X.

AREAS.

CHAPTER XI.

AREAS OF SURFACES.

CHAPTER XII.

VOLUMES.

CHAPTER XIII.

CENTRES OF GRAVITY.

CHAPTER XIV.

LINE, SURFACE, AND SPACE INTEGRALS.

CHAPTER XV.

MEAN VALUE AND PROBABILITY.

CHAPTER XVI.

ELLIPTIC INTEGRALS.

CHAPTER XVII.

INTRODUCTION TO THE THEORY OF FUNCTIONS.

CHAPTER XVIII.

KEY TO THE SOLUTION OF DIFFERENTIAL EQUATIONS.

INTEGRAL CALCULUS.

CHAPTER I.

SYMBOLS OF OPERATION.

1. It is often convenient to regard a functional symbol as indicating *an operation to be performed upon the expression which is written after the symbol.* From this point of view the symbol is called a *symbol of operation*, and the expression written after the symbol is called the *subject* of the operation.

Thus the symbol D_x in $D_x(x^2y)$ indicates that the *operation* of differentiating with respect to x is to be performed upon the *subject* (x^2y).

2. If the *result* of one operation is taken as the *subject* of a second, there is formed what is called a *compound function.*

Thus $\log\sin x$ is a *compound function*, and we may speak of the taking of the log sin as a *compound operation.*

3. When two operations are so related that the compound operation, in which the result of performing the first on any subject is taken as the subject of the second, leads to the same result as the compound operation, in which the result of performing the second on the same subject is taken as the subject of the first, the two operations are *commutative* or *relatively free.*

Or to formulate; if

$$fFu = Ffu,$$

the operations indicated by f and F are *commutative.*

For example; the operations of partial differentiation with respect to two independent variables x and y are commutative, for we know that

$$D_x D_y u = D_y D_x u. \qquad \text{(I. Art. 197).}$$

The operations of taking the sine and of taking the logarithm are not commutative, for $\log \sin u$ is not equal to $\sin \log u$.

4. If
$$f(u \pm v) = fu \pm fv$$
where u and v are any subjects, the operation f is *distributive* or *linear*.

The operation indicated by d and the operation indicated by D_x are distributive, for we know that

$$d(u \pm v) = du \pm dv,$$

and that
$$D_x(u \pm v) = D_x u \pm D_x v.$$

The operation sin is not distributive, for $\sin(u + v)$ is not equal to $\sin u + \sin v$.

5. *The compounds of distributive operations are distributive.*

Let f and F indicate distributive operations, then fF will be distributive; for

$$F(u \pm v) = Fu \pm Fv,$$

therefore
$$fF(u \pm v) = f(Fu \pm Fv) = fFu \pm fFv.$$

6. The *repetition* of any operation is indicated by writing an *exponent, equal to the number of times the operation is performed*, after the symbol of the operation.

Thus $\log^3 x$ means $\log \log \log x$; $d^3 u$ means $dddu$.

In the single case of the trigonometric functions a different use of the exponent is sanctioned by custom, and $\sin^2 u$ means $(\sin u)^2$ and not $\sin \sin u$.

7. If m and n are *whole numbers* it is easily proved that

$$f^m f^n u = f^{m+n} u.$$

This formula is assumed for all values of m *and* n, *and negative and fractional exponents are interpreted by its aid.* It is called the *law of indices.*

8. To find what interpretation must be given to a *zero exponent*, let

$$m = 0 \qquad \text{in the formula of Art. 7.}$$

$$f^0 f^n u = f^{0+n} u = f^n u,$$

or, denoting $f^n u$ by v, $\qquad f^0 v = v.$

That is; *a symbol of operation with the exponent zero has no effect on the subject*, and may be regarded as multiplying it by unity.

9. To interpret a *negative exponent*, let

$$m = -n \qquad \text{in the formula of Art. 7.}$$

$$f^{-n} f^n u = f^{-n+n} u = f^0 u = u.$$

If we call $\qquad f^n u = v$, then $f^{-n} v = u.$

If $\qquad n = 1$

we get $\qquad f^{-1} fu = u,$

and the exponent -1 indicates what we have called the anti-function of fu. (I. Art. 72.)

The exponent -1 is used in this sense even with trigonometric functions.

10. When two operations are *commutative* and *distributive*, the symbols which represent them may be combined precisely as if they were algebraic quantities.

For they obey the laws,

$$a(m + n) = am + an,$$

$$am = ma,$$

on which all the operations of arithmetic and algebra are founded.

For example; if the operation $(D_x + D_y)$ is to be performed n times in succession on a subject u, we can expand $(D_x + D_y)^n$ precisely as if it were a binominal, and then perform on u the operations indicated by the expanded expression.

$$\begin{aligned}(D_x + D_y)^3 u &= (D_x^3 + 3D_x^2 D_y + 3D_x D_y^2 + D_y^3)u \\ &= D_x^3 u + 3D_x^2 D_y u + 3D_x D_y^2 u + D_y^3 u.\end{aligned}$$

CHAPTER II.

IMAGINARIES.

11. An *imaginary* is usually defined in algebra as *the indicated even root of a negative quantity*, and although it is clear that there can be no *quantity* that raised to an even power will be negative, the assumption is made that an imaginary can be treated like any algebraic quantity.

Imaginaries are first forced upon our notice in connection with the subject of quadratic equations. Considering the typical quadratic

$$x^2 + ax + b = 0,$$

we find that it has two roots, and that these roots possess certain important properties. For example; their sum is $-a$ and their product is b. We are led to the conclusion that every quadratic has two roots whose sum and whose product are simply related to the coefficients of the equation.

On trial, however, we find that there are quadratics having but one root, and quadratics having no root.

For example; if we solve the equation

$$x^2 - 2x + 1 = 0,$$

we find that the only value of x which will satisfy it is *unity;* and if we attempt to solve

$$x^2 - 2x + 2 = 0,$$

we find that there is no value of x which will satisfy the equation.

As these results are apparently inconsistent with the conclusion to which we were led on solving the general equation, we naturally endeavor to reconcile them with it.

The difficulty in the case of the equation which has but one root is easily overcome by regarding it as having two equal roots. Thus we can say that each of the two roots of the equation

$$x^2 - 2x + 1 = 0$$

is equal to 1; and there is a decided advantage in looking at the question from this point of view, for the roots of this equation will possess the same properties as those of a quadratic having unequal roots. The sum of the roots 1 and 1 is minus the coefficient of x in the equation, and their product is the constant term.

To overcome the difficulty presented by the equation which has no root we are driven to the conception of *imaginaries.*

12. *An imaginary is not a quantity, and the treatment of imaginaries is purely arbitrary and conventional.* We begin by laying down a few arbitrary rules for our imaginary expressions to obey, which must not involve any contradiction; and we must perform all our operations upon imaginaries, and must interpret all our results by the aid of these rules.

Since imaginaries occur as roots of equations, they bear a close analogy with ordinary algebraic quantities, and they have to be subjected to the same operations as ordinary quantities; therefore our rules ought to be so chosen that the results may be comparable with the results obtained when we are dealing with real quantities.

13. By adopting the convention that

$$\sqrt{-a^2} = a\sqrt{-1},$$

where a is supposed to be *real*, we can reduce all our imaginary algebraic expressions to forms where $\sqrt{-1}$ is the only peculiar symbol. This symbol $\sqrt{-1}$ we shall define and use as *the symbol of some operation, at present unknown, the repetition of which has the effect of changing the sign of the subject of the operation.* Thus in $a\sqrt{-1}$ the symbol $\sqrt{-1}$ indicates that an operation is performed upon a which, if repeated, will change the sign of a. That is,

$$a(\sqrt{-1})^2 = -a.$$

From this point of view it would be more natural to write the symbol before instead of after the subject on which it operates, $(\sqrt{-1})a$ instead of $a\sqrt{-1}$, and this is sometimes done; but as the usage of mathematicians is overwhelmingly in favor of the second form, we shall employ it, merely as a matter of convenience, and remembering that a is the *subject* and the $\sqrt{-1}$ the *symbol of operation*.

14. The rules in accordance with which we shall use our new symbol are, first,

$$a\sqrt{-1}+b\sqrt{-1}=(a+b)\sqrt{-1}. \qquad [1]$$

In other words, the operation indicated by $\sqrt{-1}$ is to be *distributive* (Art. 4); and second,

$$a\sqrt{-1}=(\sqrt{-1})a, \qquad [2]$$

or our symbol is to be *commutative* with the symbols of quantity (Art. 3).

These two conventions will enable us to use our symbol in algebraic operations precisely as if it were a quantity (Art. 10).

When no coefficient is written before $\sqrt{-1}$ the coefficient 1 will be understood, or unity will be regarded as the subject of the operation.

15. Let us see what interpretation we can get for powers of $\sqrt{-1}$; that is, for repetitions of the operation indicated by the symbol.

$$(\sqrt{-1})^0=1 \qquad \text{(Art. 8)},$$
$$(\sqrt{-1})^1=\sqrt{-1},$$
$$(\sqrt{-1})^2=-1, \qquad \text{by definition (Art. 13)},$$
$$(\sqrt{-1})^3=(\sqrt{-1})^2\sqrt{-1}=-\sqrt{-1}, \quad \text{by definition},$$
$$(\sqrt{-1})^4=-(\sqrt{-1})^2 \quad =1,$$
$$(\sqrt{-1})^5=1\sqrt{-1} \quad =\sqrt{-1},$$
$$(\sqrt{-1})^6=(\sqrt{-1})^2 \quad =-1,$$

and so on, the values $\sqrt{-1}$, -1, $-\sqrt{-1}$, 1, occurring in cycles of four.

16. The definition we have given for the square root of a negative quantity, and the rules we have adopted concerning its use, enable us to remove entirely the difficulty felt in dealing with a quadratic which does not have real roots. Take the equation

$$x^2 - 2x + 5 = 0. \qquad (1)$$

Solving by the usual method, we get

$$x = 1 \pm \sqrt{-4};$$

$$\sqrt{-4} = 2\sqrt{-1}, \text{ by Art. 13 [1]};$$

hence $$x = 1 + 2\sqrt{-1} \text{ or } 1 - 2\sqrt{-1}.$$

On substituting these results in turn in the equation (1), performing the operations by the aid of our conventions (Art. 14 [1] and [2]), and interpreting $(\sqrt{-1})^2$ by Art. 15, we find that they both satisfy the equation, and that they can therefore be regarded as entirely analogous to real roots. We find, too, that their sum is 2 and that their product is 5, and consequently that they bear the same relations to the coefficients of the equation as real roots.

17. An imaginary root of a quadratic can always be reduced to the form $a + b\sqrt{-1}$ where a and b are real, and this is taken as the general type of an imaginary; and part of our work will be to show that when we subject imaginaries to the ordinary functional operations, all our results are reducible to this typical form.

If two imaginaries $a + b\sqrt{-1}$ and $c + d\sqrt{-1}$ are equal, a must be equal to c, and b must be equal to d.

For we have $$a + b\sqrt{-1} = c + d\sqrt{-1}.$$

Therefore $$a - c = (d - b)\sqrt{-1},$$

or a real is equal to an imaginary, unless $a - c = 0 = d - b$.

Since obviously a real and an imaginary cannot be equal, it follows that $a = c$ and $b = d$.

18. We have defined $\sqrt{-1}$ as the symbol of an operation whose repetition changes the sign of the subject.

Several different interpretations of this operation have been suggested, and the following one, in which every imaginary is graphically represented by the position of a point in a plane, is commonly adopted, and is found exceedingly useful in suggesting and interpreting relations between different imaginaries and between imaginaries and reals.

In the *Calculus of Imaginaries*, $a+b\sqrt{-1}$ is taken as the general symbol of quantity. If b is equal to zero, $a+b\sqrt{-1}$ reduces to a, and is *real;* if a is equal to zero, $a+b\sqrt{-1}$ reduces to $b\sqrt{-1}$, and is called a *pure imaginary*.

$a+b\sqrt{-1}$ is represented by the position of a point referred to a pair of rectangular axes, as in analytic geometry, a being taken as the abscissa of the point and b as its ordinate. Thus in the figure the position of the point P represents the imaginary $a+b\sqrt{-1}$.

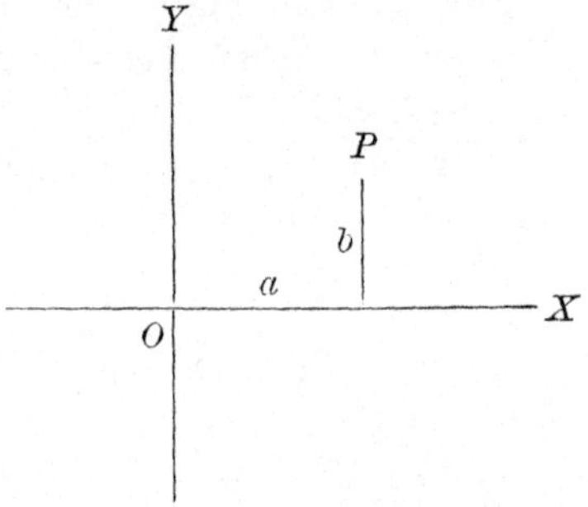

If $b=0$, and our quantity is real, P will lie on the axis of X, which on that account is called the *axis of reals;* if $a=0$, and we have a *pure imaginary*, P will lie on the axis of Y, which is called the *axis of pure imaginaries*.

It follows from Art. 17 that if two imaginaries are equal, the points representing them will coincide.

Since a and $a\sqrt{-1}$ are represented by points equally distant from the origin, and lying on the *axis of reals* and the *axis of pure imaginaries* respectively, we may regard the operation indicated by $\sqrt{-1}$ as causing the point representing the subject of the operation to rotate about the origin through an angle of 90°. A repetition of the operation ought to cause the point to rotate 90° further, and it does; for

$$a(\sqrt{-1})^2=-a,$$

and is represented by a point at the same distance from the

origin as a, and lying on the opposite side of the origin; again repeat the operation,

$$a(\sqrt{-1})^3 = -a\sqrt{-1},$$

and the point has rotated 90° further; repeat again,

$$a(\sqrt{-1})^4 = a,$$

and the point has rotated through 360°. We see, then, that if the subject is a *real* or a *pure imaginary* the effect of performing on it the operation indicated by $\sqrt{-1}$ is to rotate it about the origin through the angle 90°. We shall see later that even when the subject is neither a real nor a pure imaginary, the effect of operating on it with $\sqrt{-1}$ is still to produce the rotation just described.

19. The *sum*, the *product*, and the *quotient* of any two imaginaries, $a+b\sqrt{-1}$ and $c+d\sqrt{-1}$, are imaginaries of the typical form.

$$a+b\sqrt{-1}+c+d\sqrt{-1} \quad = a+c+(b+d)\sqrt{-1}. \qquad [1]$$

$$(a+b\sqrt{-1})\,(c+d\sqrt{-1}) = ac-bd+(bc+ad)\sqrt{-1}. \qquad [2]$$

$$\frac{a+b\sqrt{-1}}{c+d\sqrt{-1}} = \frac{(a+b\sqrt{-1})\,(c-d\sqrt{-1})}{(c+d\sqrt{-1})\,(c-d\sqrt{-1})} = \frac{ac+bd+(bc-ad)\sqrt{-1}}{c^2+d^2}$$

$$= \frac{ac+bd}{c^2+d^2} + \frac{bc-ad}{c^2+d^2}\sqrt{-1}. \qquad [3]$$

All these results are of the form $A+B\sqrt{-1}$.

20. The graphical representation we have suggested for imaginaries suggests a second typical form for an imaginary. Given the imaginary $x+y\sqrt{-1}$, let the *polar coördinates* of the point P which represents $x+y\sqrt{-1}$ be r and ϕ.

r is called the *modulus* and ϕ the *argument* of the imaginary.

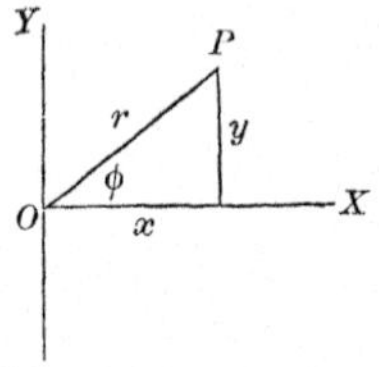

The figure enables us to establish very simple relations between x, y, r, and ϕ.

$$\left.\begin{aligned} x &= r\cos\phi, \\ y &= r\sin\phi; \end{aligned}\right\} \qquad [1]$$

$$\left.\begin{aligned} r &= \sqrt{x^2+y^2}, \\ \phi &= \tan^{-1}\frac{y}{x}. \end{aligned}\right\} \qquad [2]$$

$$x + y\sqrt{-1} = r\cos\phi + (\sqrt{-1})r\sin\phi$$
$$= r(\cos\phi + \sqrt{-1}.\sin\phi), \qquad [3]$$

where the imaginary is expressed in terms of its modulus and argument.

The value of r given by our formulas [2] is ambiguous in sign; and ϕ may have any one of an infinite number of values differing by multiples of π. In practice we always take the positive value of r, and a value of ϕ which will bring the point in question into the right quadrant. In the case of any given imaginary then, r can have but one value, while ϕ may have any one of an infinite number of values differing by multiples of 2π.

The modulus r is sometimes called the *absolute value* of the imaginary.

EXAMPLES.

(1) Find the modulus and argument of 1; of $\sqrt{-1}$; of -4; of $-2\sqrt{-1}$; of $3+3\sqrt{-1}$; of $2+4\sqrt{-1}$; and express each of these quantities in the form $r(\cos\phi + \sqrt{-1}.\sin\phi)$.

(2) Show that every positive real has the argument zero; every negative real the argument π; every positive pure imaginary the argument $\frac{\pi}{2}$; and every negative pure imaginary the argument $\frac{3\pi}{2}$.

21. If we add two imaginaries, the *modulus of the sum* is never greater than the *sum of the moduli* of the given imaginaries.

The sum of $a+b\sqrt{-1}$ and $c+d\sqrt{-1}$ is $a+c+(b+d)\sqrt{-1}$. The modulus of this sum is $\sqrt{(a+c)^2+(b+d)^2}$; the sum of the moduli of $a+b\sqrt{-1}$ and $c+d\sqrt{-1}$ is $\sqrt{a^2+b^2}+\sqrt{c^2+d^2}$. We wish to show that

$$\sqrt{(a+c)^2+(b+d)^2} \mathrel{-\!\!<} \sqrt{a^2+b^2}+\sqrt{c^2+d^2};$$

the sign $-\!\!<$ meaning "*equal to or less than.*"

Now $\sqrt{(a+c)^2+(b+d)^2} \mathrel{-\!\!<} \sqrt{a^2+b^2}+\sqrt{c^2+d^2}$,

if $(a+c)^2+(b+d)^2 \mathrel{-\!\!<} a^2+b^2+2\sqrt{(a^2+b^2)(c^2+d^2)}+c^2+d^2$,

that is, if $ac+bd \mathrel{-\!\!<} \sqrt{a^2c^2+a^2d^2+b^2c^2+b^2d^2}$;

or, squaring, if

$$a^2c^2+2abcd+b^2d^2 \mathrel{-\!\!<} a^2c^2+a^2d^2+b^2c^2+b^2d^2;$$

or, if $0 \mathrel{-\!\!<} (ad-bc)^2$.

This last result is necessarily true, as no real can have a square less than zero; hence our proposition is established.

22. *The modulus of the product of two imaginaries is the product of the moduli of the given imaginaries, and the argument of the product is the sum of the arguments of the imaginaries.*

Let us multiply

$$r_1(\cos\phi_1+\sqrt{-1}.\sin\phi_1) \quad \text{by} \quad r_2(\cos\phi_2+\sqrt{-1}.\sin\phi_2);$$

we get

$$r_1r_2[\cos\phi_1\cos\phi_2-\sin\phi_1\sin\phi_2+\sqrt{-1}(\sin\phi_1\cos\phi_2+\cos\phi_1\sin\phi_2)],$$

$$\cos\phi_1\cos\phi_2-\sin\phi_1\sin\phi_2=\cos(\phi_1+\phi_2),$$

$$\sin\phi_1\cos\phi_2+\cos\phi_1\sin\phi_2=\sin(\phi_1+\phi_2)$$

by Trigonometry; hence

$$r_1(\cos\phi_1+\sqrt{-1}.\sin\phi_1)\,r_2(\cos\phi_1+\sqrt{-1}.\sin\phi_2)$$
$$=r_1r_2[\cos(\phi_1+\phi_2)+\sqrt{-1}.\sin(\phi_1+\phi_2)],$$

and our result is in the typical form, $r_1 r_2$ being the modulus and $\phi_1 + \phi_2$ the argument of the product.

If each factor has the modulus unity, this theorem enables us to construct very easily the product of the imaginaries; it also enables us to show that the interpretation of the operation $\sqrt{-1}$, suggested in Art. 18, is perfectly general.

Let us operate on any imaginary subject,

$$r(\cos\phi + \sqrt{-1}.\sin\phi), \qquad \text{with } \sqrt{-1},$$

that is, with

$$1\left(\cos\frac{\pi}{2} + \sqrt{-1}.\sin\frac{\pi}{2}\right).$$

The modulus r will be unchanged, the argument ϕ will be increased by $\frac{\pi}{2}$, and the effect will be to cause the point representing the given imaginary to rotate about the origin through an angle of 90°.

23. Since division is the inverse of multiplication,

$$r_1(\cos\phi_1 + \sqrt{-1}.\sin\phi_1) \div r_2(\cos\phi_2 + \sqrt{-1}.\sin\phi_2)$$

will be equal to

$$\frac{r_1}{r_2}[\cos(\phi_1 - \phi_2) + \sqrt{-1}.\sin(\phi_1 - \phi_2)],$$

since if we multiply this by $r_2(\cos\phi_2 + \sqrt{-1}.\sin\phi_2)$, according to the method established in Art. 22, we must get

$$r_1(\cos\phi_1 + \sqrt{-1}.\sin\phi_1).$$

To divide one imaginary by another, we have then to take the quotient obtained by dividing the modulus of the first by the modulus of the second as our required modulus, and the argument of the first minus the argument of the second as our new argument.

24. If we are dealing with the product of n equal factors, or, in other words, if we are raising $r(\cos\phi + \sqrt{-1}.\sin\phi)$ to the

nth power, n being a positive whole number, we shall have, by Art. 22,

$$[r(\cos\phi + \sqrt{-1}.\sin\phi)]^n = r^n(\cos n\phi + \sqrt{-1}.\sin n\phi). \quad [1]$$

If r is unity, we have merely to multiply the argument by n, without changing the modulus; so that in this case increasing the exponent by unity amounts to rotating the point representing the imaginary through an angle equal to ϕ without changing its distance from the origin.

25. Since extracting a root is the inverse of raising to a power,

$$\sqrt[n]{[r(\cos\phi + \sqrt{-1}.\sin\phi)]} = \sqrt[n]{r}\left(\cos\frac{\phi}{n} + \sqrt{-1}.\sin\frac{\phi}{n}\right); \quad [1]$$

for, by Art. 24,

$$\left[\sqrt[n]{r}\left(\cos\frac{\phi}{n} + \sqrt{-1}.\sin\frac{\phi}{n}\right)\right]^n = r(\cos\phi + \sqrt{-1}.\sin\phi).$$

EXAMPLE.

Show that Art. 24 [1] holds even when n is negative or fractional.

26. As the *modulus of every quantity*, positive, negative, real, or imaginary, *is positive*, it is always possible to find the modulus of any required root; and as this modulus must be real and positive, *it can never*, in any given example, *have more than one value*. We know from algebra, however, that every equation of the nth degree containing one unknown has n roots, and that consequently every number must have n nth roots. Our formula, Art. 25 [1], appears to give us but one nth root for any given quantity. It must then be incompletè.

We have seen (Art. 20) that while the modulus of a given imaginary has but one value, its argument is indeterminate and may have any one of an infinite number of values which differ by multiples of 2π. If ϕ_0 is one of these values, the full form of the imaginary is not $r(\cos\phi_0 + \sqrt{-1}.\sin\phi_0)$, as we have hitherto written it, but is

$$r[\cos(\phi_0 + 2m\pi) + \sqrt{-1}.\sin(\phi_0 + 2m\pi)],$$

where m is zero or any whole number positive or negative. Since angles differing by multiples of 2π have the same trigonometric functions, it is easily seen that the introduction of the term $2m\pi$ into the argument of an imaginary will not modify any of our results except that of Art. 25, which becomes

$$\sqrt[n]{r[\cos(\phi_0 + 2m\pi) + \sqrt{-1}.\sin(\phi_0 + 2m\pi)]}$$
$$= \sqrt[n]{r}\left[\cos\left(\frac{\phi_0}{n} + m\frac{2\pi}{n}\right) + \sqrt{-1}.\sin\left(\frac{\phi_0}{n} + m\frac{2\pi}{n}\right)\right]. \quad [1]$$

Giving m the values 0, 1, 2, 3 ...., $n-1$, n, $n+1$, successively, we get

$$\frac{\phi_0}{n},\ \frac{\phi_0}{n} + \frac{2\pi}{n},\ \frac{\phi_0}{n} + 2\frac{2\pi}{n},\ \frac{\phi_0}{n} + 3\frac{2\pi}{n}\ \ldots\ldots,\ \frac{\phi_0}{n} + (n-1)\frac{2\pi}{n},$$

$$\frac{\phi_0}{n} + 2\pi,\ \frac{\phi_0}{n} + \frac{2\pi}{n} + 2\pi,$$

as arguments of our nth root.

Of these values the first n, that is, all except the last two, correspond to different points, and therefore to different roots; the next to the last gives the same point as the first, and the last the same point as the second, and it is easily seen that if we go on increasing m we shall get no new points. The same thing is true of negative values of m.

Hence we see that *every quantity, real or imaginary, has* n *distinct* n*th roots, all having the same modulus*, but with arguments differing by multiples of $\frac{2\pi}{n}$.

27. Any *positive real* differs from unity only in its modulus, and any *negative real* differs from -1 only in its modulus. All the nth roots of any number or of its negative may be obtained

by multiplying the nth roots of 1 or of -1 by the real positive nth root of the number.

Let us consider some of the roots of 1 and of -1; for example, the cube roots of 1 and of -1. The modulus of 1 is 1, and its argument is 0. The modulus of each of the cube roots of 1 is 1, and their arguments are 0, $\frac{2\pi}{3}$, and $\frac{4\pi}{3}$; that is, 0°, 120°, and 240°. The roots in question, then, are represented by the points P_1, P_2, P_3, in the figure. Their values are

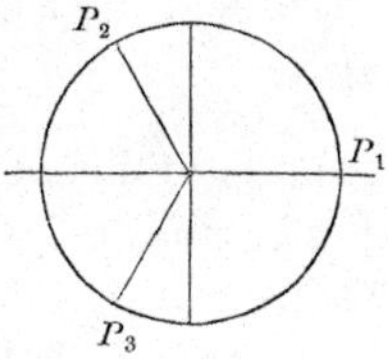

$$1(\cos 0 + \sqrt{-1}.\sin 0),$$
$$1(\cos 120° + \sqrt{-1}.\sin 120°),$$
$$\text{and } 1(\cos 240° + \sqrt{-1}.\sin 240°),$$
$$\text{or } 1,\ -\tfrac{1}{2} + \tfrac{\sqrt{3}}{2}\sqrt{-1},\ -\tfrac{1}{2} - \tfrac{\sqrt{3}}{2}\sqrt{-1}.$$

The modulus of -1 is 1, and its argument is π. The modulus of the cube roots of -1 is 1, and their arguments are $\frac{\pi}{3}$, $\frac{\pi}{3} + \frac{2\pi}{3}$, $\frac{\pi}{3} + \frac{4\pi}{3}$, that is, 60°, 180°, 300°. The roots in question, then, are represented by the points P_1, P_2, P_3, in the figure. Their values are $\frac{1}{2} + \frac{\sqrt{3}}{2}\sqrt{-1}$, -1, $\frac{1}{2} - \frac{\sqrt{3}}{2}\sqrt{-1}$.

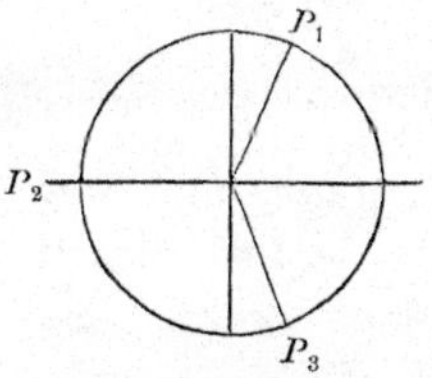

EXAMPLES.

(1) What are the square roots of 1 and -1? the 4th roots? the 5th roots? the 6th roots?

(2) Find the cube roots of -8; the 5th roots of 32.

(3) Show that an imaginary can have no real nth root; that a positive real has two real nth roots if n is even, one if n is odd; that a negative real has one real nth root if n is odd, none if n is even.

28. Imaginaries having equal moduli, and arguments differing only in sign, are called *conjugate imaginaries*.

$r(\cos\phi + \sqrt{-1}.\sin\phi)$, and $r[\cos(-\phi) + \sqrt{-1}.\sin(-\phi)]$, or $r(\cos\phi - \sqrt{-1}.\sin\phi)$ are *conjugate*.

They can be written $x + y\sqrt{-1}$ and $x - y\sqrt{-1}$, and we see that the points corresponding to them have the same abscissa, and ordinates which are equal with opposite signs.

EXAMPLES.

(1) Prove that *conjugate imaginaries* have a real sum and a real product.

(2) Prove, by considering in detail the substitution of $a + b\sqrt{-1}$ and $a - b\sqrt{-1}$ in turn for x in any algebraic polynomial in x with real coefficients, that if any algebraic equation with real coefficients has an imaginary root the *conjugate* of that root is also a root of the equation.

(3) Prove that if in any fraction where the numerator and denominator are rational algebraic polynomials in x, we substitute $a + b\sqrt{-1}$ and $a - b\sqrt{-1}$ in turn for x, the results are conjugate.

Transcendental Functions of Imaginaries.

29. We have adopted a definition of an *imaginary* and laid down rules to govern its use, that enable us to deal with it, in all expressions involving only algebraic operations, precisely as if it were a quantity. If we are going further, and are to subject it to *transcendental* operations, we must carefully define each function that we are going to use, and establish the rules which the function must obey.

The principal *transcendental* functions are e^x, $\log x$, and $\sin x$, and we wish to define and study these when x is replaced by an imaginary variable z.

As our conception and treatment of imaginaries have been entirely algebraic, we naturally wish to define our transcendental functions by the aid of algebraic functions; and since we know that the transcendental functions of a *real* variable can be expressed in terms of algebraic functions only by the aid of infinite series, we are led to use such series in defining transcendental functions of an *imaginary* variable; but we must first establish a proposition concerning the convergency of a series containing imaginary terms.

30. *If the moduli of the terms of a series containing imaginary terms form a convergent series, the given series is convergent.*

Let $u_0 + u_1 + u_2 + \cdots\cdots + u_n + \cdots\cdots$ be a series containing imaginary terms.

Let

$$u_0 = R_0(\cos\Phi_0 + \sqrt{-1}.\sin\Phi_0),\ u_1 = R_1(\cos\Phi_1 + \sqrt{-1}.\sin\Phi_1),\ \&c.,$$

and suppose that the series $R_0 + R_1 + R_2 + \cdots\cdots + R_n + \cdots\cdots$ is convergent; then will the series $u_0 + u_1 + u_2 + \cdots\cdots$ be convergent.

The series $R_0 + R_1 + \cdots\cdots$ is a convergent series composed of positive terms; if then we break up this series into parts in any way, each part will have a definite sum or will approach a definite limit as the number of terms considered is increased indefinitely.

The series $u_0 + u_1 + u_2 + \cdots\cdots u_n + \cdots\cdots$ can be broken up into the two series

$$R_0\cos\Phi_0 + R_1\cos\Phi_1 + R_2\cos\Phi_2 + \cdots\cdots + R_n\cos\Phi_n + \cdots\cdots \quad (1)$$

and

$$\sqrt{-1}(R_0\sin\Phi_0 + R_1\sin\Phi_1 + R_2\sin\Phi_2 + \cdots\cdots + R_n\sin\Phi_n + \cdots\cdots). \quad (2)$$

(1) can be separated into two parts, the first made up only of positive terms, the second only of negative terms, and can therefore be regarded as the difference between two series, each consisting of positive terms. Each term in either series will be a term of the modulus series $R_0 + R_1 + R_2 + \cdots\cdots$ multiplied by a quantity less than one, and the sum of n terms of each series will therefore approach a definite limit, as n increases indefinitely. The series (1), then, which is the abscissa of the point representing the given imaginary series, has a finite sum.

In the same way it may be shown that the coefficient of $\sqrt{-1}$ in (2) has a finite sum, and this is the ordinate of the point representing the given series. The sum of n terms of the given series, then, approaches a definite limit as n is increased indefinitely, and the series is convergent.

31. We have seen (I. Art. 133 [2]) that

$$e^x = 1 + \frac{x}{1} + \frac{x^2}{2\,!} + \frac{x^3}{3\,!} + \frac{x^4}{4\,!} + \cdots\cdots \qquad [1]$$

when x is real, and that this series is convergent for all values of x.

Let us define e^z, where $z = x + y\sqrt{-1}$, by the series

$$e^z = 1 + \frac{z}{1} + \frac{z^2}{2\,!} + \frac{z^3}{3\,!} + \frac{z^4}{4\,!} + \cdots\cdots \qquad [2]$$

This series is convergent, for if $z = r(\cos\phi + \sqrt{-1}.\sin\phi)$ the series

$$1 + \frac{r}{1} + \frac{r^2}{2\,!} + \frac{r^3}{3\,!} + \frac{r^4}{4\,!} + \cdots\cdots$$

made up of the moduli of the terms of [2] is convergent by I. Art. 133, and therefore the value we have chosen for e^z is a determinate finite one.

Write $x + y\sqrt{-1}$ for z, and we get

$$e^{x+y\sqrt{-1}} = 1 + \frac{x+y\sqrt{-1}}{1} + \frac{(x+y\sqrt{-1})^2}{2\,!} + \frac{(x+y\sqrt{-1})^3}{3\,!} + \cdots\cdots \qquad [3]$$

The terms of this series can be expanded by the Binomial Theorem. Consider all the resulting terms containing any given power of x, say x^p; we have

$$\frac{x^p}{p\,!}\left(1 + \frac{y\sqrt{-1}}{1} + \frac{(y\sqrt{-1})^2}{2\,!} + \frac{(y\sqrt{-1})^3}{3\,!} + \cdots\cdots + \frac{(y\sqrt{-1})^n}{n\,!} + \cdots\cdots\right);$$

or, separating the real terms and the imaginary terms,

$$\frac{x^p}{p\,!}\left(1 - \frac{y^2}{2\,!} + \frac{y^4}{4\,!} - \frac{y^6}{6\,!} + \cdots\cdots\right)$$
$$+ \frac{x^p}{p\,!}\sqrt{-1}\left(y - \frac{y^3}{3\,!} + \frac{y^5}{5\,!} - \frac{y^7}{7\,!} + \cdots\cdots\right),$$

or $$\frac{x^p}{p\,!}(\cos y + \sqrt{-1}.\sin y), \qquad \text{by I. Art. 134.}$$

Giving p all values from 1 to ∞ we get

$$e^{x+y\sqrt{-1}} = (\cos y + \sqrt{-1}.\sin y)\left(1 + \frac{x}{1} + \frac{x^2}{2\,!} + \frac{x^3}{3\,!} + \frac{x^4}{4\,!} + \cdots\cdots\right)$$

$$= e^x(\cos y + \sqrt{-1}.\sin y), \qquad [4]$$

which, by the way, is in one of our typical imaginary forms.

If $x = 0$, in [4],

we get $$e^{y\sqrt{-1}} = \cos y + \sqrt{-1}.\sin y, \qquad [5]$$

which suggests a new way of writing our typical imaginary; namely,

$$r(\cos\phi + \sqrt{-1}.\sin\phi) = re^{\phi\sqrt{-1}}.$$

32. We have seen that

$$e^{x+y\sqrt{-1}} = e^x e^{y\sqrt{-1}};$$

let us see if all imaginary powers of e obey the *law of indices;* that is, if the equation

$$e^u e^v = e^{u+v} \qquad [1]$$

is universally true.

Let $u = x_1 + y_1\sqrt{-1}$ and $v = x_2 + y_2\sqrt{-1}$,

then $e^u = e^{x_1 + y_1\sqrt{-1}} = e^{x_1}(\cos y_1 + \sqrt{-1}.\sin y_1)$,

$$e^v = e^{x_2 + y_2\sqrt{-1}} = e^{x_2}(\cos y_2 + \sqrt{-1}.\sin y_2),$$

$$e^u e^v = e^{x_1}e^{x_2}[\cos(y_1 + y_2) + \sqrt{-1}.\sin(y_1 + y_2)]$$

$$= e^{x_1 + x_2}[\cos(y_1 + y_2) + \sqrt{-1}.\sin(y_1 + y_2)]$$

$$= e^{x_1 + x_2 + (y_1 + y_2)\sqrt{-1}}$$

$$= e^{u+v},$$

and the *fundamental property of exponential functions holds for imaginaries as well as for reals.*

EXAMPLE.

Prove that $a^u a^v = a^{u+v}$ when u and v are imaginary.

Logarithmic Functions.

33. As a logarithm is the inverse of an exponential, we ought to be able to obtain the logarithm of an imaginary from the formula for $e^{x+y\sqrt{-1}}$. We see readily that

$$z = r(\cos\phi + \sqrt{-1}.\sin\phi) = e^{\log r + \phi\sqrt{-1}},$$

whence
$$\log z = \log r + \phi\sqrt{-1};$$

or, more strictly, since

$$z = r[\cos(\phi_0 + 2n\pi) + \sqrt{-1}.\sin(\phi_0 + 2n\pi)],$$

$$\log z = \log r + (\phi_0 + 2n\pi)\sqrt{-1} \qquad [1]$$

where n is any integer.

If $z = x + y\sqrt{-1}$, $r = \sqrt{x^2 + y^2}$, and $\phi = \tan^{-1}\frac{y}{x}$;

whence
$$\log z = \tfrac{1}{2}\log(x^2 + y^2) + \sqrt{-1}.\tan^{-1}\frac{y}{x}. \qquad [2]$$

Each of the expressions for $\log z$ is indeterminate, and represents an infinite number of values, differing by multiples of $2\pi\sqrt{-1}$.

This indeterminateness in the logarithm might have been expected *a priori*, for

$$e^{2\pi\sqrt{-1}} = \cos 2\pi + \sqrt{-1}.\sin 2\pi = 1, \qquad \text{by Art. 31.}$$

Hence, adding $2\pi\sqrt{-1}$ to the logarithm of any quantity will have the effect of multiplying the quantity by 1, and therefore will not change its value.

EXAMPLE.

Show that if an expression is imaginary, all its logarithms are imaginary; if it is real and positive, one logarithm is real and the rest imaginary; if it is real and negative, all are imaginary.

Trigonometric Functions.

34. If z is real,

$$\sin z = z - \frac{z^3}{3\,!} + \frac{z^5}{5\,!} - \frac{z^7}{7\,!} + \cdots\cdots \qquad [1]$$

$$\cos z = 1 - \frac{z^2}{2\,!} + \frac{z^4}{4\,!} - \frac{z^6}{6\,!} + \cdots\cdots \qquad [2]$$

by I. Art. 134.

If
$$z = r(\cos\phi + \sqrt{-1}.\sin\phi),$$

the series of the moduli,

$$r + \frac{r^3}{3\,!} + \frac{r^5}{5\,!} + \frac{r^7}{7\,!} + \cdots\cdots,$$

$$1 + \frac{r^2}{2\,!} + \frac{r^4}{4\,!} + \frac{r^6}{6\,!} + \cdots\cdots,$$

are easily seen to be convergent; therefore if z is imaginary, the series [1] and [2] are convergent. We shall take them as definitions of the sine and cosine of an imaginary.

EXAMPLE.

From the formulas of Art. 31, and from Art. 34 [1] and [2], show that

$$e^{z\sqrt{-1}} = \cos z + \sqrt{-1}.\sin z,$$

and
$$e^{-z\sqrt{-1}} = \cos z - \sqrt{-1}.\sin z, \quad \text{for all values of } z.$$

35. From the relations

$$e^{z\sqrt{-1}} = \cos z + \sqrt{-1}.\sin z,$$

$$e^{-z\sqrt{-1}} = \cos z - \sqrt{-1}.\sin z,$$

we get
$$\cos z = \frac{e^{z\sqrt{-1}} + e^{-z\sqrt{-1}}}{2}, \qquad [1]$$

$$\sin z = \frac{e^{z\sqrt{-1}} - e^{-z\sqrt{-1}}}{2\sqrt{-1}}, \qquad [2]$$

for all values of z.

Let $$z = x + y\sqrt{-1}.$$

$$\cos(x+y\sqrt{-1}) = \frac{e^{x\sqrt{-1}-y} + e^{-x\sqrt{-1}+y}}{2}$$

$$= \frac{(\cos x + \sqrt{-1}.\sin x)e^{-y} + (\cos x - \sqrt{-1}.\sin x)e^{y}}{2},$$

by Art. 34, Ex.,

$$= \cos x \frac{e^{y}+e^{-y}}{2} - \sqrt{-1}.\sin x \frac{e^{y}-e^{-y}}{2}. \qquad [3]$$

In the same way it may be shown that

$$\sin(x+y\sqrt{-1}) = \frac{(\cos x + \sqrt{-1}.\sin x)e^{-y} - (\cos x - \sqrt{-1}.\sin x)e^{y}}{2\sqrt{-1}}$$

$$= \sin x \frac{e^{y}+e^{-y}}{2} + \sqrt{-1}.\cos x \frac{e^{y}-e^{-y}}{2}. \qquad [4]$$

If z is real in [1] and [2], we have

$$\cos x = \frac{e^{x\sqrt{-1}} + e^{-x\sqrt{-1}}}{2},$$

$$\sin x = -\frac{e^{x\sqrt{-1}} - e^{-x\sqrt{-1}}}{2}\sqrt{-1}.$$

If $z = y\sqrt{-1}$, and is a pure imaginary,

$$\cos y\sqrt{-1} = \frac{e^{y}+e^{-y}}{2}, \qquad [5]$$

$$\sin y\sqrt{-1} = \frac{e^{y}-e^{-y}}{2}\sqrt{-1}; \qquad [6]$$

whence we see that the cosine of a pure imaginary is real, while its sine is imaginary.

By the aid of [5] and [6], [3] and [4] can be written:

$$\cos(x+y\sqrt{-1}) = \cos x \cos y\sqrt{-1} - \sin x \sin y\sqrt{-1}, \qquad [7]$$

$$\sin(x+y\sqrt{-1}) = \sin x \cos y\sqrt{-1} + \cos x \sin y\sqrt{-1}. \qquad [8]$$

EXAMPLES.

(1) From [1] and [2] show that $\sin^2 z + \cos^2 z = 1$.

(2) Prove that

$$\cos(u+v) = \cos u \cos v - \sin u \sin v,$$

$$\sin(u+v) = \sin u \cos v + \cos u \sin v,$$

where u and v are imaginary.

The relations to be proved in examples (1) and (2) are the fundamental formulas of Trigonometry, and they enable us to use trigonometric functions of imaginaries precisely as we use trigonometric functions of reals.

Differentiation of Functions of Imaginaries.

36. A function of an imaginary variable,

$$z = x + y\sqrt{-1},$$

is, strictly speaking, a function of two independent variables, x and y; for we can change z by changing either x or y, or both x and y. Its differential will usually contain dx and dy, and not necessarily dz; and if we divide its differential by dz to get its derivative with respect to z, the result will generally contain $\frac{dy}{dx}$, which will be wholly indeterminate, since x and y are entirely independent in the expression $x + y\sqrt{-1}$. It may happen, however, in the case of some simple functions, that dz will appear as a factor in the differential of the function, which in that case will have a single derivative.

37. *In differentiating, the* $\sqrt{-1}$ *may be treated like a constant;* for the operation of finding the differential of a function is an algebraic operation, and in all algebraic operations $\sqrt{-1}$ obeys the same laws as any constant.

EXAMPLE.

Prove that $d(x^2\sqrt{-1}) = 2x\sqrt{-1}.\,dx$;

and that $d\sqrt{-1}.\,\sin x = \sqrt{-1}.\,\cos x.dx$.

We have, by the aid of this principle,

if $$z = x + y\sqrt{-1},$$

$$dz = dx + \sqrt{-1}.\,dy; \qquad [1]$$

if $$z = r(\cos\phi + \sqrt{-1}.\,\sin\phi),$$

$$dz = dr(\cos\phi + \sqrt{-1}.\,\sin\phi) + rd\phi(-\sin\phi + \sqrt{-1}.\,\cos\phi)$$

$$= (dr + r\sqrt{-1}.\,d\phi)(\cos\phi + \sqrt{-1}.\,\sin\phi). \qquad [2]$$

38. Let us now consider the differentiation of z^m, e^z, $\log z$, $\sin z$, and $\cos z$.

Let $$z = r(\cos\phi + \sqrt{-1}.\,\sin\phi),$$

then

$$z^m = r^m(\cos m\phi + \sqrt{-1}.\,\sin m\phi), \qquad \text{by Art. 24 [1];}$$

$$dz^m = mr^{m-1}dr(\cos m\phi + \sqrt{-1}.\,\sin m\phi) + mr^m d\phi\,(-\sin m\phi + \sqrt{-1}.\,\cos m\phi),$$

$$dz^m = mr^{m-1}[\cos(m-1)\phi + \sqrt{-1}.\,\sin(m-1)\phi](\cos\phi + \sqrt{-1}.\,\sin\phi)dr$$
$$+ mr^m[\cos(m-1)\phi + \sqrt{-1}.\,\sin(m-1)\phi](\cos\phi + \sqrt{-1}.\,\sin\phi)\sqrt{-1}.\,d\phi,$$

$$dz^m = mr^{m-1}[\cos(m-1)\phi + \sqrt{-1}.\,\sin(m-1)\phi](dr + r\sqrt{-1}.\,d\phi)(\cos\phi + \sqrt{-1}.\,\sin\phi),$$

$$dz^m = mz^{m-1}dz, \qquad [1] \quad \text{by Art. 37 [2],}$$

$$\frac{dz^m}{dz} = mz^{m-1}, \qquad [2]$$

and a power of an imaginary variable has a single derivative.

39. If $z = x + y\sqrt{-1}$,

$$e^z = e^x(\cos y + \sqrt{-1}.\sin y), \qquad \text{by Art. 31 [4]},$$

$$de^z = e^x dx(\cos y + \sqrt{-1}.\sin y) + e^x(-\sin y + \sqrt{-1}.\cos y)\,dy,$$

$$de^z = e^x(\cos y + \sqrt{-1}.\sin y)\,(dx + \sqrt{-1}.dy),$$

$$de^z = e^z dz, \qquad [1]$$

$$\frac{de^z}{dz} = e^z. \qquad [2]$$

EXAMPLE.

Show that $da^z = a^z \log a.dz$.

40. If $z = r(\cos\phi + \sqrt{-1}.\sin\phi)$,

$$\log z = \log r + \phi\sqrt{-1}, \qquad \text{by Art. 33},$$

$$d\log z = \frac{dr}{r} + \sqrt{-1}.\,d\phi = \frac{dr + r\sqrt{-1}.\,d\phi}{r},$$

$$d\log z = \frac{(dr + r\sqrt{-1}.\,d\phi)(\cos\phi + \sqrt{-1}.\sin\phi)}{r(\cos\phi + \sqrt{-1}.\sin\phi)},$$

$$d\log z = \frac{dz}{z}, \qquad [1]$$

$$\frac{d\log z}{dz} = \frac{1}{z}. \qquad [2]$$

41. $$\sin z = \frac{e^{z\sqrt{-1}} - e^{-z\sqrt{-1}}}{2\sqrt{-1}}, \qquad \text{by Art. 35 [2]},$$

$$d\sin z = \frac{e^{z\sqrt{-1}} + e^{-z\sqrt{-1}}}{2\sqrt{-1}}\sqrt{-1}.\,dz$$

$$= \frac{e^{z\sqrt{-1}} + e^{-z\sqrt{-1}}}{2}\,dz, \qquad \text{by Art. 35 [1]},$$

$$d\sin z = \cos z.dz. \qquad [1]$$

$$\cos z = \frac{e^{z\sqrt{-1}} + e^{-z\sqrt{-1}}}{2},$$

$$d\cos z = \frac{e^{z\sqrt{-1}} - e^{-z\sqrt{-1}}}{2}\sqrt{-1}.dz = -\frac{e^{z\sqrt{-1}} - e^{-z\sqrt{-1}}}{2\sqrt{-1}}dz,$$

$$d\cos z = -\sin z.dz. \qquad [2]$$

42. We see, then, that we get the same formulas for the differentiation of simple functions of imaginaries as for the differentiation of the corresponding functions of reals. It follows that our formulas for direct integration (I. Art. 74) hold when x is imaginary.

Hyperbolic Functions.

43. We have (Art. 35 [5] and [6])

$$\cos x\sqrt{-1} = \frac{e^x + e^{-x}}{2},$$

and

$$\sin x\sqrt{-1} = \frac{e^x - e^{-x}}{2}\sqrt{-1},$$

where x is real. $\frac{e^x + e^{-x}}{2}$ is called the hyperbolic cosine of x, and is written $\cosh x$; and $\frac{e^x - e^{-x}}{2}$ is called the hyperbolic sine of x, and is written $\sinh x$;

$$\sinh x = \frac{e^x - e^{-x}}{2} = -\sqrt{-1}.\sin x\sqrt{-1}, \qquad [1]$$

$$\cosh x = \frac{e^x + e^{-x}}{2} = \cos x\sqrt{-1}. \qquad [2]$$

The hyperbolic tangent is defined as the ratio of sinh to cosh; and the hyperbolic cotangent, secant, and cosecant are the reciprocals of the tanh, cosh, and sinh respectively.

These functions, which are real when x is real, resemble in their properties the ordinary trigonometric functions.

44. For example,

$$\cosh^2 x - \sinh^2 x = 1; \qquad [1]$$

for
$$\cosh^2 x = \frac{e^{2x} + 2 + e^{-2x}}{4},$$

and
$$\sinh^2 x = \frac{e^{2x} - 2 + e^{-2x}}{4}.$$

EXAMPLES.

(1) Prove that $1 - \tanh^2 x = \text{sech}^2 x.$

(2) Prove that $1 - \text{ctnh}^2 x = -\text{csch}^2 x.$

(3) Prove that $\sinh(x + y) = \sinh x \cosh y + \cosh x \sinh y.$

(4) Prove that $\cosh(x + y) = \cosh x \cosh y + \sinh x \sinh y.$

45.
$$d \sinh x = d\frac{e^x - e^{-x}}{2} = \frac{e^x + e^{-x}}{2} dx,$$

$$d \sinh x = \cosh x . dx.$$

EXAMPLES.

(1) Prove
$$d \cosh x = \sinh x . dx.$$

$$d \tanh x = \text{sech}^2 x . dx.$$

$$d \,\text{ctnh}\, x = -\text{csch}^2 x . dx.$$

$$d \,\text{sech}\, x = -\text{sech}\, x \tanh x . dx.$$

$$d \,\text{csch}\, x = -\text{csch}\, x \,\text{ctnh}\, x . dx.$$

46. We can deal with anti-hyperbolic functions just as with anti-trigonometric functions.

To find $d \sinh^{-1} x$.

Let
$$u = \sinh^{-1} x,$$

then
$$x = \sinh u,$$

$$dx = \cosh u . du,$$

$$du = \frac{dx}{\cosh u},$$

$$\cosh u = \sqrt{1 + \sinh^2 u}, \qquad \text{by Art. 44 [1]},$$

$$\cosh u = \sqrt{1 + x^2},$$

$$d \sinh^{-1} x = \frac{dx}{\sqrt{1 + x^2}}. \qquad [1]$$

EXAMPLES.

Prove the formulas

$$d \cosh^{-1} x = \frac{dx}{\sqrt{x^2 - 1}}.$$

$$d \tanh^{-1} x = \frac{dx}{1 - x^2}.$$

$$d \operatorname{sech}^{-1} x = - \frac{dx}{x \sqrt{1 - x^2}}.$$

$$d \operatorname{csch}^{-1} x = - \frac{dx}{x \sqrt{x^2 + 1}}.$$

47. The anti-hyperbolic functions are easily expressed as logarithms.

Let $$u = \sinh^{-1} x,$$

then $$x = \sinh u = \frac{e^u - e^{-u}}{2},$$

$$2x = e^u - \frac{1}{e^u},$$

$$2xe^u = e^{2u} - 1,$$

$$e^{2u} - 2xe^u = 1,$$

$$e^{2u} - 2xe^u + x^2 = 1 + x^2,$$

$$e^u - x = \pm \sqrt{1 + x^2},$$

$$e^u = x \pm \sqrt{1 + x^2};$$

as e^u is necessarily positive, we may reject the negative value in the second member as impossible, and we have

$$e^u = x + \sqrt{1+x^2},$$

$$u = \log(x + \sqrt{1+x^2}),$$

or

$$\sinh^{-1}x = \log(x + \sqrt{1+x^2}). \qquad [1]$$

EXAMPLES.

Prove the formulas

$$\cosh^{-1}x = \log(x + \sqrt{x^2-1}).$$

$$\tanh^{-1}x = \tfrac{1}{2}\log\frac{1+x}{1-x}.$$

$$\operatorname{sech}^{-1}x = \log\left(\frac{1}{x} + \sqrt{\frac{1}{x^2} - 1}\right).$$

$$\operatorname{csch}^{-1}x = \log\left(\frac{1}{x} + \sqrt{\frac{1}{x^2} + 1}\right).$$

48. The principal advantage arising from the use of hyperbolic functions is that they bring to light some curious analogies between the integrals of certain irrational functions.

From I. Art. 71 we obtain the formulas for direct integration.

$$\int\frac{dx}{\sqrt{1-x^2}} = \sin^{-1}x. \qquad [1]$$

$$\int\frac{dx}{1+x^2} = \tan^{-1}x. \qquad [2]$$

$$\int\frac{dx}{x\sqrt{x^2-1}} = \sec^{-1}x. \qquad [3]$$

From Art. 46 we obtain the allied formulas:

$$\int\frac{dx}{\sqrt{1+x^2}} = \sinh^{-1}x = \log(x + \sqrt{1+x^2}). \qquad [4]$$

$$\int\frac{dx}{\sqrt{x^2-1}} = \cosh^{-1}x = \log(x + \sqrt{x^2-1}). \qquad [5]$$

$$\int\frac{dx}{1-x^2} = \tanh^{-1}x = \tfrac{1}{2}\log\frac{1+x}{1-x}. \qquad [6]$$

$$-\int\frac{dx}{x\sqrt{1-x^2}} = \text{sech}^{-1}x = \log\left(\frac{1}{x}+\sqrt{\frac{1}{x^2}-1}\right). \qquad [7]$$

$$-\int\frac{dx}{x\sqrt{x^2+1}} = \text{csch}^{-1}x = \log\left(\frac{1}{x}+\sqrt{\frac{1}{x^2}+1}\right). \qquad [8]$$

CHAPTER III.

GENERAL METHODS OF INTEGRATING.

49. We have defined the *integral* of any function of a single variable as *the function which has the given function for its derivative* (I. Art. 53); we have defined a *definite integral* as the *limit of the sum of a set of differentials;* and we have shown that a definite integral is the *difference between two values of an ordinary integral* (I. Art. 183).

Now that we have adopted the differential notation in place of the derivative notation, it is better to regard an integral as the inverse of a *differential* instead of as the inverse of a *derivative.* Hence the integral of $fx.dx$ will be the function whose differential is $fx.dx$; and we shall indicate it by $\int fx.dx$. In our old notation we should have indicated precisely the same function by $\int_x fx$; for if the derivative of a function is fx we know that its differential is $fx.dx$.

50. If fx is any function whatever of x, fx.dx *has an integral.* For if we construct the curve whose equation is $y = fx$, we know that the area included by the curve, the axis of X, any fixed ordinate, and the ordinate corresponding to the variable x, has for its differential ydx, or, in other words, $fx.dx$ (I. Art. 51). Such an area always exists, and it is a determinate function of x, except that, as the position of the initial ordinate is wholly arbitrary, the expression for the area will contain an arbitrary constant. Thus, if Fx is the area in question for some one position of the initial ordinate, we shall have

$$\int fx.dx = Fx + C,$$

where C is an arbitrary constant.

Moreover, $Fx + C$ is a complete expression for $\int fx.dx$; for if two functions of x have the same differential, they have the same derivative with respect to x, and therefore they change at the same rate when x changes (I. Art. 38); they can differ, then, at any instant only by the difference between their initial values, which is some constant.

Hence we see that *every expression of the form* fx.dx *has an integral, and, except for the presence of an arbitrary constant, but one integral.*

51. We have shown in I. Art. 183 that a *definite integral* is the difference between two values of an ordinary integral, and therefore contains no constant. Thus, if $Fx + C$ is the integral of $fx.dx$,

$$\int_a^b fx.dx = Fb - Fa.$$

In the same way we shall have

$$\int_a^b fz.dz = Fb - Fa;$$

and we see that a *definite integral is a function of the values between which the sum is taken* and not of the variable with respect to which we integrate.

Since
$$\int_b^a fx.dx = Fa - Fb,$$

$$\int_b^a fx.dx = -\int_a^b fx.dx.$$

EXAMPLE.

Show that $\int_a^c fx.dx + \int_c^b fx.dx = \int_a^b fx.dx.$

52. In what we have said concerning definite integrals we have tacitly assumed that the integral is a *continuous function* between the values between which the sum in question is taken. If it is not, we cannot regard the whole increment of Fx as equal

to the limit of the sum of the partial infinitesimal increments, and the reasoning of I. Art. 183 ceases to be valid.

Take, for example, $\int_{-1}^{1} \frac{dx}{x^2}$.

$$\int \frac{dx}{x^2} = \int x^{-2}\,dx = \frac{x^{-1}}{-1} = -\frac{1}{x}, \quad \text{by I. Art. 55 (7);}$$

and apparently

$$\int_{-1}^{1} \frac{dx}{x^2} = \left(-\frac{1}{x}\right)_{x=1} - \left(-\frac{1}{x}\right)_{x=-1} = -2.$$

But $\int_{-1}^{1} \frac{dx}{x^2}$ ought to be the area between the curve $y = \frac{1}{x^2}$, the axis of x, and the ordinates corresponding to $x = 1$ and $x = -1$, which evidently is not -2; and we see that the function $\frac{1}{x^2}$ is discontinuous between the values $x = -1$ and $x = 1$.

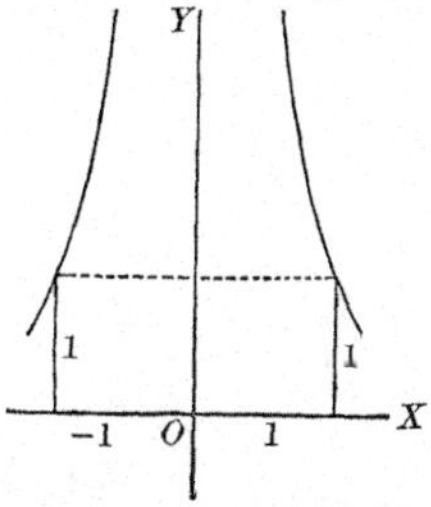

The area in question which the definite integral should represent is easily seen to be infinite, for

$$\int_{-1}^{-\epsilon} \frac{dx}{x^2} = \frac{1}{\epsilon} - 1, \text{ and } \int_{\epsilon}^{1} \frac{dx}{x^2} = \frac{1}{\epsilon} - 1,$$

and each of these expressions increases without limit as ϵ approaches zero.

53. Since a definite integral is the difference between two values of an indefinite integral, what we have to find first in any problem is the indéfinite integral. This may be found by inspection if the function to be integrated comes under any of the forms we have already obtained by differentiation, and we are then said to integrate directly. Direct integration has been illustrated, and the most important of the forms which can be integrated directly have been given in I. Chapter V. For the sake of convenience we rewrite these forms, using the differential notation, and adding one or two new forms from our sections on hyperbolic functions.

$$\int x^n\,dx = \frac{x^{n+1}}{n+1}.$$

$$\int \frac{dx}{x} = \log x.$$

$$\int a^x\,dx = \frac{a^x}{\log a}.$$

$$\int e^x\,dx = e^x.$$

$$\int \sin x.dx = -\cos x.$$

$$\int \cos x.dx = \sin x.$$

$$\int \tan x.dx = -\log\cos x.$$

$$\int \operatorname{ctn} x.dx = \log\sin x.$$

$$\int \frac{dx}{\sqrt{1-x^2}} = \sin^{-1}x.$$

$$\int \frac{dx}{\sqrt{1+x^2}} = \sinh^{-1}x = \log(x+\sqrt{1+x^2}).$$

$$\int \frac{dx}{\sqrt{x^2-1}} = \cosh^{-1}x = \log(x+\sqrt{x^2-1}).$$

$$\int \frac{dx}{1+x^2} = \tan^{-1}x.$$

$$\int \frac{dx}{1-x^2} = \tanh^{-1}x = \tfrac{1}{2}\log\frac{1+x}{1-x}.$$

$$\int \frac{dx}{x\sqrt{x^2-1}} = \sec^{-1}x.$$

$$\int \frac{dx}{x\sqrt{1-x^2}} = -\operatorname{sech}^{-1}x = -\log\left(\frac{1}{x}+\sqrt{\frac{1}{x^2}-1}\right).$$

$$\int \frac{dx}{x\sqrt{x^2+1}} = -\operatorname{csch}^{-1}x = -\log\left(\frac{1}{x}+\sqrt{\frac{1}{x^2}+1}\right).$$

$$\int \frac{dx}{\sqrt{2x-x^2}} = \operatorname{vers}^{-1}x.$$

54. We took up in I. Chap. V. the principal devices used in preparing a function for integration when it cannot be integrated directly.

The first of these methods, that of *integration by substitution*, is simplified by the use of the differential notation, because the formula for *change of variable* (I. Art. 75 [1]),

$$\int_x u = \int_y u D_y x \quad \text{becoming} \quad \int u dx = \int u \frac{dx}{dy} dy,$$

reduces to an identity and is no longer needed, and all that is required is a simple substitution.

(*a*) For example, let us find $\int \frac{dx}{x}\sqrt{1+\log x}$.

Let $1+\log x = z$; then $\frac{dx}{x} = dz$,

and $$\int \frac{dx}{x}\sqrt{1+\log x} = \int z^{\frac{1}{2}} dz = \tfrac{2}{3} z^{\frac{3}{2}} = \tfrac{2}{3}(1+\log x)^{\frac{3}{2}}.$$

When, as in this example, a factor of the quantity to be integrated is equal or proportional to the differential of some function occurring in the expression, the substitution of a new variable for the function in question will generally simplify the problem.

(*b*) Required $\int \frac{dx}{e^x + e^{-x}}$.

Let $e^x = y$; then $e^x dx = dy$,

$$\frac{dx}{e^x + e^{-x}} = \frac{e^x dx}{e^{2x}+1} = \frac{dy}{y^2+1},$$

and $$\int \frac{dx}{e^x + e^{-x}} = \int \frac{dy}{1+y^2} = \tan^{-1} y = \tan^{-1} e^x.$$

(*c*) Required $\int \sec x . dx$.

$$\sec x = \frac{1}{\cos x} = \frac{\cos x}{\cos^2 x}.$$

Let $z = \sin x$; then $dz = \cos x.dx$,

$$\cos^2 x = 1 - z^2,$$

$$\int \frac{\cos x.dx}{\cos^2 x} = \int \frac{dz}{1 - z^2} = \tfrac{1}{2} \log \frac{1 + z}{1 - z}, \qquad \text{by Art. 53,}$$

$$\int \sec x.dx = \tfrac{1}{2} \log \frac{1 + \sin x}{1 - \sin x} = \log \tan\left(\frac{\pi}{4} + \frac{x}{2}\right).$$

EXAMPLES.

Prove that (1) $\int \csc x.dx = \tfrac{1}{2} \log \frac{1 - \cos x}{1 + \cos x} = \log \tan \frac{x}{2}$.

(2) $\int \frac{x^2 dx}{\sqrt{1 - x^2}} = -\tfrac{1}{2} \cos^{-1} x - \frac{x\sqrt{1 - x^2}}{2}$.

Suggestion: Let $x = \cos z$.

55. The formula for *integration by parts* (I. Art. 79 [1]) becomes

$$\int u dv = uv - \int v du, \qquad [1]$$

when we use the differential notation. It is used as in I. Chap. V.

(*a*) For example, let us find $\int x^n \log x.dx$.

Let $u = \log x$, and $dv = x^n dx$;

then $$du = \frac{dx}{x},$$

and $$v = \frac{x^{n+1}}{n+1},$$

$$\int x^n \log x.dx = \frac{x^{n+1}}{n+1} \log x - \int \frac{x^n}{n+1} dx = \frac{x^{n+1}}{n+1}\left(\log x - \frac{1}{n+1}\right).$$

(*b*) Required $\int x \sin^{-1} x.dx$.

Let $u = \sin^{-1} x$, and $dv = x dx$;

then $$du = \frac{dx}{\sqrt{1 - x^2}},$$

and $$v = \frac{x^2}{2},$$

$$\int x \sin^{-1} x . dx = \frac{x^2}{2} \sin^{-1} x - \tfrac{1}{2} \int \frac{x^2 \, dx}{\sqrt{1 - x^2}},$$

$$\int x \sin^{-1} x . dx = \frac{x^2}{2} \sin^{-1} x + \tfrac{1}{4} (\cos^{-1} x + x \sqrt{1 - x^2}).$$

(c) Required $\int \frac{xe^x \, dx}{(1 + x)^2}$.

Let $$u = xe^x,$$

and $$dv = \frac{dx}{(1 + x)^2};$$

then $$du = (xe^x + e^x) dx = e^x (1 + x) \, dx,$$

and $$v = -\frac{1}{1 + x},$$

$$\int \frac{xe^x \, dx}{(1 + x)^2} = -\frac{xe^x}{1 + x} + \int e^x \, dx = -\frac{xe^x}{1 + x} + e^x = \frac{e^x}{1 + x}.$$

EXAMPLES.

(1) $$\int \frac{dx}{\sqrt{1 - 3x - x^2}} = \sin^{-1} \frac{3 + 2x}{\sqrt{13}}.$$

(2) $$\int x \tan^{-1} x . dx = \frac{1 + x^2}{2} \tan^{-1} x - \tfrac{1}{2} x.$$

(3) $$\int \frac{x dx}{(1 - x)^3} = -\frac{1}{1 - x} + \frac{1}{2(1 - x)^2}.$$

(4) $$\int \frac{x dx}{\sqrt{2ax - x^2}} = -\sqrt{2ax - x^2} + a \operatorname{vers}^{-1} \frac{x}{a}.$$

(5) $$\int \sqrt{2ax - x^2} . dx = \frac{x - a}{2} \sqrt{2ax - x^2} + \frac{a^2}{2} \sin^{-1} \frac{x - a}{a}.$$

Suggestion: Throw $2ax - x^2$ into the form $a^2 - (x - a)^2$.

(6) $$\int \frac{1 + \cos x}{x + \sin x} dx = \log (x + \sin x).$$

(7) $\int \frac{x + \sin x}{1 + \cos x}\,dx = x \tan \frac{x}{2}.$

Suggestion: Introduce $\frac{x}{2}$ in place of x.

(8) $\int \frac{dx}{x(\log x)^n} = -\frac{1}{(n-1)(\log x)^{n-1}}.$

(9) $\int \frac{\log(\log x)}{x}\,dx = \log x\,[\log(\log x) - 1].$

(10) $\int \frac{\sin^{-1} x . dx}{(1 - x^2)^{\frac{3}{2}}} = z \tan z + \log \cos z$, where $z = \sin^{-1} x$.

(11) $\int \frac{dx}{\sin x + \cos x} = \frac{1}{\sqrt{2}} \log \tan \left(\frac{x}{2} + \frac{\pi}{8}\right).$

(12) $\int \frac{\sin x\, dx}{a + b \cos x} = -\frac{\log(a + b \cos x)}{b}.$

(13) $\int \frac{dx}{x^2 + 4x + 5} = \tan^{-1}(x + 2).$

(14) $\int \frac{x^2\, dx}{1 - x^6} = \frac{1}{6} \log \left(\frac{1 + x^3}{1 - x^3}\right).$

(15) $\int \frac{x^3\, dx}{x^8 - x^4 - 6} = \frac{1}{20} \log \left(\frac{x^4 - 3}{x^4 + 2}\right).$

(16) $\int \frac{dx}{a^2 \cos^2 x + b^2 \sin^2 x} = \frac{1}{ab} \tan^{-1} \left(\frac{b}{a} \tan x\right).$

CHAPTER IV.

RATIONAL FRACTIONS.

56. We shall now attempt to consider systematically the methods of integrating various functions; and to this end we shall begin with *rational algebraic expressions*. *Any rational algebraic polynomial can be integrated immediately* by the aid of the formula

$$\int x^n\,dx = \frac{x^{n+1}}{n+1}.$$

Take next a *rational fraction*, that is, a fraction whose numerator and denominator are rational algebraic polynomials. A rational fraction is *proper* if its numerator is of lower degree than its denominator; *improper* if the degree of the numerator is equal to or greater than the degree of the denominator. Since an improper fraction can always be reduced to a polynomial plus a proper fraction, by actually dividing the numerator by the denominator, we need only consider the treatment of proper fractions.

57. *Every proper rational fraction can be reduced to the sum of a set of simpler fractions each of which has a constant for a numerator and some power of a binomial for its denominator;* that is, a set of fractions any one of which is of the form $\frac{A}{(x-a)^m}$.

Let our given fraction be $\frac{fx}{Fx}$.

If a, b, c, &c., are the roots of the equation,

$$Fx = 0, \tag{1}$$

we have, from the Theory of Equations,

$$Fx = A(x-a)(x-b)(x-c)\cdots\cdots \tag{2}$$

The equation (1) may have some equal roots, and then some of the factors in (2) will be repeated. Suppose a occurs p times as a root of (1), b occurs q times, c occurs r times, &c.,

then $$Fx = A(x-a)^p(x-b)^q(x-c)^r \cdots\cdots \tag{3}$$

Call $$A(x-b)^q(x-c)^r \cdots\cdots = \phi x;$$

then $$Fx = (x-a)^p \phi x,$$

and $$\frac{fx}{Fx} = \frac{fx}{(x-a)^p \phi x} = \frac{fx - \dfrac{fa}{\phi a}\phi x}{(x-a)^p \phi x} + \frac{\dfrac{fa}{\phi a}\phi x}{(x-a)^p \phi x}$$

$$= \frac{\dfrac{fa}{\phi a}}{(x-a)^p} + \frac{fx - \dfrac{fa}{\phi a}\phi x}{(x-a)^p \phi x}.$$

$\dfrac{fx - \dfrac{fa}{\phi a}\phi x}{(x-a)^p \phi x}$ is a new proper fraction, but it can be reduced to a simpler form by dividing numerator and denominator by $x-a$, which is an exact divisor of the numerator because a is a root of the equation

$$fx - \frac{fa}{\phi a}\phi x = 0.$$

If we represent by $f_1 x$ the quotient arising from the division of $fx - \dfrac{fa}{\phi a}\phi x$ by $x-a$, we shall have

$$\frac{fx}{Fx} = \frac{\dfrac{fa}{\phi a}}{(x-a)^p} + \frac{f_1 x}{(x-a)^{p-1}\phi x},$$

where $\dfrac{f_1 x}{(x-a)^{p-1}\phi x}$ is a proper fraction, and may be treated precisely as we have treated the original fraction.

Hence $$\frac{f_1 x}{(x-a)^{p-1}\phi x} = \frac{\dfrac{f_1 a}{\phi a}}{(x-a)^{p-1}} + \frac{f_2 x}{(x-a)^{p-2}\phi x}.$$

By continuing this process we shall get

$$\frac{fx}{Fx} = \frac{\dfrac{fa}{\phi a}}{(x-a)^p} + \frac{\dfrac{f_1 a}{\phi a}}{(x-a)^{p-1}} + \frac{\dfrac{f_2 a}{\phi a}}{(x-a)^{p-2}} + \cdots\cdots + \frac{\dfrac{f_{p-1} a}{\phi a}}{x-a} + \frac{f_p x}{\phi x}.$$

In the same way $\frac{f_p x}{\phi x}$ can be broken up into a set of fractions having $(x-b)^q$, $(x-b)^{q-1}$, &c., for denominators, plus a fraction which can be broken up into fractions having $(x-c)^r$, $(x-c)^{r-1}$ ....., &c., for denominators; and we shall have, in the end,

$$\frac{fx}{Fx}=\frac{A_1}{(x-a)^p}+\frac{A_2}{(x-a)^{p-1}}+\;.....\;+\frac{A_p}{x-a}+\frac{B_1}{(x-b)^q}$$
$$+\frac{B_2}{(x-b)^{q-1}}+\;.....\;+\frac{B_q}{x-b}+\;.....\;+K, \qquad [1]$$

where K is the quotient obtained when we divide out the last factor of the denominator, and is consequently a constant. More than this, K must be zero, for as (1) is identically true, it must be true when $x=\infty$; but when $x=\infty$, $\frac{fx}{Fx}$ becomes zero, because its denominator is of higher degree than its numerator, and each of the fractions in the second member also becomes zero; whence $K=0$.

58. Since we now know the form into which any given rational fraction can be thrown, we can determine the numerators by the aid of known properties of an identical equation.

Let it be required to break up $\frac{3x-1}{(x-1)^2(x+1)}$ into simpler fractions.

By Art. 57,

$$\frac{3x-1}{(x-1)^2(x+1)}=\frac{A}{(x-1)^2}+\frac{B}{x-1}+\frac{C}{x+1},$$

and we wish to determine A, B, and C. Clearing of fractions, we have

$$3x-1=A(x+1)+B(x-1)(x+1)+C(x-1)^2. \qquad (1)$$

As this equation is identically true, the coefficients of like powers of x in the two members must be equal; and we have

$$B+C=0,$$
$$A-2C=3,$$
$$A-B+C=-1;$$

whence we find
$$A = 1,$$
$$B = 1,$$
$$C = -1;$$

and
$$\frac{3x-1}{(x-1)^2(x+1)} = \frac{1}{(x-1)^2} + \frac{1}{x-1} - \frac{1}{x+1}. \qquad (2)$$

The labor of determining the required constants can often be lessened by simple algebraic devices.

For example; since the identical equation we start with is true for all values of x, we have a right to substitute for x values that will make terms of the equation disappear. Take equation [1]:

$$3x - 1 = A(x+1) + B(x+1)(x-1) + C(x-1)^2. \qquad [1]$$

Let $x = 1$,
$$2 = 2A,$$
$$A = 1,$$

then
$$2x - 2 = B(x+1)(x-1) + C(x-1)^2;$$

divide by $x - 1$,
$$2 = B(x+1) + C(x-1).$$

Let $x = 1$,
$$2 = 2B,$$
$$B = 1,$$

then
$$-x + 1 = C(x-1),$$
$$C = -1.$$

EXAMPLES.

(1) Show that when we equate the coefficients of the same powers of x on the two sides of our identical equation, we shall always have equations enough to determine all our required numerators.

(2) Break up $\dfrac{9x^2 + 9x - 128}{(x-3)^2(x+1)}$ into simpler fractions.

59. The partial fractions corresponding to any given factor of the denominator can be determined directly.

Let us suppose that the factor in question is of the first degree and occurs but once; represent it by $x - a$.

$$\frac{fx}{Fx} = \frac{A}{x-a} + \frac{f_1 x}{\phi x}, \tag{1}$$

by Art. 57, where

$$\phi x = \frac{Fx}{x-a},$$

so that

$$Fx = (x-a)\phi x.$$

Clear (1) of fractions.

$$fx = A\phi x + (x-a) f_1 x. \tag{2}$$

As (1) is an identical equation, (2) will be true for any value of x. Let $x = a$,

$$fa = A\phi a,$$

$$A = \frac{fa}{\phi a}, \tag{3}$$

a result agreeing with Art. 57.

Hence, *to find the numerator of the fraction corresponding to a factor* (x − a) *of the first degree, we have merely to strike out from the denominator of our original fraction the factor in question, and then substitute* a *for* x *in the result.*

If the factor of the denominator is of the nth degree, there are n partial fractions corresponding to it. Let $(x-a)^n$ be the factor in question.

$$\frac{fx}{Fx} = \frac{A_1}{(x-a)^n} + \frac{A_2}{(x-a)^{n-1}} + \frac{A_3}{(x-a)^{n-2}} + \cdots\cdots + \frac{A_n}{x-a} + \frac{\psi x}{\phi x}, \tag{4}$$

where

$$Fx = (x-a)^n \phi x.$$

Multiply (4) by $(x-a)^n$, and represent $(x-a)^n \dfrac{fx}{Fx}$ by Φx.

$$\Phi x = A_1 + A_2(x-a) + A_3(x-a)^2 + \cdots\cdots + A_n(x-a)^{n-1}$$
$$+ \frac{\psi x}{\phi x}(x-a)^n.$$

Differentiate successively both members of this identity, and put $x = a$ after differentiation, and we get

$$A_1 = \Phi a,$$

$$A_2 = \Phi' a,$$

$$A_3 = \frac{1}{2\,!}\Phi'' a,$$

$$A_4 = \frac{1}{3\,!}\Phi''' a,$$

$$\ldots\ldots,$$

$$A_n = \frac{1}{(n-1)\,!}\Phi^{(n-1)} a.$$

Although these results form a complete solution of the problem, and one exceedingly neat in theory, the labor of getting the successive derivatives of Φx is so great that it is usually easier in practice to use the methods of Art. 58 when we have to deal with factors of higher degree than the first. So far as the fractions corresponding to factors of the first degree and to the highest powers of factors not of the first degree are concerned, the method of this article can be profitably combined with that of Art. 58.

60. As an example where the method of the last article applies well, consider

$$\frac{3x-1}{x(x-2)(x+1)} = \frac{A}{x} + \frac{B}{x-2} + \frac{C}{x+1},$$

$$A = \left[\frac{3x-1}{(x-2)(x+1)}\right]_{x=0} = \frac{1}{2},$$

$$B = \left[\frac{3x-1}{x(x+1)}\right]_{x=2} = \frac{5}{6},$$

$$C = \left[\frac{3x-1}{x(x-2)}\right]_{x=-1} = -\frac{4}{3},$$

$$\frac{3x-1}{x(x-2)(x+1)} = \frac{1}{2}\cdot\frac{1}{x} + \frac{5}{6}\cdot\frac{1}{x-2} - \frac{4}{3}\cdot\frac{1}{x+1}. \qquad [1]$$

Again, consider

$$\frac{1}{1+x^2}=\frac{1}{(x+\sqrt{-1})\,(x-\sqrt{-1})}=\frac{A}{x+\sqrt{-1}}+\frac{B}{x-\sqrt{-1}},$$

$$A=\left[\frac{1}{x-\sqrt{-1}}\right]_{x=-\sqrt{-1}}=-\frac{1}{2\sqrt{-1}}=\frac{\sqrt{-1}}{2},$$

$$B=\left[\frac{1}{x+\sqrt{-1}}\right]_{x=\sqrt{-1}}=\frac{1}{2\sqrt{-1}}=-\frac{\sqrt{-1}}{2},$$

$$\frac{1}{1+x^2}=\frac{\sqrt{-1}}{2}\cdot\frac{1}{x+\sqrt{-1}}-\frac{\sqrt{-1}}{2}\cdot\frac{1}{x-\sqrt{-1}}. \qquad [2]$$

61. Let us now consider a more difficult example, where it is worth while to combine our methods.

To break up $\dfrac{x^2+1}{(x-1)^4\,(x^3+1)}$.

$$x^3+1=(x+1)\,(x^2-x+1)$$

$$=(x+1)\,(x-\tfrac{1}{2}-\tfrac{1}{2}\sqrt{-3})\,(x-\tfrac{1}{2}+\tfrac{1}{2}\sqrt{-3}),$$

$$\frac{x^2+1}{(x-1)^4\,(x^3+1)}=\frac{x^2+1}{(x-1)^4\,(x+1)\,(x^2-x+1)}=\frac{A_1}{(x-1)^4}$$

$$+\frac{A_2}{(x-1)^3}+\frac{A_3}{(x-1)^2}+\frac{A_4}{x-1}+\frac{B}{x+1}+\frac{C}{x-\frac{1}{2}-\frac{1}{2}\sqrt{-3}}$$

$$+\frac{D}{x-\frac{1}{2}+\frac{1}{2}\sqrt{-3}}. \qquad (1)$$

$$B=\left[\frac{x^2+1}{(x-1)^4\,(x^2-x+1)}\right]_{x=-1}=\frac{1}{24},$$

$$A_1=\left[\frac{x^2+1}{(x+1)\,(x^2-x+1)}\right]_{x=1}=1,$$

$$C=\left[\frac{x^2+1}{(x-1)^4\,(x+1)\,(x-\frac{1}{2}+\frac{1}{2}\sqrt{-3})}\right]_{x=\frac{1}{2}+\frac{1}{2}\sqrt{-3}}=-\frac{1}{3},$$

$$D=\left[\frac{x^2+1}{(x-1)^4\,(x+1)\,(x-\frac{1}{2}-\frac{1}{2}\sqrt{-3})}\right]_{x=\frac{1}{2}-\frac{1}{2}\sqrt{-3}}=-\frac{1}{3}.$$

$$\frac{-\frac{1}{3}}{x-\frac{1}{2}-\frac{1}{2}\sqrt{-3}}+\frac{-\frac{1}{3}}{x-\frac{1}{2}+\frac{1}{2}\sqrt{-3}}=-\frac{1}{3}\cdot\frac{(2x-1)}{x^2-x+1}.$$

Substitute these values and clear (1) of fractions.

$$24(x^2+1)=24(x+1)(x^2-x+1)+24A_2(x-1)(x+1)(x^2-x+1)$$
$$+24A_3(x-1)^2(x+1)(x^2-x+1)+24A_4(x-1)^3(x+1)$$
$$(x^2-x+1)+(x-1)^4(x^2-x+1)-8(2x-1)(x-1)^4(x+1);$$

$$15x^6-51x^5+45x^4+6x^3-51x^2+45x-9=24A_2(x-1)(x+1)$$
$$(x^2-x+1)+24A_3(x-1)^2(x+1)(x^2-x+1)+24A_4$$
$$(x-1)^3(x+1)(x^2+x-1).$$

The second member of this equation is divisible by

$$(x-1)(x+1)(x^2-x+1),$$

therefore the first member must be divisible by the same quantity.

Dividing, we have

$$15x^2-36x+9=24A_2+24A_3(x-1)+24A_4(x-1)^2.$$

Let $x=1$, $$-12=24A_2,$$

$$A_2=-\frac{1}{2},$$

and we get

$$15x^2-36x+21=24A_3(x-1)+24A_4(x-1)^2.$$

Divide by $x-1$;

$$15x-21=24A_3+24A_4(x-1).$$

Let $x=1$, $$-6=24A_3,$$

$$A_3=-\frac{1}{4},$$

$$15x-15=24A_4(x-1).$$

Divide by $x-1$; $$15=24A_4,$$

$$A_4=\frac{5}{8}.$$

Hence

$$\frac{x^2+1}{(x-1)^4(x^3+1)} = \frac{1}{(x-1)^4} - \frac{1}{2}\cdot\frac{1}{(x-1)^3} - \frac{1}{4}\cdot\frac{1}{(x-1)^2} + \frac{5}{8}\cdot\frac{1}{x-1}$$

$$+\frac{1}{24}\cdot\frac{1}{x+1} - \frac{1}{3}\cdot\frac{1}{x-\frac{1}{2}-\frac{1}{2}\sqrt{-3}} - \frac{1}{3}\cdot\frac{1}{x-\frac{1}{2}+\frac{1}{2}\sqrt{-3}}. \qquad (2)$$

62. Having shown that any rational fraction can be reduced to a sum of fractions which always come under one of the two forms $\frac{A}{(x-a)^n}$ and $\frac{A}{x-a}$, it remains to show that these forms can be integrated.

To find $\int \frac{Adx}{(x-a)^n}$,

let $$z = x - a,$$

then $$dz = dx,$$

and

$$\int \frac{Adx}{(x-a)^n} = A\int \frac{dz}{z^n} = -\frac{1}{(n-1)}\cdot\frac{A}{z^{n-1}} = -\frac{1}{(n-1)}\cdot\frac{A}{(x-a)^{n-1}}. \qquad [1]$$

To find $\int \frac{Adx}{x-a}$,

let $$z = x - a,$$

then $$dz = dx,$$

and $$\int \frac{Adx}{x-a} = A\int \frac{dz}{z} = A \log z = A \log(x-a). \qquad [2]$$

Turning back to Art. 58 (2), we find

$$\int \frac{(3x-1)\,dx}{(x-1)^2(x+1)} = \int \frac{dx}{(x-1)^2} + \int \frac{dx}{x-1} - \int \frac{dx}{x+1} = -\frac{1}{x-1}$$

$$+\log(x-1) - \log(x+1) = -\frac{1}{x-1} + \log\frac{x-1}{x+1}.$$

Turning to Art. 60 (1), we have

$$\int \frac{(3x-1)\,dx}{x(x-2)(x-1)} = \tfrac{1}{2}\int \frac{dx}{x} + \tfrac{5}{6}\int \frac{dx}{x-2} - \tfrac{4}{3}\int \frac{dx}{x+1}$$

$$= \tfrac{1}{2}\log x + \tfrac{5}{6}\log(x-2) - \tfrac{4}{3}\log(x+1).$$

63. If imaginary values come in when we break up our given fraction, they will disappear if we combine our results properly after integrating.

We know (Art. 28, Ex. 2) that if the denominator of our given fraction contains an imaginary factor, $(x-a-b\sqrt{-1})^n$, it will also contain the conjugate of that factor, namely, $(x-a+b\sqrt{-1})^n$. Moreover, since by Art. 59 the numerator of the partial fraction corresponding to $(x-a-b\sqrt{-1})^n$ will be the same rational algebraic function of $a+b\sqrt{-1}$ that the numerator of the partial fraction corresponding to $(x-a+b\sqrt{-1})^n$ is of $a-b\sqrt{-1}$, these two numerators must be conjugate imaginaries by Art. 28, Ex. 3. Hence, for every fraction of the form $\dfrac{A+B\sqrt{-1}}{(x-a-b\sqrt{-1})^n}$ we shall have a second of the form $\dfrac{A-B\sqrt{-1}}{(x-a+b\sqrt{-1})^n}$.

$$\int\frac{A+B\sqrt{-1}}{(x-a-b\sqrt{-1})^n}dx=-\frac{1}{(n-1)}\cdot\frac{(A+B\sqrt{-1})}{(x-a-b\sqrt{-1})^{n-1}},$$

by Art. 62 [1].

$$\int\frac{A-B\sqrt{-1}}{(x-a+b\sqrt{-1})^n}dx=-\frac{1}{(n-1)}\cdot\frac{(A-B\sqrt{-1})}{(x-a+b\sqrt{-1})^{n-1}}.$$

Let $$(x-a+b\sqrt{-1})^{n-1}=X+Y\sqrt{-1},$$

X and Y being real functions of x;

then $$(x-a-b\sqrt{-1})^{n-1}=X-Y\sqrt{-1}.$$

$$\int\frac{A+B\sqrt{-1}}{(x-a-b\sqrt{-1})^n}dx+\int\frac{A-B\sqrt{-1}}{(x-a+b\sqrt{-1})^n}dx$$

$$=-\frac{1}{(n-1)}\cdot\frac{(A+B\sqrt{-1})}{X-Y\sqrt{-1}}-\frac{1}{(n-1)}\cdot\frac{(A-B\sqrt{-1})}{X+Y\sqrt{-1}}$$

$$=-\frac{1}{(n-1)}\cdot\frac{(2AX-2BY)}{(x^2-2ax+a^2+b^2)^{n-1}}, \qquad [1]$$

a result which is free from imaginaries.

If
$$n=1,$$
we have the pair of fractions, $\dfrac{A+B\sqrt{-1}}{x-a-b\sqrt{-1}}$ and $\dfrac{A-B\sqrt{-1}}{x-a+b\sqrt{-1}}$.

$$\int\frac{A+B\sqrt{-1}}{x-a-b\sqrt{-1}}dx=(A+B\sqrt{-1})\log(x-a-b\sqrt{-1}),$$

by Art. 62 [2],

$$\int\frac{A-B\sqrt{-1}}{x-a+b\sqrt{-1}}dx=(A-B\sqrt{-1})\log(x-a+b\sqrt{-1});$$

$$\log(x-a-b\sqrt{-1})=\tfrac{1}{2}\log[(x-a)^2+b^2]-\sqrt{-1}.\tan^{-1}\frac{b}{x-a},$$

$$\log(x-a+b\sqrt{-1})=\tfrac{1}{2}\log[(x-a)^2+b^2]+\sqrt{-1}.\tan^{-1}\frac{b}{x-a}.$$

Hence
$$\int\frac{A+B\sqrt{-1}}{x-a-b\sqrt{-1}}dx+\int\frac{A-B\sqrt{-1}}{x-a+b\sqrt{-1}}dx$$
$$=A\log[(x-a)^2+b^2]+2B\tan^{-1}\frac{b}{x-a}, \qquad [2]$$
which is real.

The form of [2] can be modified by adding a constant.

$$\frac{\pi}{2}+\tan^{-1}\frac{b}{x-a}=\frac{\pi}{2}+\text{ctn}^{-1}\frac{x-a}{b}=\frac{\pi}{2}-\text{ctn}^{-1}\frac{a-x}{b}=\tan^{-1}\frac{a-x}{b}.$$

Hence
$$A\log[(x-a)^2+b^2]+2B\tan^{-1}\frac{a-x}{b} \qquad [3]$$
differs from [2] by the constant $B\pi$, and therefore is a true value of $\displaystyle\int\frac{A+B\sqrt{-1}}{x-a-b\sqrt{-1}}dx+\int\frac{A-B\sqrt{-1}}{x-a+b\sqrt{-1}}dx.$

Turning back to Art. 61 (2) we find

$$\int\frac{x^2+1}{(x-1)^4(x^3+1)}dx=\int\frac{dx}{(x-1)^4}-\tfrac{1}{2}\int\frac{dx}{(x-1)^3}-\tfrac{1}{4}\int\frac{dx}{(x-1)^2}$$
$$+\tfrac{5}{8}\int\frac{dx}{x-1}+\tfrac{1}{24}\int\frac{dx}{x+1}-\tfrac{1}{3}\int\frac{dx}{x-\frac{1}{2}-\frac{1}{2}\sqrt{-3}}-\tfrac{1}{3}\int\frac{dx}{x-\frac{1}{2}+\frac{1}{2}\sqrt{-3}}$$
$$=-\frac{1}{3}\cdot\frac{1}{(x-1)^3}+\frac{1}{4}\cdot\frac{1}{(x-1)^2}+\frac{1}{4}\cdot\frac{1}{x-1}+\tfrac{5}{8}\log(x-1)$$
$$+\tfrac{1}{24}\log(x+1)-\tfrac{1}{3}\log(x^2-x+1).$$

EXAMPLES.

(1) $\int \frac{x^2 - 3x + 3}{(x-1)(x-2)}\,dx = x + \log\frac{x-2}{x-1}.$

(2) $\int \frac{x^2 - 1}{x^2 - 4}\,dx = x + \tfrac{3}{4}\log\frac{x-2}{x+2}.$

(3) $\int \frac{dx}{x^2 + 1} = \tan^{-1}x.$

(4) $\int \frac{dx}{x^3 - 1} = \tfrac{1}{6}\log\frac{(x-1)^2}{x^2+x+1} - \frac{1}{\sqrt{3}}\tan^{-1}\frac{2x+1}{\sqrt{3}}.$

(5) $\int \frac{dx}{a^4 - x^4} = \frac{1}{2a^3}\tan^{-1}\frac{x}{a} + \frac{1}{4a^3}\log\frac{a+x}{a-x}.$

(6) $\int \frac{dx}{(x^2+1)(x^2+x+1)} = \tfrac{1}{2}\log\frac{x^2+x+1}{x^2+1} + \frac{1}{\sqrt{3}}\tan^{-1}\frac{2x+1}{\sqrt{3}}.$

(7) $\int \frac{x^2\,dx}{x^4 + x^2 - 2} = \tfrac{1}{6}\log\frac{x-1}{x+1} + \frac{\sqrt{2}}{3}\tan^{-1}\frac{x}{\sqrt{2}}.$

(8) $\int \frac{x^2 - 1}{x^4 + x^2 + 1}\,dx = \tfrac{1}{2}\log\frac{x^2 - x + 1}{x^2 + x + 1}.$

(9) $\int \frac{dx}{(x-1)^2(x^2+1)^2} = -\frac{1}{4(x-1)} - \tfrac{1}{2}\log(x-1)$

$+ \tfrac{1}{4}\tan^{-1}x - \frac{1}{4(x^2+1)} + \tfrac{1}{4}\log(x^2+1).$

(10) $\int \frac{x^2\,dx}{x^4+1} = \frac{1}{4\sqrt{2}}\log\frac{x^2 - x\sqrt{2} + 1}{x^2 + x\sqrt{2} + 1}$

$+ \frac{1}{2\sqrt{2}}[\tan^{-1}(x\sqrt{2}+1) + \tan^{-1}(x\sqrt{2}-1)].$

(11) $\int \frac{dx}{x^4+1} = \frac{1}{4\sqrt{2}}\log\frac{x^2 + x\sqrt{2} + 1}{x^2 - x\sqrt{2} + 1} + \frac{1}{2\sqrt{2}}\tan^{-1}\left(\frac{x\sqrt{2}}{1-x^2}\right).$

CHAPTER V.

REDUCTION FORMULAS.

64. The method given in the last chapter for the integration of rational fractions is open to the practical objection that it is often exceedingly laborious. In many cases much of the labor can be saved by making the required integration depend upon the integration of a simpler form. This is usually done by the aid of what is called a *reduction formula.*

Let the function to be integrated be of the form $x^{m-1}(a+bx^n)^p$, where m, n, and p may be positive or negative. If they are integers, the function in question is either an *algebraic polynomial* or a *rational fraction;* if they are fractions, the expression is irrational. The formulas we shall obtain will apply to either case.

Denote $a + bx^n$ by z; then we want $\int x^{m-1} z^p\, dx$.

Let $$z^p = u$$

and $$x^{m-1}\,dx = dv,$$ and *integrate by parts.*

$$du = pz^{p-1}\,dz = bnpx^{n-1}z^{p-1}\,dx,$$

$$v = \frac{x^m}{m},$$

$$\int x^{m-1} z^p\,dx = \frac{x^m z^p}{m} - \frac{bnp}{m}\int x^{m+n-1} z^{p-1}\,dx. \qquad [1]$$

This formula makes our integral depend upon the integral of an expression like the given one, except that the exponent of x has been increased while that of z has been decreased.

We get from [1], by transposition,

$$\int x^{m+n-1} z^{p-1}\,dx = \frac{x^m z^p}{bnp} - \frac{m}{bnp}\int x^{m-1} z^p\,dx.$$

Change $m+n$ into m and $p-1$ into p, whence m is changed into $m-n$ and p into $p+1$, and we get

$$\int x^{m-1}z^p\,dx = \frac{x^{m-n}z^{p+1}}{bn(p+1)} - \frac{m-n}{bn(p+1)}\int x^{m-n-1}z^{p+1}\,dx, \qquad [2]$$

a formula that lowers the exponent of x while it raises that of z.

Since $$z = a + bx^n,$$

$$z^p = z^{p-1}(a+bx^n),$$

hence

$$\int x^{m-1}z^p\,dx = \int x^{m-1}z^{p-1}(a+bx^n)\,dx = a\int x^{m-1}z^{p-1}\,dx$$
$$+ b\int x^{m+n-1}z^{p-1}\,dx;$$

therefore, by [1],

$$\frac{x^m z^p}{m} - \frac{bnp}{m}\int x^{m+n-1}z^{p-1}\,dx = a\int x^{m-1}z^{p-1}\,dx + b\int x^{m+n-1}z^{p-1}\,dx,$$

$$\int x^{m-1}z^{p-1}\,dx = \frac{x^m z^p}{am} - \frac{b(m+np)}{am}\int x^{m+n-1}z^{p-1}\,dx.$$

Change p into $p+1$.

$$\int x^{m-1}z^p\,dx = \frac{x^m z^{p+1}}{am} - \frac{b(m+np+n)}{am}\int x^{m+n-1}z^p\,dx. \qquad [3]$$

Change m into $m-n$, and transpose.

$$\int x^{m-1}z^p\,dx = \frac{x^{m-n}z^{p+1}}{b(m+np)} - \frac{a(m-n)}{b(m+np)}\int x^{m-n-1}z^p\,dx. \qquad [4]$$

We have seen that

$$\int x^{m-1}z^p\,dx = a\int x^{m-1}z^{p-1}\,dx + b\int x^{m+n-1}z^{p-1}\,dx,$$

and, from [1],

$$b\int x^{m+n-1}z^{p-1}\,dx = \frac{x^m z^p}{np} - \frac{m}{np}\int x^{m-1}z^p\,dx;$$

hence

$$\int x^{m-1}z^p\,dx = a\int x^{m-1}z^{p-1}\,dx + \frac{x^m z^p}{np} - \frac{m}{np}\int x^{m-1}z^p\,dx,$$

$$\int x^{m-1}z^p\,dx = \frac{x^m z^p}{m+np} + \frac{anp}{m+np}\int x^{m-1}z^{p-1}\,dx. \qquad [5]$$

Change p into $p+1$, and transpose.

$$\int x^{m-1}z^p\,dx = -\frac{x^m z^{p+1}}{an(p+1)} + \frac{m+np+n}{an(p+1)}\int x^{m-1}z^{p+1}\,dx. \qquad [6]$$

Formula [3] enables us to raise, and formula [4] to lower, the exponent of x by n without affecting the exponent of z; while formula [5] enables us to lower, and formula [6] to raise, the exponent of z by unity without affecting the exponent of x.

Formulas [1] and [3] cannot be used when $m=0$;

formulas [2] and [6] cannot be used when $p=-1$;

formulas [4] and [5] cannot be used when $m=-np$;

for in all these cases infinite values will be brought into the second member of the formula.

65. If $n=1$, $\qquad z = a + bx,$

and our last four reduction formulas become

$$\int x^{m-1}z^p dx = \frac{x^m z^{p+1}}{am} - \frac{b(m+p+1)}{am}\int x^m z^p dx. \qquad [3]$$

$$\int x^{m-1}z^p dx = \frac{x^{m-1}z^{p+1}}{b(m+p)} - \frac{a(m-1)}{b(m+p)}\int x^{m-2}z^p dx. \qquad [4]$$

$$\int x^{m-1}z^p dx = \frac{x^m z^p}{m+p} + \frac{ap}{m+p}\int x^{m-1}z^{p-1}dx. \qquad [5]$$

$$\int x^{m-1}z^p dx = -\frac{x^m z^{p+1}}{a(p+1)} + \frac{m+p+1}{a(p+1)}\int x^{m-1}z^{p+1}dx. \qquad [6]$$

If m and p are integers, and $m>0$ and $p>0$, a repeated use of [5] will reduce p to zero, and we shall have to find merely the $\int x^{m-1}dx$.

If $m<0$ and $p>0$, [3] will enable us to raise m to 0, and then [5] will enable us to lower p to 0, and we shall need only $\int\frac{dx}{x}$.

If $m>0$ and $p<0$, [6] will raise p to -1, and [4] will then lower m to 1, and we shall need $\int\frac{dx}{z}$.

If $m<0$ and $p<0$, [6] will raise p to -1, and [3] will raise m to 0, and we shall need $\int\frac{dx}{xz}$.

$$\int x^{m-1}\,dx=\frac{x^m}{m},$$

$$\int\frac{dx}{x}=\log x,$$

$$\int\frac{dx}{z}=\int\frac{dx}{a+bx}=\frac{1}{b}\log(a+bx),$$

$$\int\frac{dx}{xz}=\int\frac{dx}{x(a+bx)}=-\frac{1}{a}\log\frac{a+bx}{x}.$$

Hence, when $n=1$, and m and p are integers, our reduction formulas always lead to the desired result.

EXAMPLES.

(1) $\displaystyle\int\frac{dx}{x^5(a+bx)}=-\frac{b^4}{a^5}\log\frac{a+bx}{x}+\frac{b^3}{a^4x}-\frac{b^2}{2a^3x^2}+\frac{b}{3a^2x^3}-\frac{1}{4ax^4}.$

(2) Consider the case where $n=2$, rewriting the reduction formulas to suit the case, and giving an exhaustive investigation.

(3) $$\int\frac{x^2\,dx}{(a+bx^2)^3}=-\frac{x}{4b(a+bx^2)^2}+\frac{x}{8ab(a+bx^2)}+\frac{1}{8(ab)^{\frac{3}{2}}}\tan^{-1}x\sqrt{\frac{b}{a}}.$$

CHAPTER VI.

IRRATIONAL FORMS.

66. We have seen that algebraic polynomials and rational fractions can always be integrated. When we come to irrational expressions, however, very few forms are integrable, and most of these have to be rationalized by ingenious substitutions.

If an algebraic function is irrational because of the presence of an expression of the first degree under the radical sign, it can be easily made rational.

Let $f(x, \sqrt[n]{a+bx})$ be the function in question.

Let $$z = \sqrt[n]{a+bx};$$

then $$z^n = a + bx.$$

$$nz^{n-1}dz = bdx,$$

$$dx = \frac{nz^{n-1}dz}{b};$$

$$x = \frac{z^n - a}{b}.$$

Hence $$\int f(x, \sqrt[n]{a+bx})\,dx = \frac{n}{b}\int f\left(\frac{z^n-a}{b}, z\right)z^{n-1}\,dz,$$

which is rational and can be treated by the methods of Chapter IV.

EXAMPLES.

(1) $\int \frac{\sqrt{x}+1}{\sqrt{x}-1}\,dx = x + 4\sqrt{x} + 4\log(\sqrt{x}-1)$.

(2) $\int \sqrt[n]{(ax+b)^m}\,dx = \frac{n\sqrt[n]{(ax+b)^{m+n}}}{a(m+n)}$.

(3) $\int [x\sqrt[n]{(x+a)} + \sqrt{(x+a)}]\,dx$

$$= \frac{n\sqrt[n]{(x+a)^{2n+1}}}{2n+1} - \frac{na\sqrt[n]{(x+a)^{n+1}}}{n+1} + \tfrac{2}{3}\sqrt{(x+a)^3}.$$

67. A case not unlike the last is $\int f(x, \sqrt[n]{c + \sqrt[m]{a + bx}})\,dx$.

Let
$$z = \sqrt[n]{c + \sqrt[m]{a + bx}};$$
$$z^n = c + \sqrt[m]{a + bx},$$
$$(z^n - c)^m = a + bx,$$
$$x = \frac{(z^n - c)^m - a}{b},$$
$$dx = \frac{mn(z^n - c)^{m-1} z^{n-1}\,dz}{b}.$$

Hence
$$\int f(x, \sqrt[n]{c + \sqrt[m]{a + bx}})\,dx$$
$$= \frac{mn}{b}\int f\left[\frac{(z^n - c)^m - a}{b}, z\right](z^n - c)^{m-1} z^{n-1}\,dz.$$

EXAMPLES.

(1) Find $\int \frac{x\,dx}{\sqrt{c + \sqrt{a + bx}}}$.

(2) Find $\int \frac{dx}{\sqrt[4]{1 + \sqrt{1 - x}}}$.

68. If the expression under the radical is of a higher degree than the first the function cannot in general be rationalized. The only important exceptional case is where the function to be integrated is irrational by reason of containing the square root of a quantity of the second degree.

Required $\int f(x, \sqrt{a + bx + cx^2})\,dx$.

First Method. Let c be positive; take out $\sqrt{c}$ as a factor, and the radical may be written $\sqrt{A + Bx + x^2}$.

Let
$$\sqrt{A + Bx + x^2} = x + z,$$
$$A + Bx + x^2 = x^2 + 2xz + z^2,$$
$$x = \frac{z^2 - A}{B - 2z},$$

$$dx = -\frac{2(z^2 - Bz + A)dz}{(B - 2z)^2},$$

$$\sqrt{A + Bx + x^2} = x + z = -\frac{z^2 - Bz + A}{B - 2z},$$

and the substitution of these values will render the given function rational.

Second Method. Let c be positive; take out $\sqrt{c}$ as a factor, and, as before, the radical may be written $\sqrt{A + Bx + x^2}$.

Let $$\sqrt{A + Bx + x^2} = \sqrt{A} + xz;$$

$$A + Bx + x^2 = A + 2\sqrt{A}\,.\,xz + x^2z^2,$$

$$x = \frac{2\sqrt{A}\,.\,z - B}{1 - z^2},$$

$$dx = \frac{2(\sqrt{A}\,.\,z^2 - Bz + \sqrt{A})dz}{(1 - z^2)^2},$$

$$\sqrt{A + Bx + x^2} = \sqrt{A} + xz = \frac{\sqrt{A}\,.\,z^2 - Bz + \sqrt{A}}{1 - z^2},$$

and the substitution of these values will render the given function rational.

If c is negative the radical can be reduced to the form $\sqrt{A + Bx - x^2}$, and the method just given will present no difficulty.

Third Method. Let c be positive; the radical will reduce to $\sqrt{A + Bx + x^2}$. Resolve the quantity under the radical into the product of two binomial factors $(x - \alpha)(x - \beta)$, α and β being the roots of the equation $A + Bx + x^2 = 0$.

Let $$\sqrt{(x - \alpha)(x - \beta)} = (x - \alpha)z;$$

$$(x - \alpha)(x - \beta) = (x - \alpha)^2z^2,$$

$$x = \frac{\beta - \alpha z^2}{1 - z^2},$$

$$dx = \frac{2z(\beta - \alpha)dz}{(1 - z^2)^2},$$

$$\sqrt{(x - \alpha)(x - \beta)} = (x - \alpha)z = \frac{(\beta - \alpha)z}{1 - z^2},$$

and the substitution of these values will make the given function rational.

If c is negative the radical will reduce to $\sqrt{A + Bx - x^2}$, and may be written $\sqrt{(\alpha - x)(x - \beta)}$ where α and β are the roots of $x^2 - Bx - A = 0$, and the method just explained will apply.

In general, that one of the three methods is preferable which will avoid introducing imaginary constants; the first, if $c > 0$; the second, if $c < 0$ and $\frac{a}{-c} > 0$; the third, if $c < 0$ and $\frac{a}{-c} < 0$. If the roots α and β are imaginary, and $A = \frac{a}{-c}$ is negative, it will be impossible to avoid imaginaries, for in that case $A + Bx - x^2$ will be negative for all real values of x.

69. Let us compare the working of the three methods just given by applying them in turn to the example $\int \frac{dx}{\sqrt{2 + 3x + x^2}}$.

1st. Let $\sqrt{2 + 3x + x^2} = x + z;$

$$\int \frac{dx}{\sqrt{2 + 3x + x^2}} = \int \frac{2(z^2 - 3z + 2)\,dz}{(3 - 2z)^2} \cdot \frac{3 - 2z}{z^2 - 3z + 2} = \int \frac{2\,dz}{3 - 2z}$$

$$= -\log(3 - 2z),$$

$$\int \frac{dx}{\sqrt{2 + 3x + x^2}} = -\log(3 + 2x - 2\sqrt{2 + 3x + x^2})$$

$$= \log \frac{1}{3 + 2x - 2\sqrt{2 + 3x + x^2}}$$

$$= \log \frac{3 + 2x + 2\sqrt{2 + 3x + x^2}}{9 + 12x + 4x^2 - 8 - 12x - 4x^2}$$

$$= \log [3 + 2x + 2\sqrt{2 + 3x + x^2}]. \qquad (1)$$

2d. Let $\sqrt{2 + 3x + x^2} = \sqrt{2} + xz;$

$$\int \frac{dx}{\sqrt{2 + 3x + x^2}} = 2\int \frac{(\sqrt{2}.z^2 - 3z + \sqrt{2})\,dz}{(1 - z^2)^2} \cdot \frac{1 - z^2}{\sqrt{2}.z^2 - 3z + \sqrt{2}}$$

$$= 2\int \frac{dz}{1 - z^2} = \log \frac{1 + z}{1 - z}. \qquad \text{(Art. 53)}$$

$$\int\frac{dx}{\sqrt{2+3x+x^2}}=\log\frac{x-\sqrt{2}+\sqrt{2+3x+x^2}}{x+\sqrt{2}-\sqrt{2+3x+x^2}}$$

$$=\log\frac{x^2+2x\sqrt{2+3x+x^2}+2+3x+x^2-2}{x^2+2\sqrt{2}.x+2-2-3x-x^2}$$

$$=\log\frac{3+2x+2\sqrt{2+3x+x^2}}{2\sqrt{2}-3}$$

$$=\log(3+2x+2\sqrt{2+3x+x^2})-\log(2\sqrt{2}-3),$$

or, dropping the constant $\log(2\sqrt{2}-3)$,

$$\int\frac{dx}{\sqrt{2+3x+x^2}}=\log(3+2x+2\sqrt{2+3x+x^2}). \qquad (2)$$

3d. Let $\sqrt{2+3x+x^2}=\sqrt{(x+1)(x+2)}=(x+1)z$;

$$\int\frac{dx}{\sqrt{2+3x+x^2}}=2\int\frac{-zdz}{(1-z^2)^2}\,\frac{1-z^2}{-z}=2\int\frac{dz}{1-z^2}=\log\frac{1+z}{1-z}.$$

$$\int\frac{dx}{\sqrt{2+3x+x^2}}=\log\frac{1+\sqrt{\dfrac{x+2}{x+1}}}{1-\sqrt{\dfrac{x+2}{x+1}}}-\log\frac{\sqrt{x+1}+\sqrt{x+2}}{\sqrt{x+1}-\sqrt{x+2}}$$

$$=\log\frac{x+1+2\sqrt{2+3x+x^2}+x+2}{x+1-x-2}$$

$$=\log(3+2x+2\sqrt{2+3x+x^2})+\log(-1),$$

or, dropping the imaginary constant $\log(-1)$,

$$\int\frac{dx}{\sqrt{2+3x+x^2}}=\log(3+2x+2\sqrt{2+3x+x^2}). \qquad (3)$$

EXAMPLES.

(1) $$\int\frac{dx}{(2+3x)\sqrt{4-x^2}}=\frac{1}{4\sqrt{2}}\log\frac{\sqrt{4+2x}-\sqrt{2-x}}{\sqrt{4+2x}+\sqrt{2-x}}.$$

(2) $$\int\frac{dx}{\sqrt{x^2+x}}=\log(\tfrac{1}{2}+x+\sqrt{x^2+x}).$$

(3) $$\int\frac{dx}{\sqrt{a+bx+cx^2}}=\frac{1}{\sqrt{c}}\log\left(\frac{b}{2\sqrt{c}}+x\sqrt{c}+\sqrt{a+bx+cx^2}\right).$$

70. If the function is irrational through the presence, under the radical sign, of a fraction whose numerator and denominator are of the first degree, it can always be rationalized.

Required $\int f\left(x, \sqrt[n]{\frac{ax+b}{lx+m}}\right) dx.$

Let
$$z = \sqrt[n]{\frac{ax+b}{lx+m}},$$

$$z^n = \frac{ax+b}{lx+m},$$

$$x = \frac{b-mz^n}{lz^n-a},$$

$$dx = \frac{n(am-bl)z^{n-1}dz}{(lz^n-a)^2},$$

and the substitution of these values will make the given function rational.

EXAMPLE.

$$\int \frac{dx}{(1+x)^2}\sqrt[3]{\frac{1-x}{1+x}} = -\tfrac{3}{8}\sqrt[3]{\left(\frac{1-x}{1+x}\right)^4}.$$

71. If the function to be integrated is of the form $x^{m-1}(a+bx^n)^p$, m, n, and p being any numbers positive or negative, and one at least of them being fractional, the reduction formulas of Art. 64 will often lead to the desired integral.

EXAMPLES.

(1) $\int \frac{x^4\,dx}{(1-x^2)^{\frac{1}{2}}} = \tfrac{3}{8}\sin^{-1}x - \frac{x\sqrt{1-x^2}}{8}(3+2x^2).$

(2) $\int \frac{dx}{x^3\sqrt{1-x^2}} = \tfrac{1}{2}\log\frac{1-\sqrt{1-x^2}}{x} - \frac{\sqrt{1-x^2}}{2x^2}.$

(3) $\int \frac{x^2\,dx}{(2ax-x^2)^{\frac{1}{2}}} = -(2ax-x^2)^{\frac{1}{2}}\left(\frac{x}{2}+\frac{3a}{2}\right)+3a^2\sin^{-1}\sqrt{\frac{x}{2a}}.$

(4) $\int \frac{x^3\,dx}{(a^2+x^2)^{\frac{5}{2}}} = -\frac{(2a^2+3x^2)}{3(a^2+x^2)^{\frac{3}{2}}}.$

72. We have said that when an irrational function contains a quantity of a higher degree than the second, under the square-root sign, it cannot *ordinarily* be integrated. It would be more correct to say that its integral cannot ordinarily be finitely expressed in terms of the functions with which we are familiar.

The integrals of a large class of such irrational expressions have been specially studied under the name of Elliptic Integrals. They have peculiar properties, and can be expressed in terms of ordinary functions only by the aid of infinite series.

CHAPTER VII.

TRANSCENDENTAL FUNCTIONS.

73. In dealing with the integration of transcendental functions the method of *integration by parts* is generally the most effective.

For example. Required $\int x(\log x)^2 dx$.

Let
$$u = (\log x)^2,$$
$$dv = x.dx;$$
$$du = \frac{2\log x.dx}{x},$$
$$v = \frac{x^2}{2},$$
$$\int x(\log x)^2 = \frac{x^2(\log x)^2}{2} - \int x\log x.dx = \frac{x^2}{2}[(\log x)^2 - \log x + \tfrac{1}{2}].$$

Again. Required $\int e^x \sin x.dx$.
$$u = \sin x,$$
$$dv = e^x\, dx;$$
$$du = \cos x.dx,$$
$$v = e^x,$$
$$\int e^x \sin x.dx = e^x \sin x - \int e^x \cos x.dx,$$
$$\int e^x \cos x.dx = e^x \cos x + \int e^x \sin x.dx;$$

whence
$$\int e^x \sin x.dx = \frac{e^x(\sin x - \cos x)}{2},$$

and
$$\int e^x \cos x.dx = \frac{e^x(\sin x + \cos x)}{2}.$$

EXAMPLES.

(1) $\int x^m (\log x)^3 dx = \frac{x^{m+1}}{m+1}\left[(\log x)^3 - \frac{3(\log x)^2}{m+1} + \frac{6\log x}{(m+1)^2} - \frac{6}{(m+1)^3}\right].$

(2) $\int \frac{\log x . dx}{(1-x)^2} = \frac{x \log x}{1-x} + \log(1-x).$

(3) $\int e^{ax}\sqrt{(1-e^{2ax})}.dx = \frac{1}{2a}\left[e^{ax}\sqrt{(1-e^{2ax})} + \sin^{-1} e^{ax}\right].$

74. The method of integration by parts gives us important reduction formulas for transcendental functions. Let us consider $\int \sin^n x.dx.$

$$u = \sin^{n-1}x,$$

$$dv = \sin x.dx;$$

$$du = (n-1)\sin^{n-2}x \cos x.dx,$$

$$v = -\cos x;$$

$$\int \sin^n x.dx = -\sin^{n-1}x \cos x + (n-1)\int \sin^{n-2}x \cos^2 x.dx$$

$$= -\sin^{n-1}x \cos x + (n-1)\int (\sin^{n-2}x - \sin^n x)dx;$$

$$\int \sin^n x.dx = -\frac{1}{n}\sin^{n-1}x \cos x + \frac{n-1}{n}\int \sin^{n-2}x.dx. \quad [1]$$

Transposing, and changing n into $n+2$, we get

$$\int \sin^n x.dx = \frac{1}{n+1}\sin^{n+1}x \cos x + \frac{n+2}{n+1}\int \sin^{n+2}x.dx. \quad [2]$$

In like manner we get

$$\int \cos^n x.dx = \frac{1}{n}\sin x \cos^{n-1}x + \frac{n-1}{n}\int \cos^{n-2}x.dx, \quad [3]$$

$$\int \cos^n x.dx = -\frac{1}{n+1}\sin x \cos^{n+1}x + \frac{n+2}{n+1}\int \cos^{n+2}x.dx. \quad [4]$$

If n is a positive integer, formulas [1] and [3] will enable us to reduce the exponent of the sine or cosine to one or to zero, and then we can integrate by inspection. If n is a negative integer, formulas [2] and [4] will enable us to raise the exponent to zero or to minus one. In the latter case we shall need $\int\frac{dx}{\cos x}$, or $\int\frac{dx}{\sin x}$, which have been found in Art. 54 (c).

EXAMPLES.

(1) $\int \sin^4 x.dx = -\frac{\sin x\cos x}{4}\left(\sin^2 x + \frac{3}{2}\right) + \frac{3}{8}x.$

(2) $\int \cos^6 x.dx = \frac{\sin x\cos^3 x}{6}\left(\cos^2 x + \frac{5}{4}\right) + \frac{5}{16}(\sin x\cos x + x).$

(3) $\int \frac{dx}{\sin^3 x} = -\frac{\cos x}{2\sin^2 x} + \frac{1}{2}\log\tan\frac{x}{2}.$

(4) Obtain the formulas

$$\int \sinh^n x.dx = \frac{1}{n}\sinh^{n-1}x\cosh x - \frac{n-1}{n}\int\sinh^{n-2}x.dx.$$

$$\int \sinh^n x.dx = \frac{1}{n+1}\sinh^{n+1}x\cosh x - \frac{n+2}{n+1}\int\sinh^{n+2}x.dx.$$

$$\int \cosh^n x.dx = \frac{1}{n}\sinh x\cosh^{n-1}x + \frac{n-1}{n}\int\cosh^{n-2}x.dx.$$

$$\int \cosh^n x.dx = -\frac{1}{n+1}\sinh x\cosh^{n+1}x + \frac{n+2}{n+1}\int\cosh^{n+2}x.dx.$$

(5) $\int \frac{dx}{\sinh^3 x} = -\frac{1}{2}\frac{\cosh x}{\sinh^2 x} - \frac{1}{4}\log\frac{\cosh x - 1}{\cosh x + 1}.$

75. The $(\sin^{-1}x)^n dx$ can be integrated by the aid of a reduction formula.

Let $$z = \sin^{-1}x;$$

then $$x = \sin z,$$

$$dx = \cos z.dz,$$

and $$\int(\sin^{-1}x)^n dx = \int z^n \cos z.dz.$$

Let
$$u = z^n,$$
$$dv = \cos z.dz;$$
$$du = nz^{n-1}dz,$$
$$v = \sin z;$$
$$\int z^n \cos z.dz = z^n \sin z - n\int z^{n-1}\sin z.dz.$$

$\int z^{n-1}\sin z.dz$ can be reduced in the same way, and is equal to $-z^{n-1}\cos z + (n-1)\int z^{n-2}\cos z.dz$;

hence

$$\int z^n \cos z.dz = z^n \sin z + nz^{n-1}\cos z - n(n-1)\int z^{n-2}\cos z.dz, \quad [1]$$

or
$$\int (\sin^{-1}x)^n dx = x(\sin^{-1}x)^n + n\sqrt{1-x^2}(\sin^{-1}x)^{n-1}$$
$$- n(n-1)\int (\sin^{-1}x)^{n-2}dx. \quad [2]$$

If n is a positive integer, this will enable us to make our required integral depend upon $\int dx$ or $\int \sin^{-1}x.dx$, the latter of which forms has been found in (I. Art. 81).

EXAMPLES.

(1) Obtain a formula for $\int (\text{vers}^{-1}x)^n dx$.

(2) $\int (\sin^{-1}x)^4 dx = x[(\sin^{-1}x)^4 - 4.3.(\sin^{-1}x)^2 + 4.3.2.1]$
$+ 4\sqrt{1-x^2}\sin^{-1}x[(\sin^{-1}x)^2 - 3.2]$.

76. Integration by substitution is sometimes a valuable method in dealing with transcendental forms, and in the case of the trigonometric functions often enables us to reduce the given form to an algebraic one. Let it be required to find $\int (f\sin x)\cos x.dx$.

Let
$$z = \sin x,$$
$$dz = \cos x.dx;$$
$$\int (f\sin x)\cos x.dx = \int fz.dz.$$

In the same way we see that

$$\int (f\cos x)\sin x.dx = -\int fz.dz \qquad \text{if } z=\cos x,$$

and

$$\int [f(\sin x,\ \cos x)]\cos x.dx = \int [f(z,\ \sqrt{1-z^2})]\,dz \ \text{ if } z=\sin x,$$

$$\int [f(\cos x,\ \sin x)]\sin x.dx = -\int [f(z,\ \sqrt{1-z^2})]\,dz \ \text{ if } z=\cos x,$$

or, more generally,

$$\int f(\sin x,\ \cos x)\,dx = \int f(z,\ \sqrt{1-z^2})\frac{dz}{\sqrt{1-z^2}} \qquad \text{if } z=\sin x,$$

$$\int f(\cos x,\ \sin x)\,dx = -\int f(z,\ \sqrt{1-z^2})\frac{dz}{\sqrt{1-z^2}} \qquad \text{if } z=\cos x.$$

Since any trigonometric function of x may be expressed in terms of $\sin x$ and $\cos x$, the formulas just given enable us to make the integration of any trigonometric function depend on the integration of an algebraic function, which, however, is frequently complicated by the presence of the radical $\sqrt{1-z^2}$.

77. A better substitution than that of the last article, when the form to be treated does not contain $\sin x$ or $\cos x$ as a factor, is $z=\tan\frac{x}{2}$.

This gives us

$$dx = \frac{2\,dz}{1+z^2},$$

$$\sin x = \frac{2z}{1+z^2},$$

$$\cos x = \frac{1-z^2}{1+z^2};$$

whence

$$\int f(\sin x,\ \cos x)\,dx = 2\int f\left(\frac{2z}{1+z^2},\ \frac{1-z^2}{1+z^2}\right)\frac{dz}{1+z^2}. \qquad [1]$$

As an example, let us find $\int \frac{dx}{a+b\cos x}$.

Here we have

$$\int\frac{dx}{a+b\cos x}=2\int\frac{dz}{(1+z^2)\left[a+b\dfrac{1-z^2}{1+z^2}\right]}=2\int\frac{dz}{a+b+(a-b)z^2}$$

$$=\frac{2}{a-b}\int\frac{dz}{\dfrac{a+b}{a-b}+z^2}=\frac{2}{\sqrt{a^2-b^2}}\tan^{-1}\left(\sqrt{\frac{a-b}{a+b}}\cdot z\right)$$

by I. Art. 77, Ex. 1.

Hence $\displaystyle\int\frac{dx}{a+b\cos x}=\frac{2}{\sqrt{a^2-b^2}}\tan^{-1}\left(\sqrt{\frac{a-b}{a+b}}\cdot\tan\frac{x}{2}\right)$, if $a>b$.

78. $\int\sin^m x\cos^n x.dx$ can be readily found by the method of Art. 76 if m and n are positive integers, and if either of them is odd. Let n be odd, then

$$\cos^n x=\cos^{n-1}x\cos x=(1-\sin^2 x)^{\frac{n-1}{2}}\cos x,$$

$$\int\sin^m x\cos^n x.dx=\int\sin^m x(1-\sin^2 x)^{\frac{n-1}{2}}\cos x.dx.$$

Let
$$z=\sin x,$$
$$dz=\cos x.dx,$$

$$\int\sin^m x\cos^n x.dx=\int z^m(1-z^2)^{\frac{n-1}{2}}dz,$$

which can be expanded into an algebraic polynomial and integrated directly.

If m and n are positive integers, and are both even,

$$\int\sin^m x\cos^n x.dx=\int\sin^m x(1-\sin^2 x)^{\frac{n}{2}}dx.$$

$\sin^m x(1-\sin^2 x)^{\frac{n}{2}}$ can be expanded and thus integrated by Art. 74 [1].

If m or n is negative, and odd, we can write

$$\cos^n x=\cos^{n-1}x\cos x,\quad\text{or}\quad\sin^m x=\sin^{m-1}x\sin x,$$

and reduce the function to be integrated to a rational fraction by the substitution of

$$z=\cos x,\quad\text{or}\quad z=\sin x.$$

$\int\sin^m x\cos^n x.dx$ can also be treated by the aid of reduction formulas easily obtained.

79. $\int \tan^n x dx$ and $\int \frac{dx}{\tan^n x}$ can be handled by the methods of Art. 78, but they can be simplified greatly by a reduction formula.

We have

$$\int \tan^n x.dx = \int \tan^{n-2} x \tan^2 x.dx = \int \tan^{n-2} x (\sec^2 x - 1)\, dx$$

$$= \int \tan^{n-2} x\, d(\tan x) - \int \tan^{n-2} x.dx,$$

whence $$\int \tan^n x.dx = \frac{\tan^{n-1} x}{n-1} - \int \tan^{n-2} x.dx; \qquad [1]$$

and $$\int \frac{dx}{\tan^n x} = \int \frac{\sec^2 x - \tan^2 x}{\tan^n x} dx = \int \frac{d(\tan x)}{\tan^n x} - \int \frac{dx}{\tan^{n-2} x},$$

whence $$\int \frac{dx}{\tan^n x} = -\frac{1}{(n-1)\tan^{n-1} x} - \int \frac{dx}{\tan^{n-2} x}. \qquad [2]$$

EXAMPLES.

(1) $\int \sin^3 x \cos^7 x.dx = \frac{\cos^{10} x}{10} - \frac{\cos^8 x}{8}.$

(2) $\int \cos^3 x \sqrt{\sin x}.dx = \frac{2 \sin^{\frac{3}{2}} x}{3} - \frac{2 \sin^{\frac{7}{2}} x}{7}.$

(3) $\int \frac{\sin^3 x.dx}{\sqrt{\cos x}} = \frac{2 \cos^{\frac{5}{2}} x}{5} - 2 \cos^{\frac{1}{2}} x.$

(4) $\int \cos^2 x \sin^4 x.dx = \frac{\sin x \cos x}{2}\left(\frac{\sin^4 x}{3} - \frac{\sin^2 x}{12} - \frac{1}{8}\right) + \frac{x}{16}.$

(5) $\int \frac{dx}{\sin x \cos^2 x} = \sec x + \log \tan \frac{x}{2}.$

(6) $\int \frac{dx}{\sin^3 x \cos^2 x} = \sec x - \frac{\cos x}{2 \sin^2 x} + \frac{3}{2} \log \tan \frac{x}{2}.$

(7) $\int \frac{dx}{\tan^5 x} = -\frac{1}{4 \tan^4 x} + \frac{1}{2 \tan^2 x} + \log \sin x.$

$$(8) \int \frac{dx}{a + b \cos x} = \frac{1}{\sqrt{b^2 - a^2}} \log \frac{\sqrt{b+a} + \sqrt{b-a} \, . \tan\frac{x}{2}}{\sqrt{b+a} - \sqrt{b-a} \, . \tan\frac{x}{2}}.$$

$$(9) \int \frac{dx}{5 + 4 \sin x} = \frac{2}{3} \tan^{-1}\left(\frac{4 + 5 \tan\frac{x}{2}}{3}\right).$$

$$(10) \int \frac{dx}{3 \sin x + \sin 2x} = \frac{1}{5} \log \sin\frac{x}{2} - \log \cos\frac{x}{2} + \frac{2}{5} \log(3 + 2 \cos x).$$

$$(11) \int \frac{\cos x dx}{(5 + 4 \cos x)^2} = \frac{5}{9} \frac{\sin x}{5 + 4 \cos x} - \frac{8}{27} \tan^{-1}\left(\frac{1}{3} \tan\frac{x}{2}\right).$$

$$(12) \int \frac{dx}{a + b \sin x + c \cos x} = \frac{2}{\sqrt{a^2 - b^2 - c^2}} \tan^{-1}\left[\frac{(a-c) \tan\frac{x}{2} + b}{\sqrt{a^2 - b^2 - c^2}}\right].$$

CHAPTER VIII.

DEFINITE INTEGRALS.

80. In I. Art. 183, a definite integral has been defined as the limit of a sum of infinitesimal terms, and has been proved equal to the difference between two values of an ordinary integral.

We are now ready to put our definition into more precise, and at the same time more general, form.

If fx is finite, continuous, and single-valued between the values $x = a$ and $x = b$, and we form the sum

$$(x_1 - a)fa + (x_2 - x_1)fx_1 + (x_3 - x_2)fx_2 + \cdots + (x_{n-1} - x_{n-2})fx_{n-2} + (b - x_{n-1})fx_{n-1},$$

where $x_1, x_2, x_3 \cdots x_{n-1}$ are $n-1$ successive values of x lying between a and b, the limit approached by this sum as n is indefinitely increased, while at the same time each of the increments $(x_1 - a)$, $(x_2 - x_1)$, etc., is made to approach zero, is the definite integral of fx from a to b, and will be denoted by $\int_a^b fx.dx$.

If we construct the curve $y = fx$ in rectangular co-ordinates, this definition clearly requires us to break up the projection on the axis of X of the portion of the curve between the points A and B into n intervals, to multiply each interval by the ordinate at its beginning, and to take the limit of the sum of these products as each interval is indefinitely decreased; that is, the limit of the sum of the small rectangles in the figure, and this is easily proved to be the area ABA_1B_1.

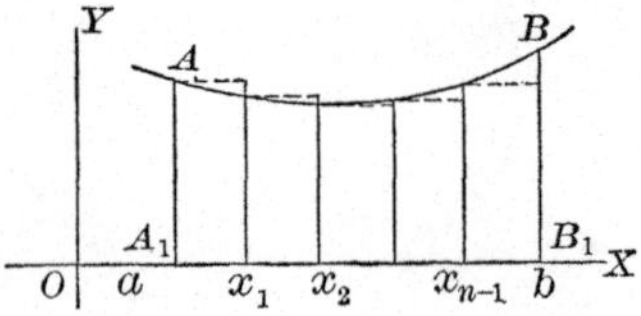

Now the area ABA_1B_1, found by the method of I. Chap. V., is

$$\left[\int fx.dx\right]_{x=b} - \left[\int fx.dx\right]_{x=a}.$$

Therefore $$\int_a^b fx.dx = \left[\int fx.dx\right]_{x=b} - \left[\int fx.dx\right]_{x=a}. \qquad [1]$$

That is, $\int_a^b fx.dx$ is the increment produced in $\int fx.dx$ by changing x from a to b.

It is to be noted that the successive increments $(x_1 - a)$, $(x_2 - x_1)$, $(x_3 - x_2)$, etc., that is, the successive values of dx, are not necessarily equal; and also, that if we multiply each interval, not by the ordinate at its beginning, but by an ordinate erected at any point of its length, the limit of our sum will be unaltered. (*v.* I. Arts. 161, 149.)

81. It is instructive to find a few definite integrals by actually performing the summation suggested in the definition (Art. 80), and then finding the limit of the sum.

(*a*) $$\int_a^b x.dx.$$

Let us divide the interval from a to b into n equal parts, and call each of them dx.

Then $$ndx = b - a.$$

Our sum is

$$S = adx + (a + dx)\,dx + (a + 2\,dx)\,dx + \cdots + (a + (n-1)\,dx)\,dx$$

$$= nadx + (1 + 2 + 3 + \cdots + (n-1))\,dx^2$$

$$= a(b-a) + \frac{n\,(n-1)}{2}\,dx^2,$$

since $ndx = b - a$, and the sum of the arithmetical progression

$$1 + 2 + 3 + \cdots + (n-1) = \frac{n(n-1)}{2}.$$

$$\frac{n\,(n-1)}{2}\,dx^2 = \tfrac{1}{2}\,(n^2dx^2 - ndx^2) = \frac{(b-a)^2}{2} - \frac{(b-a)dx}{2}.$$

Hence $$S = \frac{b^2 - a^2}{2} - \frac{(b-a)}{2}\,dx.$$

As we increase n indefinitely, dx approaches zero, and

$$\int_a^b x.dx = \underset{dx \doteq 0}{\text{limit}}\left[\frac{b^2-a^2}{2} - \frac{(b-a)\,dx}{2}\right] = \frac{b^2}{2} - \frac{a^2}{2}.$$

(b) $$\int_a^b e^x dx.$$

Let $$dx = \frac{b-a}{n}.$$

$$S = e^a dx + e^{a+dx} dx + e^{a+2dx} dx + \cdots + e^{a+(n-1)dx} dx$$
$$= e^a dx\,[1 + e^{dx} + e^{2dx} + e^{3dx} + \cdots + e^{(n-1)dx}];$$

but $1 + e^{dx} + e^{2dx} + \cdots + e^{(n-1)dx}$ is a geometrical progression, and its sum is

$$\frac{e^{ndx}-1}{e^{dx}-1} = \frac{e^{b-a}-1}{e^{dx}-1}.$$

Hence $$S = \frac{e^{b-a}-1}{e^{dx}-1}\cdot e^a dx = (e^b - e^a)\frac{dx}{e^{dx}-1},$$

and $$\int_a^b e^x dx = (e^b - e^a)\underset{dx \doteq 0}{\text{limit}}\left[\frac{dx}{e^{dx}-1}\right];$$

but as dx approaches zero, $\dfrac{dx}{e^{dx}-1}$ approaches the indeterminate form $\dfrac{0}{0}$; but since the *true value* of

$$\left[\frac{a}{e^a-1}\right]_{a=0} = \left[\frac{1}{e^a}\right]_{a=0} = 1,$$

$$\int_a^b e^x dx = e^b - e^a.$$

(c) $$\int_0^\pi \cos^3 x.dx.$$

Let $dx = \dfrac{\pi}{n}$, and let n be an odd number.

Then

$$S = dx + \cos^3 dx\cdot dx + \cos^3 2\,dx\cdot dx + \cdots + \cos^3(n-2)\,dx\cdot dx$$
$$+ \cos^3(n-1)\,dx\cdot dx$$
$$= dx + \cos^3 dx\cdot dx + \cos^3 2\,dx\cdot dx + \cdots + \cos^3(\pi - 2\,dx)\cdot dx$$
$$+ \cos^3(\pi - dx)\cdot dx$$
$$= dx + \cos^3 dx\cdot dx + \cos^3 2\,dx\cdot dx + \cdots - \cos^3 2\,dx\cdot dx$$
$$- \cos^3 dx\cdot dx,$$

since $\cos(\pi - \phi) = -\cos\phi.$

Hence the terms cancel in pairs, and we have left

$$S = dx$$

and $$\int_0^{\pi} \cos^3 x.dx = \underset{dx \doteq 0}{\text{limit}} \left[dx \right] = 0.$$

(*d*) $$\int_0^{\frac{\pi}{2}} \sin^2 x.dx.$$

Let $dx = \dfrac{\pi}{2n}$, and let n be an odd number.

$$S = \sin^2 0 \cdot dx + \sin^2 dx \cdot dx + \sin^2 2\,dx \cdot dx + \cdots + \sin^2 (n-2)\,dx \cdot dx + \sin^2 (n-1)\,dx \cdot dx$$

$$= \sin^2 dx \cdot dx + \sin^2 2dx \cdot dx + \cdots + \sin^2\left(\frac{\pi}{2} - 2dx\right)dx + \sin^2\left(\frac{\pi}{2} - dx\right)dx$$

$$= \sin^2 dx \cdot dx + \sin^2 2dx \cdot dx + \cdots + \cos^2 2dx \cdot dx + \cos^2 dx \cdot dx,$$

since $\sin\left(\dfrac{\pi}{2} - \phi\right) = \cos\phi.$

Then $S = dx + dx + dx \cdots = \dfrac{n-1}{2} dx,$

since $\sin^2\phi + \cos^2\phi = 1.$

Therefore $S = \dfrac{\pi}{4} - \dfrac{dx}{2},$

and $$\int_0^{\frac{\pi}{2}} \sin^2 x.dx = \frac{\pi}{4}.$$

(*e*) $$\int_a^b \frac{dx}{x}.$$

Here it is best to divide the interval between a and b into unequal parts.

Let the values $x_1, x_2, x_3 \cdots x_{n-1}$ be such as to form with a and b a geometrical progression.

For this purpose take $q = \sqrt[n]{\dfrac{b}{a}}$, so that $aq^n = b$.

Then the values in question are aq, aq^2, $aq^3 \cdots aq^{n-1}$, and the intervals are $a(q-1)$, $aq(q-1)$, $aq^2(q-1) \cdots aq^{n-1}(q-1)$, and the sum

$$S = \frac{a(q-1)}{a} + \frac{aq(q-1)}{aq} + \frac{aq^2(q-1)}{aq^2} + \cdots + \frac{aq^{n-1}(q-1)}{aq^{n-1}}$$

$$= n(q-1).$$

To prove our division legitimate we have only to show that each of our intervals, $a(q-1)$, $aq(q-1) \cdots aq^{n-1}(q-1)$, approaches the limit zero as n increases indefinitely. Since

$$q^n = \frac{b}{a}$$

the limiting value of q as n increases must be 1, as otherwise $\underset{n=\infty}{\text{limit}}\, q^n$ would not be finite.

Therefore $\underset{n=\infty}{\text{limit}}[aq^k(q-1)] = \underset{q \doteq 1}{\text{limit}}[aq^k(q-1)] = 0.$

We have then

$$\int_a^b \frac{dx}{x} = \underset{n=\infty}{\text{limit}}[S] = \underset{n=\infty}{\text{limit}}[n(q-1)] = \underset{q \doteq 1}{\text{limit}}[n(q-1)]$$

$$= \underset{q \doteq 1}{\text{limit}}\left[\frac{\log \frac{b}{a}}{\log q}(q-1)\right],$$

since $$n \log q = \log \frac{b}{a}.$$

But $$\underset{q \doteq 1}{\text{limit}}\left[\frac{\log \frac{b}{a}}{\log q}(q-1)\right] = \log \frac{b}{a} \underset{q \doteq 1}{\text{limit}}\left[\frac{q-1}{\log q}\right] = \log \frac{b}{a}.$$

For $$\left[\frac{q-1}{\log q}\right]_{q=1} = \left[\frac{1}{\frac{1}{q}}\right]_{q=1} = 1.$$

Therefore $$\int_a^b \frac{dx}{x} = \log b - \log a.$$

EXAMPLES.

(1) Prove by the methods of this article that

$$\int_b^c a^x\,dx = \frac{a^c - a^b}{\log a}.$$

(2) By the aid of the trigonometric formulas

$$\cos\theta + \cos 2\theta + \cos 3\theta + \cdots + \cos(n-1)\theta$$
$$= \tfrac{1}{2}\left[\sin n\theta \operatorname{ctn}\frac{\theta}{2} - 1 - \cos n\theta\right],$$

$$\sin\theta + \sin 2\theta + \sin 3\theta + \cdots + \sin(n-1)\theta$$
$$= \tfrac{1}{2}\left[(1 - \cos n\theta)\operatorname{ctn}\frac{\theta}{2} - \sin n\theta\right],$$

prove that $$\int_a^b \cos x . dx = \sin b - \sin a,$$

and $$\int_a^b \sin x . dx = \cos a - \cos b.$$

(3) Show that $$\int_0^{2\pi} \sin^5 x . dx = 0,$$

and that $$\int_0^{\pi} \cos^2 x . dx = \frac{\pi}{2}.$$

(4) Show that $\displaystyle\int_a^b x^m\,dx = \frac{b^{m+1} - a^{m+1}}{m+1}$, using the method of Art. 81 (*e*).

82. When the indefinite integral can be found, the definite integral $\int_a^b fx.dx$ can usually be most easily obtained by employing the formula [1] Art. 80, and this can always be done with safety when fx is finite, continuous, and single-valued between $x = a$ and $x = b$.

Of course, if the indefinite integral is a multiple-valued function, we must choose the values of the indefinite integral corresponding to $x = a$ and $x = b$, so that they may be ordinates of the same branch of the curve $y = \int fx.dx$.

Consider, for example, $\int_{-1}^{1}\frac{dx}{1+x^2}$. The indefinite integral $\int\frac{dx}{1+x^2}=\tan^{-1}x$ and $\tan^{-1}x$ is a multiple-valued function. Indeed, $y=\tan^{-1}x$ is a curve consisting of an infinite number of separate branches so related that ordinates corresponding to the same value of x differ by multiples of π. On the branch which passes through the origin, when $x=-1$, $y=\tan^{-1}x=-\frac{\pi}{4}$; on the same branch, when $x=1$, $y=\tan^{-1}x=\frac{\pi}{4}$. On the next branch above, when $x=-1$, $y=\tan^{-1}x=\frac{3\pi}{4}$; and when $x=1$, $y=\frac{5\pi}{4}$. On any branch, when $x=-1$, $y=\tan^{-1}x=-\frac{\pi}{4}+n\pi$; and on the same branch, when $x=1$, $y=\frac{\pi}{4}+n\pi$.

Hence $$\int_{-1}^{1}\frac{dx}{1+x^2}=\tan^{-1}(1)-\tan^{-1}(-1)=\frac{\pi}{4}+\frac{\pi}{4}=\frac{\pi}{2};$$

or $$\int_{-1}^{1}\frac{dx}{1+x^2}=\frac{5\pi}{4}-\frac{3\pi}{4}=\frac{\pi}{2},$$

or $$\int_{-1}^{1}\frac{dx}{1+x^2}=\frac{\pi}{4}+n\pi-\left(-\frac{\pi}{4}+n\pi\right)=\frac{\pi}{2}.$$

By $\int_a^{\infty}fx.dx$ we mean the limit approached by $\int_a^b fx.dx$ as b is indefinitely increased.

EXAMPLES.

(1) Work the examples of Art. 81 by the method of Art. 82.

(2) $\int_0^{\frac{\pi}{4}}\frac{\sin x.dx}{\cos^2 x}=\sqrt{2}-1.$

(3) $\int_0^{a}\frac{dx}{\sqrt{x+a}+\sqrt{x}}=\frac{4}{3}\sqrt{a}\,(\sqrt{2}-1).$

(4) $\int_0^{\infty}\frac{dx}{a^2+x^2}=\frac{\pi}{2a}.$

(5) $\int_0^\infty \frac{a\,dx}{a^2+x^2} = \frac{\pi}{2}$ if $a>0$, and $-\frac{\pi}{2}$ if $a<0$, and 0 if $a=0$.

(6) $\int_0^\infty e^{-ax}\,dx = \frac{1}{a}$ if $a>0$.

(7) $\int_0^\infty e^{-ax}\sin mx.dx \quad = \frac{m}{a^2+m^2}$ if $a>0$.

(8) $\int_0^\infty e^{-ax}\cos mx.dx \quad = \frac{a}{a^2+m^2}$ if $a>0$.

(9) $\int_0^1 \frac{dx}{1+2x\cos\phi+x^2} = \frac{\phi}{2\sin\phi}$.

(10) $\int_0^\infty \frac{dx}{1+2x\cos\phi+x^2} = \frac{\phi}{\sin\phi}$.

83. When fx is finite and single-valued between $x=a$ and $x=b$, but has a *finite* discontinuity at some intermediate value $x=c$

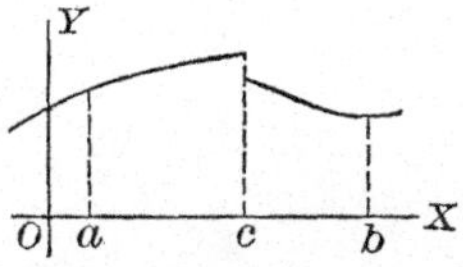

$$\int_a^b fx.dx = \int_a^c fx.dx + \int_c^b fx.dx,$$

and therefore $\int_a^b fx.dx$ can be found by Art. 82 when the indefinite integral $\int fx.dx$ can be obtained; but when fx becomes *infinite* for $x=a$, or for $x=b$, or for some intermediate value $x=c$, special care must be exercised, and some special investigation is usually required.

If fx is infinite when $x=a$ and $\int_{a+\epsilon}^b fx.dx$ approaches a finite limit as ϵ approaches zero, this limit is what we shall mean by $\int_a^b fx.dx$; if $\int_{a+\epsilon}^b fx.dx$ increases indefinitely as ϵ approaches zero, we shall say that $\int_a^b fx.dx$ is infinite; and if $\int_{a+\epsilon}^b fx.dx$ neither approaches a finite limit nor increases indefinitely as ϵ

approaches zero, we shall say that $\int_a^b fx.dx$ is indeterminate. It is in the first case only that $\int_a^b fx.dx$ can be safely employed in mathematical work.

If fx is infinite when $x = b$ and $\int_a^{b-\epsilon} fx.dx$ approaches a finite limit as ϵ approaches zero, that limit is the value of $\int_a^b fx.dx$.

If fx is infinite when $x = c$, and each of the expressions $\int_a^{c-\epsilon} fx.dx$ and $\int_{c+\epsilon}^b fx.dx$ approaches a finite limit as ϵ approaches zero, the sum of these limits is $\int_a^b fx.dx$. Should either or both of the expressions,

$$\int_a^{c-\epsilon} fx.dx, \qquad \int_{c+\epsilon}^b fx.dx,$$

fail to approach a finite limit as ϵ approaches zero, $\int_a^b fx.dx$ is either infinite or indeterminate, and cannot be safely used.

When the indefinite integral of $fx.dx$ can be obtained there is little difficulty in deciding on the nature of $\int_a^b fx.dx$ in any of the cases just considered, or in getting its value when that value is finite and determinate.

For example,

(*a*) $\int_0^1 \frac{dx}{x}$ is infinite, since

$$\int \frac{dx}{x} = \log x \text{ and } \int_\epsilon^1 \frac{dx}{x} = \log(1) - \log \epsilon = \log \frac{1}{\epsilon},$$

and increases indefinitely as ϵ approaches zero.

(*b*) $\int_0^1 \frac{dx}{1-x^2}$ is not finite and determinate, for

$$\int \frac{dx}{1-x^2} = \tfrac{1}{2} \log \frac{1+x}{1-x},$$

$$\int_0^{1-\epsilon} \frac{dx}{1-x^2} = \tfrac{1}{2} \log \left(\frac{2-\epsilon}{\epsilon} \right),$$

and increases indefinitely as ϵ approaches zero.

(c) $\int_0^a \frac{dx}{\sqrt{a^2 - x^2}}$ is finite and determinate, for

$$\int \frac{dx}{\sqrt{a^2 - x^2}} = \sin^{-1}\frac{x}{a},$$

$$\int_0^{a-\epsilon} \frac{dx}{\sqrt{a^2 - x^2}} = \sin^{-1}\frac{a-\epsilon}{a} - \sin^{-1}0 = \sin^{-1}\frac{a-\epsilon}{a},$$

and its limiting value as ϵ approaches zero is $\sin^{-1}(1)$ or $\frac{\pi}{2}$.

(d) $\int_0^2 \frac{xdx}{(1-x)^{\frac{1}{3}}}$ is finite and determinate, for

$$\int \frac{xdx}{(1-x)^{\frac{1}{3}}} = \tfrac{3}{5}(1-x)^{\frac{5}{3}} - \tfrac{3}{2}(1-x)^{\frac{2}{3}},$$

$$\int_0^{1-\epsilon} \frac{xdx}{(1-x)^{\frac{1}{3}}} = \tfrac{3}{5}\epsilon^{\frac{5}{3}} - \tfrac{3}{2}\epsilon^{\frac{2}{3}} - \tfrac{3}{5} + \tfrac{3}{2},$$

and its limiting value as ϵ approaches zero is $\frac{3}{2} - \frac{3}{5}$.

$$\int_{1+\epsilon}^2 \frac{xdx}{(1-x)^{\frac{1}{3}}} = -\tfrac{3}{5} - \tfrac{3}{2} + \tfrac{3}{5}\epsilon^{\frac{5}{3}} + \tfrac{3}{2}\epsilon^{\frac{2}{3}},$$

and its limiting value as ϵ approaches zero is $-\frac{3}{5} - \frac{3}{2}$, and consequently

$$\int_0^2 \frac{xdx}{(1-x)^{\frac{1}{3}}} = \tfrac{3}{2} - \tfrac{3}{5} - \tfrac{3}{5} - \tfrac{3}{2} = -\tfrac{6}{5}.$$

84. When, as is sometimes the case, the indefinite integral cannot be obtained, and the function to be integrated becomes infinite at or between the limits of integration, it is only necessary to investigate the limiting value of $\epsilon f(a+\epsilon)$ as ϵ approaches zero if fx becomes infinite when $x = a$; of $\epsilon f(b-\epsilon)$ if fx becomes infinite when $x = b$; and of both $\epsilon f(c-\epsilon)$ and $\epsilon f(c+\epsilon)$ if fx becomes infinite when $x = c$. If each of the values in question has zero for its limit, $\int_a^b fx.dx$ is finite and determinate, otherwise it is infinite or indeterminate.

For, in order that the area $EE'B'B$ should not increase indefinitely as ϵ approaches zero, the rectangle $AA'E'E$, whose area is $\epsilon f(a+\epsilon)$, must approach zero as its limit; and the same reasoning holds good for the other cases considered above.

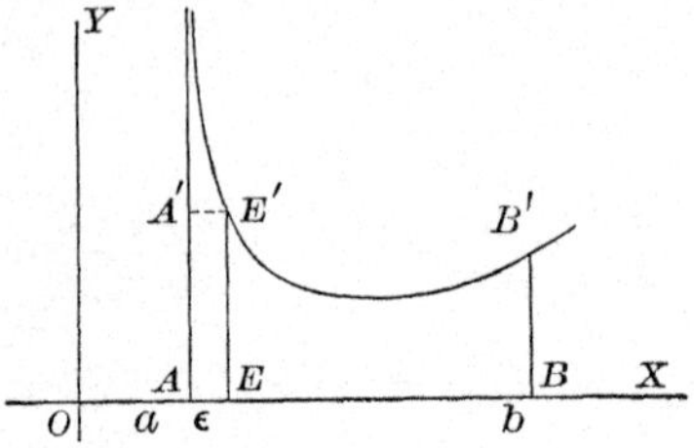

Let us apply this test to the examples considered in Art. 83.

$$(a)\quad \int_0^1 \frac{dx}{x} = \infty \quad \text{because} \quad \underset{\epsilon \doteq 0}{\text{limit}} \left[\frac{\epsilon}{\epsilon}\right] = 1.$$

$$(b)\quad \int_0^\infty \frac{dx}{1-x^2} \text{ is indeterminate, for}$$

$$\underset{\epsilon \doteq 0}{\text{limit}} \left[\frac{\epsilon}{1-(1-\epsilon)^2}\right] = \underset{\epsilon \doteq 0}{\text{limit}} \left[\frac{\epsilon}{2\epsilon-\epsilon^2}\right] = \underset{\epsilon \doteq 0}{\text{limit}} \left[\frac{1}{2-\epsilon}\right] = \tfrac{1}{2},$$

and

$$\underset{\epsilon \doteq 0}{\text{limit}} \left[\frac{\epsilon}{1-(1+\epsilon)^2}\right] = -\tfrac{1}{2}.$$

$$(c)\quad \int_0^a \frac{dx}{\sqrt{a^2-x^2}} \text{ is finite and determinate, for}$$

$$\underset{\epsilon \doteq 0}{\text{limit}} \left[\frac{\epsilon}{\sqrt{a^2-(a-\epsilon)^2}}\right] = \underset{\epsilon \doteq 0}{\text{limit}} \left[\frac{\epsilon}{\sqrt{2a\epsilon-\epsilon^2}}\right] = \underset{\epsilon \doteq 0}{\text{limit}} \left[\frac{1}{\sqrt{\frac{2a}{\epsilon}-1}}\right] = 0.$$

$$(d)\quad \int_0^2 \frac{xdx}{(1-x)^{\frac{1}{3}}} \text{ is finite and determinate, for}$$

$$\underset{\epsilon \doteq 0}{\text{limit}} \left[\frac{\epsilon(1-\epsilon)}{(1-(1-\epsilon))^{\frac{1}{3}}}\right] = \underset{\epsilon \doteq 0}{\text{limit}} \left[\epsilon^{\frac{2}{3}}(1-\epsilon)\right] = 0,$$

and

$$\underset{\epsilon \doteq 0}{\text{limit}} \left[\frac{\epsilon(1+\epsilon)}{(1-(1+\epsilon))^{\frac{1}{3}}}\right] = -\underset{\epsilon \doteq 0}{\text{limit}} \left[\epsilon^{\frac{2}{3}}(1+\epsilon)\right] = 0.$$

Even when, as in the examples just given, the indefinite integral can be obtained, there is a decided advantage in using the very simple method of this article. For if the application of the test shows that the definite integral in question is infinite or indeterminate, the labor of finding the indefinite integral is saved; and if the application of the test proves the definite integral finite and determinate, it follows that the indefinite integral does not become infinite for the value of x which makes the given function infinite, and consequently when the indefinite integral has been obtained, the method of Art. 82 can be used without hesitation.

As an example, where the indefinite integral cannot be obtained, let us consider at some length

$$\int_0^1 \left(\log\frac{1}{x}\right)^n dx.$$

If n is positive, $\left(\log\frac{1}{x}\right)^n$ is continuous and single-valued between $x=0$ and $x=1$, but becomes infinite when $x=0$. We must then investigate the limiting value of $\epsilon\left(\log\frac{1}{\epsilon}\right)^n$ as ϵ approaches zero.

$\epsilon\left(\log\frac{1}{\epsilon}\right)^n$ is indeterminate when $\epsilon=0$, but its *true value* is easily found to be zero if n is positive, whether n is whole or fractional. For positive values of n, $\int_0^1 \left(\log\frac{1}{x}\right)^n dx$ is, then, finite and determinate.

If n is negative, call $n=-m.$

Then $$\int_0^1 \left(\log\frac{1}{x}\right)^n dx = \int_0^1 \frac{dx}{\left(\log\frac{1}{x}\right)^m}.$$

$$\frac{1}{\left(\log\frac{1}{x}\right)^m}$$

is continuous and single-valued from $x=0$ to $x=1$, but becomes infinite when $x=1$.

We must, then, find $\underset{\epsilon \doteq 0}{\text{limit}}\left[\dfrac{\epsilon}{\left(\log\dfrac{1}{1-\epsilon}\right)^m}\right]$,

which proves to be $\underset{\epsilon \doteq 0}{\text{limit}}\left[\dfrac{1-\epsilon}{m\left(\log\dfrac{1}{1-\epsilon}\right)^{m-1}}\right]$:

if $m<1$, this is zero; if $m=1$, it is 1; and if $m>1$, it is infinite.

$$\int_0^1 \frac{dx}{\left(\log\frac{1}{x}\right)^m}$$

is, then, finite and determinate if $m<1$, but infinite if $m=1$ or $m>1$; and we reach the result that

$$\int_0^1 \left(\log\frac{1}{x}\right)^n dx$$

is finite and determinate if $n>-1$, but infinite if $n=-1$ or $n<-1$.

EXAMPLES.

(1) Prove that

$$\int_0^1 \frac{\log x}{1-x}\,.\,dx,\quad \int_0^1 \frac{\log x}{1-x^2}\,.\,dx,\quad \int_0^1 \frac{dx}{x}\log\left(\frac{1+x}{1-x}\right),$$

are finite and determinate.

(2) Prove that

$\displaystyle\int_0^\infty \frac{dx}{1-x^4}, \int_0^\infty \frac{x^{2m}dx}{1-x^{2n}}$, where m and n are integers, and

$\displaystyle\int_0^\infty \frac{x^{a-1}}{1-x}\,.\,dx$, are not finite and determinate.

(3) Find for what values of n $\displaystyle\int_0^1 (\log x)^n dx$ is finite and determinate.

(4) Find for what values of m and n $\int_0^1 x^m\left(\log\frac{1}{x}\right)^n dx$ is finite and determinate.

(5) Show that $\int_0^1 x^{m-1}(1-x)^{n-1}dx$ is finite and determinate if m and n are positive.

(6) Prove that $\int_0^{\frac{\pi}{2}} \log\sin x.dx$ is finite and determinate.

(7) Show that the following integrals are finite and determinate, and obtain their values:

$$\int_0^a \frac{dx}{\sqrt{a^2-x^2}} = \frac{\pi}{2}.$$

$$\int_0^a \frac{dx}{\sqrt{ax-x^2}} = \pi.$$

$$\int_1^2 \frac{dx}{x\sqrt{x^2-1}} = \frac{\pi}{3}.$$

85. It was stated in Art. 82 that by $\int_a^\infty fx.dx$ we mean the limit approached by $\int_a^b fx.dx$ as b is indefinitely increased, and, as we have seen, if the indefinite integral $\int fx.dx$ can be found, there is no difficulty in investigating the nature of $\int_a^\infty fx.dx$ and in obtaining its value if it is finite and determinate. There are, however, many exceedingly important definite integrals of the form $\int_a^\infty fx.dx$ whose values are obtained by ingenious devices without employing the indefinite integral, and these devices are valid only provided that the integral in question is finite and determinate, since an infinite value not recognized and treated as such, or a value absolutely indeterminate, renders inconclusive any piece of mathematical reasoning into which it enters.

If we construct the curve $y = fx$, $\int_a^\infty fx.dx$ is the limiting value approached by the area ABB_1A_1, as OB_1 is indefinitely increased; and in order that this area should be finite and determinate, it is clearly necessary that the area BCC_1B_1 should approach zero as its limit as OB_1 and OC_1 are indefinitely increased, however great the amount by which OC_1 exceeds OB_1.

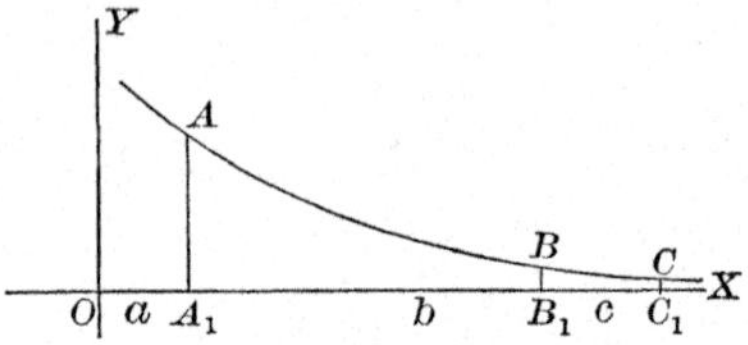

That is,
$$\lim_{\substack{b=\infty \\ c=\infty}} \left[\int_b^c fx.dx\right]$$

must be equal to zero no matter how much more rapidly c increases than b.

86. The investigation of the limiting value of $\int_b^c fx.dx$, as b and c are indefinitely increased, is usually made with the aid of the following important theorem known as the *Maximum-Minimum Theorem*.

If a given function of x *is the product of two functions, one of which* v *does not change its sign between* x = a *and* x = b, *and if* M *is algebraically the greatest and* N *the least value of the other factor* u *between* x = a *and* x = b, $\int_a^b uv.dx$ *lies between* $M\int_a^b v.dx$ *and* $N\int_a^b v.dx$.

To prove this theorem, let us first suppose that v is positive between $x = a$ and $x = b$. Now, $M - u$ is positive for the values of x considered, $(M - u)v$ is positive, and therefore

$$\int_a^b (M-u)v.dx > 0 \quad \text{and} \quad M\int_a^b v.dx > \int_a^b uv.dx. \qquad (1)$$

$u - N$ also is positive for all values of x between $x = a$ and $x = b$, $(u - N)v$ is positive, and therefore

$$\int_a^b (u - N)v.dx > 0 \quad \text{and} \quad \int_a^b uv.dx > N\int_a^b v.dx. \quad (2)$$

Hence $\int_a^b uv.dx$ lies between $M\int_a^b v.dx$ and $N\int_a^b v.dx$.

It is easy to modify this proof to meet the case where v is negative.

(*a*) As an example of the use of this theorem, we will prove $\int_0^\infty e^{-x^2}dx$ finite and determinate.

e^{-x^2} is single-valued, finite, and continuous for all positive values of x; if, then, we can show that $\int_b^c e^{-x^2}dx$ has the limit zero as b and c are indefinitely increased, c remaining greater than b, our proof is complete.

e^{-x^2} can be written $x^2e^{-x^2}\cdot\frac{1}{x^2}$, and $\frac{1}{x^2}$ never changes its sign.

As x increases $x^2e^{-x^2}$ eventually decreases, and continues to decrease toward zero as x increases indefinitely, as may be proved by determining its value for $x = \infty$.

We have $$\left[\frac{x^2}{e^{x^2}}\right]_{x=\infty} = \left[\frac{2x}{2xe^{x^2}}\right]_{x=\infty} = \left[\frac{1}{e^{x^2}}\right]_{x=\infty} = 0.$$

Hence, eventually the greatest value of $x^2e^{-x^2}$ between $x = b$ and $x = c$ is $b^2e^{-b^2}$, and the least value is $c^2e^{-c^2}$.

Therefore, by the Maximum-Minimum Theorem,

$$b^2e^{-b^2}\int_b^c \frac{dx}{x^2} > \int_b^c x^2e^{-x^2}\cdot\frac{dx}{x^2} > c^2e^{-c^2}\int_b^c \frac{dx}{x^2},$$

$$b^2e^{-b^2}\left(\frac{1}{b} - \frac{1}{c}\right) > \int_b^c e^{-x^2}dx > c^2e^{-c^2}\left(\frac{1}{b} - \frac{1}{c}\right).$$

As b and c are indefinitely increased, the first and third members of our last inequality approach the same limit, zero. Consequently the limiting value of $\int_b^c e^{-x^2}dx$ is zero, and

$$\int_0^\infty e^{-x^2}dx \text{ is finite and determinate.}$$

(*b*) Let us consider $\int_0^\infty \frac{\sin ax}{x}.dx.$

$\frac{\sin ax}{x}$ is equal to a when $x=0$, and is single-valued, finite, and continuous for all positive values of x.

By integration by parts,

$$\int \frac{\sin ax}{x}dx = -\frac{\cos ax}{ax} - \frac{1}{a}\int \frac{\cos ax}{x^2}.dx.$$

Therefore

$$\int_b^c \frac{\sin ax}{x}dx = \frac{\cos ab}{ab} - \frac{\cos ac}{ac} - \frac{1}{a}\int_b^c \frac{\cos ax}{x^2}.dx.$$

As b and c are indefinitely increased, $\frac{\cos ab}{ab}$ and $\frac{\cos ac}{ac}$ approach the limiting value zero.

$\frac{1}{x^2}$ does not change sign between $x=b$ and $x=c$, and the greatest value of $\cos ax$ is 1, and its least value is -1.

Hence $1\int_b^c \frac{dx}{x^2} > \int_b^c \frac{\cos ax}{x^2}dx > -1\int_b^c \frac{dx}{x^2},$

or $$\frac{1}{b}-\frac{1}{c} > \int_b^c \frac{\cos ax}{x^2}dx > -\left(\frac{1}{b}-\frac{1}{c}\right);$$

and the limiting value of $\int_b^c \frac{\cos ax}{x^2}dx$ as b and c increase is zero. The same thing is true of $\int_b^c \frac{\sin ax}{x}dx$; and therefore

$$\int_0^\infty \frac{\sin ax}{x}dx \text{ is finite and determinate.}$$

(c) $\int_0^\infty \cos(x^2)dx$ is finite and determinate.

For $\cos(x^2)$ is single-valued, finite, and continuous for all positive values of x; and it is not difficult to prove that $\int_b^c \cos(x^2)dx$ approaches the limit zero as b and c are indefinitely increased.

We have

$$\int_b^c \cos(x^2)\,dx = \int_b^c \frac{2x\cos(x^2)\,dx}{2x} = \left[\frac{\sin(x^2)}{2x}\right]_b^c + \tfrac{1}{2}\int_b^c \frac{\sin(x^2)\,dx}{x^2}.$$

The limiting value of $\dfrac{\sin(c^2)}{2c} - \dfrac{\sin(b^2)}{2b}$ as b and c are increased indefinitely is zero; and the limiting value of $\displaystyle\int_b^c \frac{\sin(x^2)}{x^2}dx$ can be proved zero by the method followed in (b).

EXAMPLES.

(1) Construct the curves $y = e^{-x^2}$; $y = \dfrac{\sin ax}{x}$; $y = \cos(x^2)$.

(2) Prove that the following integrals are finite and determinate:

$$\int_0^\infty \frac{\sin^2 x}{x^2}.\,dx, \qquad \int_0^\infty \frac{\sin x}{\sqrt{x}}.\,dx, \qquad \int_0^\infty \frac{e^{-ax}\sin mx}{x}.\,dx,$$

$$\int_0^\infty e^{-a^2x^2}\cos bx.\,dx, \qquad \int_0^\infty e^{-ax^2}x^{2n}.\,dx, \qquad \int_0^\infty e^{-x^2-\frac{a^2}{x^2}}.\,dx,$$

$$\int_0^\infty \log\left(\frac{e^x+1}{e^x-1}\right).\,dx.$$

(3) Show that $\displaystyle\int_0^\infty x^n e^{-x}.\,dx$ is finite and determinate for all values of n greater than -1.

87. When we have occasion to use a reduction formula in finding the value of a definite integral, it is often worth while to substitute the limits of integration in the general formula before attempting to apply it to the particular problem.

For example, let us find $\displaystyle\int_0^a \frac{x^6.\,dx}{\sqrt{a^2-x^2}}$.

We can reduce the exponent of x by [4], Art. 64,

$$\int x^{m-1}z^p\,dx = \frac{x^{m-n}z^{p+1}}{b(m+np)} - \frac{a(m-n)}{b(m+np)}\int x^{m-n-1}z^p\,dx.$$

For our example this becomes

$$\int x^{m-1}(a^2-x^2)^{-\frac{1}{2}}dx = \frac{x^{m-2}(a^2-x^2)^{\frac{1}{2}}}{-m+1} - \frac{a^2(m-2)}{-m+1}\int x^{m-3}(a^2-x^2)^{-\frac{1}{2}}dx.$$

When $x=0$, and also when $x=a$, $\dfrac{x^{m-2}(a^2-x^2)^{\frac{1}{2}}}{-m+1} = 0.$

Hence

$$\int_0^a x^{m-1}(a^2-x^2)^{-\frac{1}{2}}\,dx = \frac{a^2(m-2)}{m-1}\int_0^a x^{m-3}(a^2-x^2)^{-\frac{1}{2}}\,dx;$$

$$\int_0^a x^6(a^2-x^2)^{-\frac{1}{2}}dx = \frac{5}{6}\cdot a^2\int_0^a x^4(a^2-x^2)^{-\frac{1}{2}}dx$$

$$= \frac{5}{6}\cdot\frac{3}{4}\cdot a^4\int_0^a x^2(a^2-x^2)^{-\frac{1}{2}}dx$$

$$= \frac{5}{6}\cdot\frac{3}{4}\cdot\frac{1}{2}\cdot a^6\int_0^a \frac{dx}{\sqrt{a^2-x^2}}.$$

Therefore $\displaystyle\int_0^a \frac{x^6\,dx}{\sqrt{a^2-x^2}} = \frac{1}{2}\cdot\frac{3}{4}\cdot\frac{5}{6}\cdot\frac{\pi a^6}{2}.$

EXAMPLES.

(1) $\displaystyle\int_0^a \frac{x^5\,dx}{\sqrt{a^2-x^2}} = \frac{2}{3}\cdot\frac{4}{5}\,a^5.$

(2) $\displaystyle\int_0^a \sqrt{a^2-x^2}\cdot dx = \frac{\pi a^2}{4}.$

(3) $\displaystyle\int_0^a x^2\sqrt{a^2-x^2}\cdot dx = \frac{1}{4}\cdot\frac{\pi a^4}{4}.$

(4) $\displaystyle\int_0^a x^2(a^2-x^2)^{\frac{3}{2}}\cdot dx = \frac{1}{6}\cdot\frac{3}{16}\cdot\pi a^6.$

(5) $\displaystyle\int_0^{\frac{\pi}{2}} \sin^n x.dx = \frac{1.3.5\ldots(n-1)}{2.4.6\ldots n}\cdot\frac{\pi}{2}$ when n is even,

$\displaystyle = \frac{2.4.6\ldots(n-1)}{3.5.7\ldots n}$ when n is odd.

(6) Show that $\int_0^{\frac{\pi}{2}} \cos^n x.dx = \int_0^{\frac{\pi}{2}} \sin^n x.dx.$

(7) $\int_0^1 \frac{x^{2n}\,dx}{\sqrt{1-x^2}} = \frac{1.3.5\ldots(2n-1)}{2.4.6\ldots 2n}\cdot\frac{\pi}{2}.$

Suggestion: let $x = \sin\theta$.

(8) $\int_0^1 \frac{x^{2n+1}\,dx}{\sqrt{1-x^2}} = \frac{2.4.6\ldots 2n}{3.5.7\ldots(2n+1)}.$

(9) From Exs. 7 and 8 obtain Wallis's formula

$$\frac{\pi}{2} = \frac{2.2.4.4.6.6.8.8\ldots}{1.3.3.5.5.7.7.9\ldots}.$$

Suggestion: $\int_0^1 \frac{x^{2n-1}\,dx}{\sqrt{1-x^2}} > \int_0^1 \frac{x^{2n}\,dx}{\sqrt{1-x^2}} > \int_0^1 \frac{x^{2n+1}\,dx}{\sqrt{1-x^2}}.$

88. When in finding $\int_a^b fx.dx$ the method of *integration by substitution* is used, and $y = Fx$ is introduced in place of x, we can regard the new integral as a definite integral, the limits of integration being Fa and Fb, and thus avoid the labor of replacing y by its value in terms of x in the result of the indefinite integration.

Let us find $$\int_{-\infty}^0 e^{ax}\sqrt{1-e^{2ax}}\,.\,dx.$$

Substitute $$y = e^{ax}.$$

$$dy = ae^{ax}\,dx.$$

Hence, $$\int e^{ax}\sqrt{1-e^{2ax}}\,.\,dx = \frac{1}{a}\int \sqrt{1-y^2}\,.\,dy.$$

When $x = -\infty$, $y = 0$, and when $x = 0$, $y = 1$.

Therefore $$\int_{-\infty}^0 e^{ax}\sqrt{1-e^{2ax}}\,.\,dx = \frac{1}{a}\int_0^1 \sqrt{1-y^2}\,.\,dy = \frac{\pi}{4a}.$$

There is one rather rare class of cases where special care is needed in using the method just described. It is when y has a maximum or a minimum value between $x = a$ and $x = b$, say for $x = c$, and x is a multiple-valued function of y.

For suppose y a maximum when $x=c$, then as x increases from a to b, y increases to the value Fc, and then decreases to the value Fb, instead of simply increasing or decreasing from Fa to Fb. If x is a single-valued function of y, and $\phi(y)dy$ is the result of substituting y for x in $fx.dx$, ϕy is a single-valued function of y, and $\int_{Fa}^{Fb}\phi y.dy=\int_{Fa}^{Fc}\phi y.dy+\int_{Fc}^{Fb}\phi y.dy$, and there is no error in using $\int_{Fa}^{Fb}\phi y.dy$ for $\int_a^b fx.dx$. But if x is a multiple-valued function of y, it will always happen that when y passes through a maximum, we pass from one set of values of x to another, and therefore from one set of values of ϕy to another, and in that case it is necessary to express our required integral as $\int_{Fa}^{Fc}\phi y.dy+\int_{Fc}^{Fb}\phi y.dy$, taking pains to select the correct set of values for ϕy in each integral.

If y is a minimum between $x=a$ and $x=b$, essentially the same reasoning holds good.

A couple of examples will make this clearer.

(*a*) Take
$$\int_0^{2a}\frac{x.dx}{\sqrt{2ax-x^2}}.$$

Let $y=2ax-x^2$. Then $\frac{dy}{dx}=2(a-x)=0$ when $x=a$.

$\frac{d^2y}{dx^2}=-2$, and y is a maximum when $x=a$.

$$x=a\pm\sqrt{a^2-y},$$

$$dx=\mp\frac{dy}{2\sqrt{a^2-y}}.$$

Since $\frac{dy}{dx}$ is positive from $x=0$ to $x=a$, and negative from $x=a$ to $x=2a$, $dx=\frac{dy}{2\sqrt{a^2-y}}$ and $x=a-\sqrt{a^2-y}$ from $x=0$ to $x=a$, and $dx=-\frac{dy}{2\sqrt{a^2-y}}$, and $x=a+\sqrt{a^2-y}$ from $x=a$ to $x=2a$.

Hence

$$\int_0^{2a}\frac{xdx}{\sqrt{2\,ax-x^2}}=\int_0^{a}\frac{xdx}{\sqrt{2\,ax-x^2}}+\int_a^{2a}\frac{xdx}{\sqrt{2\,ax-x^2}}$$

$$=\frac{1}{2}\int_0^{a^2}\frac{a-\sqrt{a^2-y}}{\sqrt{a^2y-y^2}}\cdot dy-\frac{1}{2}\int_{a^2}^{0}\frac{a+\sqrt{a^2-y}}{\sqrt{a^2y-y^2}}\cdot dy$$

$$=\frac{1}{2}\int_0^{a^2}\frac{a-\sqrt{a^2-y}}{\sqrt{a^2y-y^2}}\cdot dy+\frac{1}{2}\int_0^{a^2}\frac{a+\sqrt{a^2-y}}{\sqrt{a^2y-y^2}}\cdot dy$$

$$=\int_0^{a^2}\frac{ady}{\sqrt{a^2y-y^2}}=\pi a. \qquad \text{(Ex. 7, Art. 84)}$$

(*b*)
$$\int_0^{\frac{\pi}{2}}\frac{dx}{(\sin x+\cos x)^2}.$$

Let $y=\sin x+\cos x$. $\dfrac{dy}{dx}=\cos x-\sin x=0$ when $x=\dfrac{\pi}{4}$. $\dfrac{d^2y}{dx^2}=-\sin x-\cos x=-\sqrt{2}$ when $x=\dfrac{\pi}{4}$. Therefore y has a maximum value $\sqrt{2}$ when $x=\dfrac{\pi}{4}$.

$$y=\sin x+\cos x=\sqrt{2}\,.\cos\left(\frac{\pi}{4}-x\right),$$

$$x=\frac{\pi}{4}-\cos^{-1}\frac{y}{\sqrt{2}}, \qquad dx=\pm\frac{dy}{\sqrt{2-y^2}}.$$

Since $\dfrac{dy}{dx}=0$ and $\dfrac{d^2y}{dx^2}<0$ when $x=\dfrac{\pi}{4}$, it follows that $\dfrac{dy}{dx}$ is positive from $x=0$ to $x=\dfrac{\pi}{4}$, and negative from $x=\dfrac{\pi}{4}$ to $x=\dfrac{\pi}{2}$. Hence we have

$$\int_0^{\frac{\pi}{2}}\frac{dx}{(\sin x+\cos x)^2}=\int_0^{\frac{\pi}{4}}\frac{dx}{(\sin x+\cos x)^2}+\int_{\frac{\pi}{4}}^{\frac{\pi}{2}}\frac{dx}{(\sin x+\cos x)^2}$$

$$=\int_1^{\sqrt{2}}\frac{dy}{y^2\sqrt{2-y^2}}-\int_{\sqrt{2}}^{1}\frac{dy}{y^2\sqrt{2-y^2}}=2\int_1^{\sqrt{2}}\frac{dy}{y^2\sqrt{2-y^2}}.$$

Let
$$\frac{y}{\sqrt{2}} = \sin\theta;$$

$$\int_1^{\sqrt{2}} \frac{dy}{y^2\sqrt{2-y^2}} = \frac{1}{2}\int_{\frac{\pi}{4}}^{\frac{\pi}{2}} \csc^2\theta d\theta = \frac{1}{2},$$

and
$$\int_0^{\frac{\pi}{2}} \frac{dx}{(\sin x + \cos x)^2} = 1.$$

EXAMPLE.

(1) Show that $$\int_0^{\pi} \frac{dx}{(\sin x + \cos x)^2} = \infty.$$

89. *Differentiation of a definite integral.*

We have seen in Art. 51 that a definite integral is a function of the *limits of integration*, and not of the variable with respect to which we integrate; that is, that $\int_a^b fx.dx$ is a function of a and b, and not a function of x. Strictly speaking, $\int_a^b fx.dx$ is a function of a and b, and of any constants that fx may contain, where by *constant* we mean any quantity that is independent of x.

If the limits a and b are variables, they are always independent of the x with respect to which the integration is performed, which must from the nature of the case disappear when the definite integral is formed, as it always may be in theory, from the indefinite integral; and this assertion holds good even when the same letter which is used for the variable with respect to which the integration is performed appears explicitly in the limits of integration.

Thus if we write $\int_0^x \sin x.dx$, the x in $\sin x.dx$ and the x which is the upper limit of integration do not represent the same variable, and are entirely unconnected. Indeed, the former x

may be replaced by any other letter without affecting the value of the integral. For

$$\int_0^x \sin x.dx = \int_0^x \sin z.dz = 1 - \cos x.$$

Let us now consider the possibility of differentiating a definite integral.

Required $D_\alpha \int_a^b f(x, \alpha)\,dx$, where α is independent of x, and a and b do not depend upon α.

We have

$$D_\alpha \int_a^b f(x,\alpha)\,dx = \underset{\Delta\alpha \doteq 0}{\text{limit}} \left[\frac{\int_a^b f(x, \alpha + \Delta\alpha)\,dx - \int_a^b f(x, \alpha)\,dx}{\Delta\alpha}\right]$$

$$= \underset{\Delta\alpha \doteq 0}{\text{limit}} \left[\int_a^b \frac{f(x, \alpha + \Delta\alpha) - f(x, \alpha)}{\Delta\alpha}\right] dx$$

$$= \int_a^b \left(\underset{\Delta\alpha \doteq 0}{\text{limit}} \left[\frac{f(x, \alpha + \Delta\alpha) - f(x, \alpha)}{\Delta\alpha}\right]\right) dx.$$

Hence $$D_\alpha \int_a^b f(x, \alpha)\,dx = \int_a^b [D_\alpha f(x, \alpha)]\,dx, \qquad [1]$$

and we find that we have merely to differentiate under the sign of integration.

The truth of the converse of the last proposition can be easily established, and we have

$$\int \left[\int_a^b f(x, \alpha)\,dx\right] d\alpha = \int_a^b \left[\int f(x, \alpha)\,d\alpha\right] dx, \qquad [2]$$

or even

$$\int_c^d \left[\int_a^b f(x, \alpha)\,dx\right] d\alpha = \int_a^b \left[\int_c^d f(x, \alpha)\,d\alpha\right] dx, \qquad [3]$$

if a, b, c, and d are entirely independent.

Suppose now that we are dealing with variable limits of integration.

Let us find first $\frac{d}{dz}\int_a^z fx.dx.$

Let $\int fx.dx = Fx$, then $\int_a^z fx.dx = Fz - Fa$; and since by definition $\frac{dFx}{dx} = fx$, it follows that $\frac{dFz}{dz} = fz$.

Hence $$\frac{d}{dz}\int_a^z fx.dx = \frac{d\,(Fz - Fa)}{dz} = fz. \qquad [4]$$

In the same way it may be shown that

$$\frac{d}{dz}\int_z^b fx.dx = -fz. \qquad [5]$$

Let us now take the most complicated case, namely, to find $\frac{d}{d\alpha}\int_a^b f(x, \alpha)\,dx$, where a and b are functions of α.

Let $$\int f(x, \alpha)\,dx = F(x, \alpha);$$

then $$u = \int_a^b f(x, \alpha)\,dx = F(b, \alpha) - F(a, \alpha),$$

and $$\frac{du}{d\alpha} = \frac{dF(b, \alpha)}{d\alpha} - \frac{dF(a, \alpha)}{d\alpha};$$

but as b and a are functions of α,

$$\frac{dF(b, \alpha)}{d\alpha} = D_b F(b, \alpha)\frac{db}{d\alpha} + D_\alpha F(b, \alpha),$$

and $$\frac{dF(a, \alpha)}{d\alpha} = D_a F(a, \alpha)\frac{da}{d\alpha} + D_\alpha F(a, \alpha),$$

by I. Art. 200.

$$D_b F(b, \alpha) = f(b, \alpha),$$

$$D_a F(a, \alpha) = f(a, \alpha).$$

Hence $$\frac{du}{d\alpha} = D_\alpha [F(b, \alpha) - F(a, \alpha)] + f(b, \alpha)\frac{db}{d\alpha} - f(a, \alpha)\frac{da}{d\alpha},$$

or

$$\frac{d}{d\alpha}\int_a^b f(x, \alpha)\,dx = \int_a^b (D_\alpha f(x, \alpha))\,dx + f(b, \alpha)\frac{db}{d\alpha} - f(a, \alpha)\frac{da}{d\alpha}. \qquad [6]$$

EXAMPLES.

(1) $\frac{d}{dy}\int_0^{xy} \sin(x+y)\,dx = (x+1)\sin(xy+y) - \sin y.$

(2) $\frac{d}{dx}\int_0^{\sqrt[3]{x}} x^2\,dx = \frac{1}{3}.$

(3) $\frac{d}{d\phi}\int_0^{e^\phi} \sqrt{1-\cos\phi}\,.\,d\phi = e^\phi\sqrt{1-\cos e^\phi}.$

90. When the indefinite integral cannot be found, the problem of obtaining the value of the definite integral usually becomes a more or less difficult mathematical puzzle, which can be solved, if solved at all, only by the exercise of great ingenuity. Some of the results arrived at, however, are so important, and some of the devices employed so interesting, that we shall present them briefly here. But we must repeat the warning that most of the methods are valid only in case the definite integral is finite and determinate; and erroneous results have more than once been obtained and published when a little attention to the precautions described in Articles 83–86 would have prevented the mistake.

91. *Integration by development in series.*

(a) $\int_0^1 \frac{\log x}{1-x}\,.\,dx.$ (v. Art. 84, Ex. 1.)

$$\frac{1}{1-x} = (1-x)^{-1} = 1 + x + x^2 + x^3 + \cdots, \text{ if } x < 1.$$

$$\int_0^1 \frac{\log x}{1-x}\,.\,dx = \int_0^1 (\log x + x\log x + x^2\log x + \cdots)\,dx.$$

$$\int_0^1 x^n \log x . dx = -\frac{1}{(n+1)^2}. \qquad \text{(v. Art. 55 (a).)}$$

Therefore

$$\int_0^1 \frac{\log x}{1-x}\,.\,dx = -\left(\frac{1}{1^2} + \frac{1}{2^2} + \frac{1}{3^2} + \frac{1}{4^2} + \cdots\right) = -\frac{\pi^2}{6}.$$

(v. Todhunter's Trigonometry, Chap. XXIII., Ex. 1.)

(b) $$\int_0^\infty \log\left(\frac{e^x+1}{e^x-1}\right)dx.$$ (v. Art. 86, Ex. 2.)

$$\log\left(\frac{e^x+1}{e^x-1}\right) = \log\left(\frac{1+e^{-x}}{1-e^{-x}}\right) = \log(1+e^{-x}) - \log(1-e^{-x})$$

$$= e^{-x} - \frac{e^{-2x}}{2} + \frac{e^{-3x}}{3} - \frac{e^{-4x}}{4} \cdots - \left(-e^{-x} - \frac{e^{-2x}}{2} - \frac{e^{-3x}}{3} - \frac{e^{-4x}}{4} \cdots\right).$$

(I. Art. 130.)

Hence

$$\int_0^\infty \log\left(\frac{e^x+1}{e^x-1}\right)dx = 2\int_0^\infty \left(e^{-x} + \frac{e^{-3x}}{3} + \frac{e^{-5x}}{5} + \frac{e^{-7x}}{7} + \cdots\right)dx$$

$$= 2\left(1 + \frac{1}{3^2} + \frac{1}{5^2} + \frac{1}{7^2} + \cdots\right).$$

But $$\frac{1}{1^2} + \frac{1}{3^2} + \frac{1}{5^2} + \frac{1}{7^2} + \cdots = \frac{\pi^2}{8}.$$

(v. Todhunter's Trig., Chap. XXIII., Ex. 1.)

Therefore $$\int_0^\infty \log\left(\frac{e^x+1}{e^x-1}\right)dx = \frac{\pi^2}{4}.$$

EXAMPLES.

(1) $$\int_0^1 \frac{\log x}{1+x}.dx = -\frac{\pi^2}{12}.$$

(2) $$\int_0^1 \frac{\log x}{1-x^2}.dx = -\frac{\pi^2}{8}.$$

(3) $$\int_0^1 \frac{dx}{x}.\log\left(\frac{1+x}{1-x}\right) = \frac{\pi^2}{4}.$$

(4) $$\int_0^{\frac{\pi}{2}} \frac{d\phi}{\sqrt{1-k^2\sin^2\phi}} = \frac{\pi}{2}\left[1 + \left(\frac{1}{2}\right)^2 k^2 + \left(\frac{1.3}{2.4}\right)^2 k^4 + \left(\frac{1.3.5}{2.4.6}\right)^2 k^6 + \cdots\right]$$ if $k^2 < 1$.

(5) $$\int_0^{\frac{\pi}{2}} \sqrt{1-k^2\sin^2\phi}\,.\,d\phi = \frac{\pi}{2}\left[1-\left(\frac{1}{2}\right)^2\frac{k^2}{1}-\left(\frac{1.3}{2.4}\right)^2\frac{k^4}{3}\right.$$
$$\left.-\left(\frac{1.3.5}{2.4.6}\right)^2\frac{k^6}{5}-\cdots\right] \qquad \text{if } k^2<1.$$

92. *Integration by ingenious devices.*

(*a*) $\int_0^{\frac{\pi}{2}} \log\sin x.dx.$ (v. Art. 84, Ex. 6.)

Let $$u=\int_0^{\frac{\pi}{2}}\log\sin x.dx.$$

Substitute $$y=\frac{\pi}{2}-x.$$

$$u=-\int_{\frac{\pi}{2}}^0 \log\cos y.dy=\int_0^{\frac{\pi}{2}}\log\cos x.dx.$$

$$2u=\int_0^{\frac{\pi}{2}}(\log\sin x+\log\cos x)\,dx=\int_0^{\frac{\pi}{2}}\log(\sin x\cos x)\,dx$$

$$=\int_0^{\frac{\pi}{2}}\log\left(\frac{\sin 2x}{2}\right).dx$$

$$=-\frac{\pi}{2}\log(2)+\int_0^{\frac{\pi}{2}}\log\sin 2x.dx$$

$$=-\frac{\pi}{2}\log(2)+\frac{1}{2}\int_0^{\pi}\log\sin x.dx.$$

$$\int_0^{\pi}\log\sin x.dx=\int_0^{\frac{\pi}{2}}\log\sin x.dx+\int_{\frac{\pi}{2}}^{\pi}\log\sin x.dx$$
$$=u+\int_{\frac{\pi}{2}}^{\pi}\log\sin x.dx.$$

Substitute $$y=\pi-x,$$

and

$$\int_{\frac{\pi}{2}}^{\pi} \log \sin x . dx = -\int_{\frac{\pi}{2}}^{0} \log \sin y . dy = \int_{0}^{\frac{\pi}{2}} \log \sin x . dx = u.$$

Hence $$2u = -\frac{\pi}{2} \log (2) + u,$$

and

$$u = \int_{0}^{\frac{\pi}{2}} \log \sin x . dx = \int_{0}^{\frac{\pi}{2}} \log \cos x . dx = -\frac{\pi}{2} \log (2). \qquad [1]$$

(*b*) $$\int_{0}^{\infty} e^{-x^2} dx. \qquad \text{(v. Art. 86 } (a).)$$

Let $$u = \int_{0}^{\infty} e^{-x^2} dx, \quad \text{and let } x = \alpha y;$$

then $$u = \int_{0}^{\infty} \alpha e^{-\alpha^2 y^2} dy = \int_{0}^{\infty} \alpha e^{-\alpha^2 x^2} dx,$$

$$u e^{-\alpha^2} = \int_{0}^{\infty} \alpha e^{-(1+x^2)\alpha^2} dx,$$

$$u \int_{0}^{\infty} e^{-\alpha^2} d\alpha = u^2 = \int_{0}^{\infty} \left(\int_{0}^{\infty} \alpha e^{-(1+x^2)\alpha^2} d\alpha \right) dx, \quad \text{by [3], Art. 89.}$$

But

$$\int_{0}^{\infty} \alpha e^{-(1+x^2)\alpha^2} d\alpha = \frac{1}{2} \frac{1}{1+x^2}.$$

Hence $$u^2 = \frac{1}{2} \int_{0}^{\infty} \frac{dx}{1+x^2} = \frac{\pi}{4},$$

and $$\int_{0}^{\infty} e^{-x^2} dx = \frac{1}{2} \sqrt{\pi}. \qquad [2]$$

(*c*) $$\int_{0}^{\infty} \frac{\sin mx}{x} . dx. \qquad \text{(v. Art. 86 } (b).)$$

We have $$\frac{1}{x} = \int_{0}^{\infty} e^{-\alpha x} d\alpha \quad \text{if } x > 0. \qquad \text{(Art. 82, Ex. 6.)}$$

Hence

$$\int_0^\infty \frac{\sin mx}{x}.dx = \int_0^\infty \left(\sin mx \int_0^\infty e^{-ax}\,da\right)dx$$

$$= \int_0^\infty \left(\int_0^\infty e^{-ax}\sin mx.da\right)dx$$

$$= \int_0^\infty \left(\int_0^\infty e^{-ax}\sin mx.dx\right)da, \text{ by [3], Art. 89.}$$

$$= \int_0^\infty \frac{mda}{a^2+m^2}. \qquad \text{(Art. 82, Ex. 7.)}$$

Therefore

$$\left.\begin{aligned}\int_0^\infty \frac{\sin mx}{x}.dx &= \frac{\pi}{2} \text{ if } m>0\\ &= -\frac{\pi}{2} \text{ if } m<0\\ &= 0 \text{ if } m=0\end{aligned}\right\}, \qquad [3]$$

by Art. 82, Ex. 5.

EXAMPLES.

(1) $\int_0^\pi x \log \sin x\,.\,dx = -\frac{\pi^2}{2}\log(2).$

(2) $\int_0^\infty \log\left(x+\frac{1}{x}\right)\frac{dx}{1+x^2} = \pi\log(2).$

Suggestion: let $x = \tan\theta$.

(3) $\int_0^\infty e^{-a^2x^2}\,dx = \frac{1}{2a}\sqrt{\pi}.$

(4) $\int_0^1 \frac{dx}{\sqrt{\log\frac{1}{x}}} = \sqrt{\pi}.$

(5) $\int_0^\infty \frac{\sin x \cos mx}{x}.dx = 0$ if $m<-1$ or $m>1$

$= \frac{\pi}{4}$ if $m=-1$ or $m=1$

$= \frac{\pi}{2}$ if $-1<m<1$.

(6) $\int_0^\infty \frac{\sin^2 x}{x^2}\,dx = \frac{\pi}{2}$. *Suggestion:* integrate by parts.

93. *Differentiation or integration with respect to a quantity which is independent of* x. (v. Art. 89.)

(*a*) We have $\int_0^\infty e^{-ax}\,dx = \frac{1}{a}$. (Art. 82, Ex. 6.)

Differentiate both members with respect to a,

$$\int_0^\infty (-xe^{-ax}dx) = -\frac{1}{a^2} \quad \text{or} \quad \int_0^\infty xe^{-ax}\,dx = \frac{1}{a^2}.$$

Differentiate again,

$$\int_0^\infty x^2 e^{-ax}\,dx = \frac{2\,!}{a^3}.$$

Differentiating n times,

$$\int_0^\infty x^n e^{-ax}\,dx = \frac{n\,!}{a^{n+1}}. \qquad [1] \quad \text{(v. Art. 86, Ex. 3.)}$$

(*b*) We have $\int_0^\infty e^{-ax^2}\,dx = \frac{1}{2}\frac{\sqrt{\pi}}{a^{\frac{1}{2}}}$. (Art. 92, Ex. 3.)

Differentiating n times with respect to a,

$$\int_0^\infty x^{2n} e^{-ax^2}\,dx = \frac{1.3.5\ldots(2n-1)}{2^{n+1}a^n}\sqrt{\frac{\pi}{a}}. \qquad [2]$$

(v. Art. 86, Ex. 2.)

(*c*) We have $\int_0^\infty e^{-cx}\,dx = \frac{1}{c}$. (Art. 82, Ex. 6.)

Multiply by dc, and integrate from a to b,

$$\int_0^\infty \left(\int_a^b e^{-cx}\,dc\right) dx = \int_a^b \frac{dc}{c}.$$

Hence $\int_0^\infty \frac{e^{-ax} - e^{-bx}}{x}\,dx = \log\frac{b}{a}$. [3]

(*d*) $$\int_0^1 x^\alpha\,dx = \frac{1}{\alpha+1}.$$

Multiply by $d\alpha$, and integrate from b to a,

$$\int_0^1\left(\int_b^a x^\alpha\,d\alpha\right)dx = \int_b^a \frac{d\alpha}{\alpha+1}.$$

Hence $$\int_0^1 \frac{x^a - x^b}{\log x}\,dx = \log\left(\frac{a+1}{b+1}\right). \qquad [4]$$

EXAMPLES.

(1) From $\int_0^\infty \frac{dx}{x^2+a} = \frac{\pi}{2}\frac{1}{\sqrt{a}}$ obtain

$$\int_0^\infty \frac{dx}{(x^2+a)^{n+1}} = \frac{\pi}{2}\,\frac{1.3.5\ldots(2n-1)}{2.4.6\ldots 2n}\cdot\frac{1}{\sqrt{a^{2n+1}}}.$$

(2) From $\int_0^1 x^n\,dx = \frac{1}{n+1}$ obtain

$$\int_0^1 x^n(\log x)^m\,dx = (-1)^m\frac{m!}{(n+1)^{m+1}}.$$

(3) From $\int_0^\infty e^{-ax}\cos mx.dx = \frac{a}{a^2+m^2}$ obtain

$$\int_0^\infty \frac{e^{-ax}-e^{-bx}}{x}\cos mx.dx = \tfrac{1}{2}\log\left(\frac{b^2+m^2}{a^2+m^2}\right).$$

(4) From $\int_0^\infty e^{-ax}\sin mx.dx = \frac{m}{a^2+m^2}$ obtain

$$\int_0^\infty \frac{e^{-ax}-e^{-bx}}{x}\sin mx.dx = \tan^{-1}\frac{b}{m} - \tan^{-1}\frac{a}{m}.$$

94. The method illustrated in Art. 93 can be applied to much more complicated forms.

(*a*) $$\int_0^\infty e^{-x^2-\frac{a^2}{x^2}}.dx. \qquad \text{(v. Art. 86, Ex. 2.)}$$

Let $$u = \int_0^\infty e^{-x^2 - \frac{a^2}{x^2}} dx;$$

then $$\frac{du}{da} = -2\int_0^\infty \frac{adx}{x^2} \cdot e^{-x^2 - \frac{a^2}{x^2}}.$$

Substitute $$z = \frac{a}{x},$$

and $$\frac{du}{da} = -2\int_0^\infty e^{-z^2 - \frac{a^2}{z^2}} dz = -2\int_0^\infty e^{-x^2 - \frac{a^2}{x^2}} dx = -2u.$$

Hence $$\frac{du}{u} = -2\,da.$$

Integrating,
$$\log u = -2a + C,$$

and $$u = C_1 e^{-2a}.$$

When $a = 0$, $$u = \int_0^\infty e^{-x^2} dx = \tfrac{1}{2}\sqrt{\pi}.$$ (Art. 92 (*b*) [2].)

Therefore $$C_1 = \tfrac{1}{2}\sqrt{\pi},$$

and $$u = \int_0^\infty e^{-x^2 - \frac{a^2}{x^2}} dx = \frac{e^{-2a}\sqrt{\pi}}{2}. \qquad [1]$$

(*b*) $$\int_0^\infty e^{-a^2x^2} \cos bx.dx.$$ (v. Art. 86, Ex. 2.)

Let $$u = \int_0^\infty e^{-a^2x^2} \cos bx.dx,$$

then $$\frac{du}{db} = -\int_0^\infty xe^{-a^2x^2} \sin bx.dx.$$

Integrating by parts,
$$\int_0^\infty xe^{-a^2x^2} \sin bx.dx = \frac{b}{2a^2}\int_0^\infty e^{-a^2x^2} \cos bx.dx = \frac{b}{2a^2} u.$$

Therefore $$\frac{du}{db} = -\frac{b}{2a^2} u,$$

or $$\frac{du}{u} = -\frac{b}{2a^2} db.$$

Integrating, we have

$$\log u = -\frac{b^2}{4a^2} + C,$$

or

$$u = C_1 e^{-\frac{b^2}{4a^2}}.$$

When $b = 0$, $\quad u = \int_0^\infty e^{-a^2x^2}\,dx = \frac{\sqrt{\pi}}{2a}$. (Art. 92, Ex. 3.)

Hence

$$u = \int_0^\infty e^{-a^2x^2} \cos bx . dx = \frac{\sqrt{\pi}}{2a} e^{-\frac{b^2}{4a^2}}. \qquad [2]$$

EXAMPLES.

(1) $\int_0^\infty \frac{e^{-ax} \sin mx}{x} . dx = \tan^{-1}\frac{m}{a}.$

(2) $\int_0^\infty \frac{\cos mx}{1+x^2} \cdot dx \quad = \frac{\pi}{2} e^{-m}.$

Suggestion: $\frac{1}{1+x^2} = 2\int_0^\infty ue^{-a^2(1+x^2)}\,da.$

95. *Introduction of imaginary constants.*

$$\int_0^\infty \cos(x^2)\,dx. \qquad \text{(v. Art. 86 (c).)}$$

We have $\quad \int_0^\infty e^{-a^2x^2}\,dx = \frac{1}{2a}\sqrt{\pi}$. (Art. 92, Ex. 3.)

Let

$$a^2 = c^2\sqrt{-1} = c^2\left(\cos\frac{\pi}{2} + \sqrt{-1}\sin\frac{\pi}{2}\right).$$

Then

$$a = c\left(\cos\frac{\pi}{4} + \sqrt{-1}\sin\frac{\pi}{4}\right) = \frac{c\sqrt{2}}{2}(1+\sqrt{-1}),$$

(Art. 25.)

and

$$\frac{1}{2a} = \frac{1}{c\sqrt{2}(1+\sqrt{-1})} = \frac{1}{2c\sqrt{2}}(1-\sqrt{-1}).$$

Hence $$\int_0^\infty e^{-c^2x^2\sqrt{-1}}\,dx = \frac{1}{2c}\sqrt{\frac{\pi}{2}}\cdot(1-\sqrt{-1}).$$

But $$e^{-c^2x^2\sqrt{-1}} = \cos(c^2x^2) - \sqrt{-1}\sin(c^2x^2).$$

([5] Art. 31.)

Therefore

$$\int_0^\infty \cos(c^2x^2)\,dx - \sqrt{-1}\int_0^\infty \sin(c^2x^2)\,dx = \frac{1}{2c}\sqrt{\frac{\pi}{2}}(1-\sqrt{-1}),$$

and

$$\int_0^\infty \cos(c^2x^2)\,dx = \int_0^\infty \sin(c^2x^2)\,dx = \frac{1}{2c}\sqrt{\frac{\pi}{2}}. \qquad [1]$$

(Art. 17.)

Let $$c = 1,$$

and $$\int_0^\infty \cos(x^2)\,dx = \int_0^\infty \sin(x^2)\,dx = \tfrac{1}{2}\sqrt{\frac{\pi}{2}}. \qquad [2]$$

If we substitute $y = x^2$ in [2], we get

$$\int_0^\infty \frac{\cos y}{\sqrt{y}}\,dy = \int_0^\infty \frac{\sin y}{\sqrt{y}}\,dy = \sqrt{\frac{\pi}{2}}. \qquad [3]$$

Gamma Functions.

96. It was shown in Art. 84 that $\int_0^1 \left(\log\frac{1}{x}\right)^n dx$ is finite and determinate for all values of n greater than -1, and infinite when n is equal to or less than -1. The substitution of $y = \log\frac{1}{x}$ reduces this integral to $\int_0^\infty y^n e^{-y}\,dy$, or, what is the same thing, to $\int_0^\infty x^n e^{-x}\,dx$; and in Art. 86, Ex. 3, the student has been required to show that this integral is finite and determinate for all values of n greater than -1.

$$\int x^n e^{-x}\,dx = -x^n e^{-x} + n\int x^{n-1}e^{-x}\,dx,$$

by integration by parts.

If n is greater than zero,

$$x^n e^{-x} = 0 \quad \text{when } x = 0,$$

and $x^n e^{-x}$ is indeterminate when $x = \infty$. Its true value when $x = \infty$, obtained by the method of I. Art. 141, is, however, zero.

Therefore $$\int_0^\infty x^n e^{-x} dx = n \int_0^\infty x^{n-1} e^{-x} dx \qquad [1]$$

for all positive values of n.

If n is an integer, a repeated use of [1] gives

$$\int_0^\infty x^n e^{-x} dx = n! \int_0^\infty e^{-x} dx;$$

but $$\int_0^\infty e^{-x} dx \quad = 1,$$

and we have $$\int_0^\infty x^n e^{-x} dx = n! \qquad [2]$$

provided that n *is a positive whole number.*

If n is not a positive integer, but is greater than -1, $\int_0^\infty x^n e^{-x} dx$ is a finite and determinate function of n, and its value can be computed to any required degree of accuracy by methods which we have not space to consider here.

$\int_0^\infty x^{n-1} e^{-x} dx$ is generally represented by $\Gamma(n)$, and has been very carefully studied under the name of the *Gamma Function.*

If n is a positive integer, we have from [2]

$$\Gamma(n+1) = n!. \qquad [3]$$

From [3], $$\Gamma(2) \quad = 1. \qquad [4]$$

Since $$\Gamma(1) \quad = \int_0^\infty x^0 e^{-x} dx = \int_0^\infty e^{-x} dx,$$

$$\Gamma(1) \quad = 1. \qquad [5]$$

We have always from [1]

$$\Gamma(n+1) = n\Gamma(n), \qquad [6]$$

if n is greater than zero.

Since $\int_0^\infty x^n e^{-x} dx$ is infinite when n is equal to or less than -1, it follows from the definition of $\Gamma(n)$ that $\Gamma(n) = \infty$ if n is equal to or less than zero. It has, however, been found convenient to adopt formula [6] as the definition of $\Gamma(n)$ when n is equal to or less than zero, and to restrict the original definition to positive values of n. The result easily deduced is that $\Gamma(n)$ is infinite when n is equal to zero or to a negative integer, but is finite and determinate for all other values of n.

97. We may regard the formula

$$\Gamma(n+1) = n\Gamma(n)$$

as a sort of *reduction formula;* and since each time we apply it we can raise or lower the value of n by unity, we can obtain any required *Gamma Function* by the aid of a table containing the values of $\Gamma(n)$ corresponding to the values of n between any two arbitrarily chosen consecutive whole numbers.

Such tables have been computed, and we give one here containing the *common* logarithms of the values of $\Gamma(n)$ from $n = 1$ to $n = 2$. The table is carried out to four decimal places, and each logarithm is printed with the characteristic 9, which, of course, is ten units too large, the true characteristic being -1.

$10 + \log \Gamma(n)$.

n	0	1	2	3	4	5	6	7	8	9
1.0	..	9975	9951	9928	9905	9883	9862	9841	9821	9802
1.1	9.9783	9765	9748	9731	9715	9699	9684	9669	9655	9642
1.2	9.9629	9617	9605	9594	9583	9573	9564	9554	9546	9538
1.3	9.9530	9523	9516	9510	9505	9500	9495	9491	9487	9483
1.4	9.9481	9478	9476	9475	9473	9473	9472	9473	9473	9474
1.5	9.9475	9477	9479	9482	9485	9488	9492	9496	9501	9506
1.6	9.9511	9517	9523	9529	9536	9543	9550	9558	9566	9575
1.7	9.9584	9593	9603	9613	9623	9633	9644	9656	9667	9679
1.8	9.9691	9704	9717	9730	9743	9757	9771	9786	9800	9815
1.9	9.9831	9846	9862	9878	9895	9912	9929	9946	9964	9982

Such a table enables us to compute with Gamma Functions as readily as with Trigonometric Functions, and consequently the problem of obtaining the value of a definite integral is practically solved if the integral in question can be expressed in terms of Gamma Functions.

For example, let us consider

(a) $$\int_0^\infty x^n e^{-ax}\,dx.$$

Let $$y = ax;$$

then $$\int_0^\infty x^n e^{-ax}\,dx = \frac{1}{a^{n+1}}\int_0^\infty y^n e^{-y}\,dy = \frac{1}{a^{n+1}}\int_0^\infty x^n e^{-x}\,dx.$$

Hence $$\int_0^\infty x^n e^{-ax}\,dx = \frac{\Gamma(n+1)}{a^{n+1}}, \qquad [1]$$

provided that a is positive and $n > -1$.

(b) $$\int_0^1 x^m \left(\log\frac{1}{x}\right)^n dx. \qquad \text{(v. Art. 84, Ex. 4.)}$$

Let $$y = -\log x.$$

then $$\int_0^1 x^m \left(\log\frac{1}{x}\right)^n dx = \int_0^\infty y^n e^{-(m+1)y}\,dy.$$

Hence, by [1],

$$\int_0^1 x^m \left(\log\frac{1}{x}\right)^n dx = \frac{\Gamma(n+1)}{(m+1)^{n+1}}, \qquad [2]$$

if $m > -1$ and $n > -1$.

(c) $$\int_0^\infty e^{-x^2}\,dx.$$

Let $$y = x^2;$$

then $$\int_0^\infty e^{-x^2}\,dx = \tfrac{1}{2}\int_0^\infty \frac{e^{-y}}{\sqrt{y}}\,dy = \tfrac{1}{2}\int_0^\infty x^{-\frac{1}{2}} e^{-x}\,dx.$$

Hence $$\int_0^\infty e^{-x^2}\,dx = \tfrac{1}{2}\Gamma(\tfrac{1}{2}). \qquad [3]$$

98. $$\int_0^1 x^{m-1}(1-x)^{n-1}dx = B(m, n) \qquad [1]$$

is an exceedingly important integral that can be expressed in terms of Gamma Functions; it is known as the *Beta Function*, or the *First Eulerian Integral*, $\Gamma(n)$ being sometimes called the *Second Eulerian Integral*.

In the Beta Function, m and n are *positive*, and $B(m, n)$ is always finite and determinate. (v. Art. 84, Ex. 5.)

In $$\int_0^1 x^{m-1}(1-x)^{n-1}dx \quad \text{let} \quad y = 1 - x,$$

and we get

$$\int_0^1 x^{m-1}(1-x)^{n-1}dx = \int_0^1 y^{n-1}(1-y)^{m-1}dy,$$

or $$B(m, n) = B(n, m). \qquad [2]$$

In $$\int_0^1 x^{m-1}(1-x)^{n-1}dx \quad \text{let} \quad x = \frac{y}{1+y},$$

and we get

$$\int_0^1 x^{m-1}(1-x)^{n-1}dx = \int_0^\infty \frac{y^{m-1}dy}{(1+y)^{m+n}} = \int_0^\infty \frac{x^{m-1}}{(1+x)^{m+n}}dx.$$

Hence $$\int_0^\infty \frac{x^{m-1}}{(1+x)^{m+n}}dx = B(m, n). \qquad [3]$$

We have seen in [1] Art. 97 (a) that

$$\int_0^\infty x^n e^{-ax}dx = \frac{\Gamma(n+1)}{a^{n+1}}.$$

Hence $$\Gamma(m) = \int_0^\infty a^m x^{m-1} e^{-ax} dx,$$

$$\Gamma(m)\, a^{n-1} e^{-a} = \int_0^\infty a^{m+n-1} x^{m-1} e^{-a(1+x)} dx,$$

$$\Gamma(m)\int_0^\infty a^{n-1} e^{-a} da = \int_0^\infty x^{m-1}\left(\int_0^\infty a^{m+n-1} e^{-a(1+x)} da\right) dx,$$

$$\Gamma(m)\,\Gamma(n) = \int_0^\infty x^{m-1} \frac{\Gamma(m+n)}{(1+x)^{m+n}} dx.$$

Therefore $$\frac{\Gamma(m)\,\Gamma(n)}{\Gamma(m+n)} = \int_0^\infty \frac{x^{m-1}}{(1+x)^{m+n}}.dx; \qquad [4]$$

or by [3],

$$B(m,n) = \int_0^1 x^{m-1}(1-x)^{n-1}dx = \frac{\Gamma(m)\,\Gamma(n)}{\Gamma(m+n)}. \qquad [5]$$

If $n = 1 - m$, then since $\Gamma(1) = 1$,

$$\int_0^1 \frac{x^{m-1}}{(1-x)^m}dx = \int_0^\infty \frac{x^{m-1}}{1+x}dx = \Gamma(m)\,\Gamma(1-m). \qquad [6]$$

Formula [6] leads to an interesting confirmation of Art. 92 (b).

Let $m = \frac{1}{2}$, and we have from [6]

$$[\Gamma(\tfrac{1}{2})]^2 = \int_0^\infty \frac{dx}{(1+x)\sqrt{x}}.$$

Substitute $$y = \sqrt{x},$$

and we have $$\int_0^\infty \frac{dx}{(1+x)\sqrt{x}} = 2\int_0^\infty \frac{dy}{1+y^2} = \pi.$$

Hence $$\Gamma(\tfrac{1}{2}) = \sqrt{\pi}; \qquad [7]$$

and since by Art. 97 (c)

$$\int_0^\infty e^{-x^2}dx = \tfrac{1}{2}\Gamma(\tfrac{1}{2}),$$

$$\int_0^\infty e^{-x^2}dx = \tfrac{1}{2}\sqrt{\pi}.$$

99. By the aid of formulas [4], [5], and [7] of Art. 98 a number of important integrals can be obtained.

For example, let us consider

$$\int_0^{\frac{\pi}{2}} \sin^n x.dx, \qquad \text{where } n \text{ is greater than } -1.$$

Let $$y = \sin x,$$

and we have $$\int_0^{\frac{\pi}{2}} \sin^n x.dx = \int_0^1 y^n(1-y^2)^{-\frac{1}{2}}dy.$$

Let, now, $z = y^2$,

and

$$\int_0^1 y^n (1-y^2)^{-\frac{1}{2}} dy = \tfrac{1}{2}\int_0^1 z^{\frac{n}{2}-\frac{1}{2}} (1-z)^{-\frac{1}{2}} dz$$

$$= \tfrac{1}{2}\int_0^1 x^{\frac{n+1}{2}-1} (1-x)^{\frac{1}{2}-1} dx = \tfrac{1}{2} B\left(\frac{n+1}{2}, \frac{1}{2}\right).$$

But

$$B\left(\frac{n+1}{2}, \frac{1}{2}\right) = \frac{\Gamma\left(\frac{n+1}{2}\right)\Gamma(\frac{1}{2})}{\Gamma\left(\frac{n}{2}+1\right)} \qquad \text{by [5] Art. 98.}$$

$$= \sqrt{\pi}\,.\frac{\Gamma\left(\frac{n+1}{2}\right)}{\Gamma\left(\frac{n}{2}+1\right)} \qquad \text{by [7] Art. 98.}$$

Hence

$$\int_0^{\frac{\pi}{2}} \sin^n x.dx = \frac{\sqrt{\pi}}{2}\,\frac{\Gamma\left(\frac{n+1}{2}\right)}{\Gamma\left(\frac{n}{2}+1\right)}. \qquad [1]$$

If n is a whole number, this will reduce to the result given in Art. 87, Ex. 5.

EXAMPLES.

(1) $$\int_0^1 \frac{x^{2n}dx}{\sqrt{1-x^2}} = \frac{\sqrt{\pi}}{2}\,\frac{\Gamma(n+\frac{1}{2})}{\Gamma(n+1)}.$$

(2) $$\int_0^{\frac{\pi}{2}} \sin^n x \cos^m x.dx = \frac{\Gamma\left(\frac{m+1}{2}\right)\Gamma\left(\frac{n+1}{2}\right)}{2\,\Gamma\left(\frac{m+n}{2}+1\right)}.$$

$$(3)\quad \int_0^1 \frac{dx}{\sqrt{1-x^n}} = \frac{\sqrt{\pi}}{n}\,\frac{\Gamma\left(\frac{1}{n}\right)}{\Gamma\left(\frac{1}{n}+\frac{1}{2}\right)}.$$

$$(4)\quad \int_0^1 x^m(1-x^n)^p\,dx = \frac{1}{n}\,\frac{\Gamma(p+1)\,\Gamma\left(\frac{m+1}{n}\right)}{\Gamma\left(p+1+\frac{m+1}{n}\right)}.$$

CHAPTER IX.

LENGTHS OF CURVES.

100. If we use rectangular coördinates, we have seen (I. Art. 27) that

$$\tan \tau = \frac{dy}{dx}, \qquad [1]$$

and (I. Arts. 52 and 181) that

$$ds^2 = dx^2 + dy^2. \qquad [2]$$

From these we get

$$\sin \tau = \frac{dy}{ds}, \qquad [3]$$

$$\cos \tau = \frac{dx}{ds}, \qquad [4]$$

by the aid of a little elementary Trigonometry.

These formulas are of great importance in dealing with all properties of curves that concern in any way the lengths of arcs.

We have already considered the use of [2] in the first volume of the Calculus, and we have worked several examples by its aid in rectification of curves. Before going on to more of the same sort we shall find it worth while to obtain the equations of two very interesting transcendental curves, the *catenary* and the *tractrix*.

The Catenary.

101. The common *catenary* is the curve in which a uniform heavy flexible string hangs when its ends are supported.

As the string is flexible, the only force exerted by one portion of the string on an adjacent portion is a pull along the string, which we shall call the tension of the string, and shall represent by T. T of course has different values at different points of the string, and is some function of the coördinates of the point in question.

The tension at any point has to support the weight of the portion of the string below the point, and a certain amount of side pull, due to the fact that the string would hang vertically were it not that its ends are forcibly held apart.

Let the origin be taken at the lowest point of the curve, and suppose the string fastened at that point.

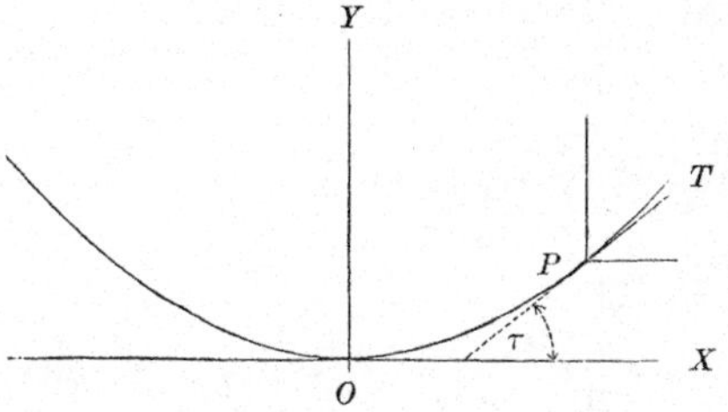

Let s be the arc OP, P being any point of the string. As the string is uniform, the weight of OP is proportional to its length; we shall call this weight ms.

This weight acts vertically downward, and must be balanced by the vertical effect of T, which, by I. Art. 112, is $T \sin \tau$.

Hence $$T \sin \tau = ms. \qquad (1)$$

As there is no *external* horizontal force acting, the horizontal effect of the tension at one end of any portion of the string must be the same as the horizontal effect at the other end. In other words, $$T \cos \tau = c \qquad (2)$$

where c is a constant. Dividing (1) by (2) we get

$$s = \frac{c}{m} \tan \tau,$$

or $$s = a \tan \tau, \qquad (3)$$

where a is some constant. From this we want to get an equation in terms of x and y.

$$\tan \tau = \sqrt{\sec^2 \tau - 1} = \sqrt{\frac{ds^2}{dx^2} - 1};$$

hence $$s^2 = a^2 \left(\frac{ds^2}{dx^2} - 1 \right),$$

or $$a^2 ds^2 = (a^2 + s^2) dx^2,$$

and $$\frac{a ds}{(a^2 + s^2)^{\frac{1}{2}}} = dx.$$ Integrate both members.

$$a \log(s + \sqrt{a^2 + s^2}) = x + C;$$

when $x = 0,\ s = 0$,

hence $$C = a \log a,$$

and $$\log(s + \sqrt{a^2 + s^2}) = \frac{x}{a} + \log a,$$

$$s + \sqrt{a^2 + s^2} = ae^{\frac{x}{a}},$$

$$\sqrt{a^2 + s^2} = ae^{\frac{x}{a}} - s,$$

$$a^2 = a^2 e^{2\frac{x}{a}} - 2ae^{\frac{x}{a}} s,$$

$$s = \frac{a}{2}(e^{\frac{x}{a}} - e^{-\frac{x}{a}}) = a \tan \tau \qquad \text{by (3).}$$

Hence $$a\frac{dy}{dx} = \frac{a}{2}(e^{\frac{x}{a}} - e^{-\frac{x}{a}}),$$

and $$y = \frac{a}{2}(e^{\frac{x}{a}} + e^{-\frac{x}{a}}) + C.$$

If we change our axes, taking the origin at a point a units below the lowest point of the curve, $y = a$ when $x = 0$, and therefore $C = 0$, and we get, as the equation of the catenary,

$$y = \frac{a}{2}(e^{\frac{x}{a}} + e^{-\frac{x}{a}}). \tag{4}$$

EXAMPLE.

Find the curve in which the cables of a suspension-bridge must hang. *Ans.* A parabola.

The Tractrix.

102. If two particles are attached to a string, and rest on a rough horizontal plane, and one, starting with the string stretched, moves in a straight line at right angles with the initial position of the string, dragging the other particle after it, the path of the second particle is called the *tractrix*.

Take as the axis of X the path of the first particle, and as the axis of Y the initial position of the string, and let a be the length of the string. From the nature of the curve the string is always a tangent, and we shall have for any point P

$$\frac{y}{a} = -\sin\tau, \qquad [1]$$

for τ lying in the fourth quadrant has a negative sine.

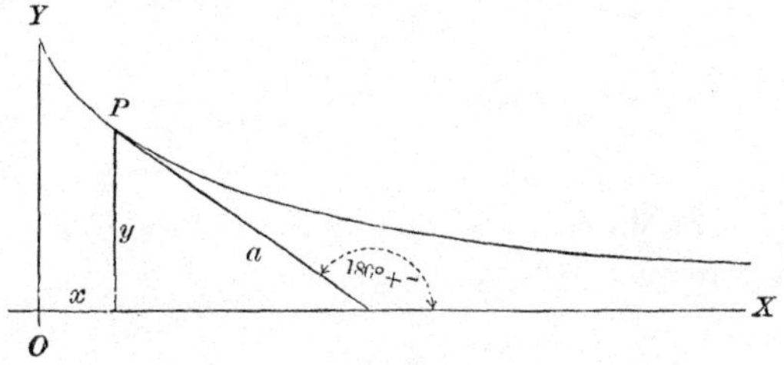

$$\frac{y^2}{a^2} = \sin^2\tau = \frac{dy^2}{ds^2} = \frac{dy^2}{dx^2 + dy^2};$$

hence
$$y^2(dx^2 + dy^2) = a^2 dy^2,$$
$$y^2 dx^2 = (a^2 - y^2)\,dy^2,$$
and
$$dx = \pm\frac{(a^2 - y^2)^{\frac{1}{2}}dy}{y}$$

is the differential equation of the *tractrix*.

On the right-hand half of the curve τ is in the fourth quadrant, $\frac{dy}{dx}$ or $\tan\tau$ is negative, and we shall write the equation

$$dx = -\frac{(a^2 - y^2)^{\frac{1}{2}}dy}{y}. \qquad [2]$$

If we allow the radical to be ambiguous in sign we shall get also the curve that would be described if the first particle went to the left instead of to the right. The tractrix curve, generally considered, includes these two portions.

Integrating both members of [2], and determining the arbitrary constant, we get

$$x = -\sqrt{a^2 - y^2} + a\log\frac{a + \sqrt{a^2 - y^2}}{y} \qquad [3]$$

as the equation of the *tractrix*.

EXAMPLES.

(1) Show by Art. 102 (1) that in the tractrix $s = a \log \frac{a}{y}$ if s is measured from the starting-point.

(2) Find the evolute of the tractrix. (I. Art. 93.)

Rectification of Curves.

103. In finding the length of an arc of a given curve we can regard it as the limit of the sum of the differentials of the arc, and express it by a definite integral.

We shall have
$$s = \int_{x=x_0}^{x=x_1} \sqrt{dx^2 + dy^2}.$$

Of course in using this formula we must express $\sqrt{dx^2 + dy^2}$ in terms of x only, or of y only, or of some single variable on which x and y depend, before we can integrate.

For example; let us find the length of an arc of the circle

$$x^2 + y^2 = a^2.$$

$$2x.dx + 2y.dy = 0,$$

$$dy = -\frac{x.dx}{y},$$

$$dx^2 + dy^2 = \frac{x^2 + y^2}{y^2} dx^2 = \frac{a^2}{y^2} dx^2 = \frac{a^2}{a^2 - x^2} dx^2,$$

$$s = a \int_{x_0}^{x_1} \frac{dx}{\sqrt{a^2 - x^2}} = a \left(\sin^{-1} \frac{x_1}{a} - \sin^{-1} \frac{x_0}{a} \right).$$

The length of a quadrant $= a \int_0^a \frac{dx}{\sqrt{a^2 - x^2}} = \frac{\pi a}{2}$;

the length of a circumference $= 2\pi a$.

Length of Arc of Cycloid.

104. For the cycloid we have

$$\left.\begin{array}{l} x = a\theta - a\sin\theta \\ y = a \quad - a\cos\theta \end{array}\right\}. \qquad \text{(I. Art. 99.)}$$

$$dx = a(1 - \cos\theta)\,d\theta = y\,d\theta,$$

$$\theta = \text{vers}^{-1}\frac{y}{a},$$

$$d\theta = \frac{1}{a}\cdot\frac{dy}{\sqrt{2\frac{y}{a} - \frac{y^2}{a^2}}} = \frac{dy}{\sqrt{2ay - y^2}},$$

$$dx = \frac{y\,dy}{\sqrt{2ay - y^2}},$$

$$ds^2 = dx^2 + dy^2 = \frac{2ay\,dy^2}{2ay - y^2} = \frac{2a\,dy^2}{2a - y},$$

$$ds = \sqrt{2a}\,.\frac{dy}{\sqrt{2a - y}},$$

$$s = \sqrt{2a}\int_{y_0}^{y_1}\frac{dy}{\sqrt{2a - y}} = 2\sqrt{2a}\left(\sqrt{2a - y_0} - \sqrt{2a - y_1}\right).$$

If the arc is measured from the cusp, $y_0 = 0$,

$$s = 4a - 2\sqrt{2a}\sqrt{2a - y_1}. \qquad [1]$$

If the arc is measured to the highest point, $y_1 = 2a$,

$$s = 4a.$$

The whole arch $= 8a$.

EXAMPLE.

Taking the origin at the vertex, and taking the direction downward as the positive direction for y, the equations become

$$\left.\begin{array}{l} x = a\theta + a\sin\theta \\ y = \;\; a - a\cos\theta \end{array}\right\}. \qquad \text{(I. Art. 100.)}$$

Show that $s = 2\sqrt{2ay}$ when the arc is measured from the summit of the curve.

105. We can rectify the cycloid without eliminating θ.

$$\left.\begin{aligned} x &= a\theta - a\sin\theta \\ y &= \ a - a\cos\theta \end{aligned}\right\},$$

$$dx = a(1-\cos\theta)\,d\theta,$$

$$dy = a\sin\theta . d\theta,$$

$$dx^2 + dy^2 = 2a^2 d\theta^2 (1-\cos\theta),$$

and
$$s = a\sqrt{2}\int_{\theta_0}^{\theta_1}(1-\cos\theta)^{\frac{1}{2}}d\theta,$$

$$s = a\sqrt{2}\int_{\theta_0}^{\theta_1}\left[2\sin^2\frac{\theta}{2}\right]^{\frac{1}{2}}d\theta = 4a\int_{\theta_0}^{\theta_1}\sin\frac{\theta}{2}d\frac{\theta}{2} = 4a\left(\cos\frac{\theta_0}{2} - \cos\frac{\theta_1}{2}\right).$$

If $\theta_0 = 0$ and $\theta_1 = 2\pi$, we get $s = 8a$ as the whole curve.

106. Let us find the *length of an arch of the epicycloid.*

$$\left.\begin{aligned} x &= (a+b)\cos\theta - b\cos\frac{(a+b)}{b}\theta \\ y &= (a+b)\sin\theta - b\sin\frac{a+b}{b}\theta \end{aligned}\right\}, \text{(I. Art.109 [1].)}$$

$$dx = \left[-(a+b)\sin\theta + (a+b)\sin\frac{a+b}{b}\theta\right]d\theta,$$

$$dy = \left[\ \ (a+b)\cos\theta - (a+b)\cos\frac{a+b}{b}\theta\right]d\theta.$$

$$ds^2 = (a+b)^2 d\theta^2\left[2 - 2\left(\cos\frac{a+b}{b}\theta\cos\theta + \sin\frac{a+b}{b}\theta\sin\theta\right)\right]$$

$$= 2(a+b)^2 d\theta^2\left(1 - \cos\frac{a}{b}\theta\right).$$

$$s = (a+b)\sqrt{2}\int_{\theta_0}^{\theta_1}\left(1 - \cos\frac{a}{b}\theta\right)^{\frac{1}{2}}d\theta,$$

$$s = \frac{4b(a+b)}{a}\left[\cos\frac{a}{2b}\theta_0 - \cos\frac{a}{2b}\theta_1\right]. \qquad [1]$$

To get a complete arch we must let $\theta_0 = 0$ and $\theta_1 = \frac{2b}{a}\pi$. Hence, for a whole arch,

$$8b(a+b)$$

EXAMPLES.

(1) Find the length of an arch of a hypocycloid.

Ans. $s = \dfrac{8b(a-b)}{a}$.

(2) Find the length of an arch of the curve $x^{\frac{2}{3}} + y^{\frac{2}{3}} = a^{\frac{2}{3}}$, and show that it agrees with the result of Ex. 1.

(v. I. Art. 109, Ex. 2.)

107. Let us attempt to find the *length of an arc of the ellipse*

$$\frac{x^2}{a^2} + \frac{y^2}{b^2} = 1.$$

We have
$$\frac{2x\,dx}{a^2} + \frac{2y\,dy}{b^2} = 0,$$

$$dy = -\frac{b^2 x}{a^2 y}\,dx,$$

$$ds^2 = \frac{b^4x^2 + a^4y^2}{a^4y^2}\,dx^2 = \frac{a^2 - \dfrac{a^2-b^2}{a^2}x^2}{a^2-x^2}\,dx^2 = \frac{a^2-e^2x^2}{a^2-x^2}\,dx^2,$$

where e is the *eccentricity* of the ellipse.

Hence
$$s = \int_{x_0}^{x_1}\left[\frac{a^2-e^2x^2}{a^2-x^2}\right]^{\frac{1}{2}} dx. \qquad [1]$$

The length of the *elliptic quadrant* is

$$s_q = \int_0^a \left[\frac{a^2-e^2x^2}{a^2-x^2}\right]^{\frac{1}{2}} dx. \qquad [2]$$

These integrals cannot be obtained directly, but

$$\left[\frac{a^2-e^2x^2}{a^2-x^2}\right]^{\frac{1}{2}} = \left[1 + \frac{(1-e^2)\,x^2}{a^2-x^2}\right]^{\frac{1}{2}}$$

can be expanded by the Binomial Theorem, and the terms of the result can be integrated separately, and we shall have the required length expressed by a series.

A more convenient way of dealing with the problem is to use an auxiliary angle. Instead of $\dfrac{x^2}{a^2} + \dfrac{y^2}{b^2} = 1$ we can use the pair of equations

$$\left.\begin{aligned} x &= a\sin\phi \\ y &= b\cos\phi \end{aligned}\right\}, \qquad \text{(I. Art. 150),}$$

$$dx = a\cos\phi . d\phi,$$
$$dy = -b\sin\phi . d\phi,$$
$$ds^2 = (a^2\cos^2\phi + b^2\sin^2\phi)\,d\phi^2 = [a^2 - (a^2 - b^2)\sin^2\phi]\,d\phi^2$$
$$= a^2\left(1 - \frac{a^2 - b^2}{a^2}\sin^2\phi\right)d\phi^2 = a^2(1 - e^2\sin^2\phi)\,d\phi^2,$$

where e is the eccentricity of the ellipse.

$$s = a\int_{\phi_0}^{\phi_1}(1 - e^2\sin^2\phi)^{\frac{1}{2}}\,d\phi \qquad [3]$$
$$= a\int_{\phi_0}^{\phi_1}[1 - \tfrac{1}{2}e^2\sin^2\phi - \tfrac{1}{2}\cdot\tfrac{1}{4}e^4\sin^4\phi - \tfrac{1}{2}\cdot\tfrac{1}{4}\cdot\tfrac{3}{6}e^6\sin^6\phi \cdots]\,d\phi.$$

For the arc of a quadrant we have

$$s_q = a\int_0^{\frac{\pi}{2}}[1 - e^2\sin^2\phi]^{\frac{1}{2}}\,d\phi. \qquad [4]$$

EXAMPLES.

(1) Obtain s_q as a series from [2], and also from [4], and compare the results with Art. 91, Ex. 5.

(2) Show that the length of an arc of the hyberbola is

$$s = b\int_{\phi_0}^{\phi_1}\left[1 + \frac{a^2e^2}{b^2}\sinh^2\phi\right]^{\frac{1}{2}}d\phi.$$

Polar Formulæ.

108. If we use polar coördinates we have

$$ds = \sqrt{dr^2 + r^2\,d\phi^2}, \qquad \text{(I. Art. 207, Ex. 2.)}$$
$$\tan\epsilon = \frac{r\,d\phi}{dr}, \qquad \text{(I. Art. 207.)}$$

From these we get, by Trigonometry,

$$\sin\epsilon = \frac{r\,d\phi}{ds}, \qquad \cos\epsilon = \frac{dr}{ds}.$$

109. Let us find the equation of the curve which crosses all its radii vectores at the same angle. Here

$$\tan\epsilon = a, \quad \text{a constant}, \qquad \frac{r\,d\phi}{dr} = a, \qquad \frac{a\,dr}{r} = d\phi,$$

$$a \log r = \phi + C, \qquad r = e^{\frac{\phi}{a} + \frac{C}{a}} = e^{\frac{C}{a}} e^{\frac{\phi}{a}},$$

$$r = b e^{\frac{\phi}{a}}. \tag{1}$$

where b is some constant depending upon the position of the origin. This curve is known as the *Logarithmic* or *Equiangular* Spiral.

110. To rectify the *Logarithmic Spiral.* We have, from 109 (1),

$$\frac{\phi}{a} = \log \frac{r}{b};$$

$$d\phi = a \frac{dr}{r},$$

$$r d\phi = a dr,$$

$$ds^2 = dr^2 + r^2 d\phi^2 = (1 + a^2) dr^2;$$

$$s = \int_{r_0}^{r_1} (1 + a^2)^{\frac{1}{2}} dr = (1 + a^2)^{\frac{1}{2}} (r_1 - r_0).$$

EXAMPLES.

(1) Find the length of an arc of the parabola from its polar equation

$$r = \frac{m}{1 + \cos \phi}.$$

(2) Find the length of an arc of the Spiral of Archimedes

$$r = a\phi.$$

111. To rectify the *Cardioide.* We have

$$r = 2a(1 - \cos \phi), \qquad \text{(I. Art. 109, Ex. 1)},$$

$$dr = 2a \sin \phi . d\phi,$$

$$ds^2 = 4a^2 \sin^2 \phi . d\phi^2 + 4a^2 (1 - \cos \phi)^2 d\phi^2$$

$$= 8a^2 d\phi^2 (1 - \cos \phi),$$

$$s = 2\sqrt{2} . a \int_{\phi_0}^{\phi_1} (1 - \cos \phi)^{\frac{1}{2}} d\phi = 8a \left[\cos \frac{\phi_0}{2} - \cos \frac{\phi_1}{2} \right]$$

$$= 16a \text{ for the whole perimeter.}$$

Involutes.

112. If we can express the length of the arc of a given curve, measured from a fixed point, in terms of the coördinates of its variable extremity, we can find the equation of the *involute* of the curve.

We have found the equations of the evolute of $y = fx$ in the form

$$\left.\begin{aligned} x' &= x - \rho \cos \nu \\ y' &= y - \rho \sin \nu \end{aligned}\right\}, \qquad \text{(I. Art. 91).}$$

We have proved that $\tan \nu = \tan \tau'$, (I. Art. 95),

and that $$\frac{ds'}{d\rho} = 1, \qquad \text{(I. Art. 96);}$$

$$\left.\begin{aligned} \sin \tau' &= \frac{dy'}{ds'} \\ \cos \tau' &= \frac{dx'}{ds'} \end{aligned}\right\}, \qquad \text{(Art. 100).}$$

Since $\tan \nu = \tan \tau'$, $\nu = \tau'$ or $\nu = 180° + \tau'$.

As normal and radius of curvature have opposite directions, we shall consider $\nu = 180° + \tau'$.

Then $\sin \nu = - \sin \tau'$ and $\cos \nu = - \cos \tau'$.

Hence $$x' = x + \rho \frac{dx'}{ds'}, \qquad (1)$$

$$y' = y + \rho \frac{dy'}{ds'}. \qquad (2)$$

Since $$d\rho = ds',$$

$$\rho = s' + l \qquad (3)$$

where l is an arbitrary constant. Since x and y are the coördinates of any point of the *involute*, it is only necessary to eliminate x', y', and ρ by combining equations (1), (2), and (3) with the equation of the evolute.

As we are supposed to start with the equation of the evolute and work towards the equation of the involute, it will be more natural to accent the letters belonging to the latter curve instead of those going with the former; and our equations may be written

$$x = x' + \rho' \frac{dx}{ds}; \qquad y = y' + \rho' \frac{dy}{ds}; \qquad \rho' = s + l. \tag{4}$$

Since $\rho' = l$ when $s = 0$, it follows that l is the free portion of the string with which we start. (I. Art. 97.) By varying l we may get different involutes of the same curve.

To test our method, let us find the involute of the curve

$$y^2 = \frac{8}{27m}(x - m)^3, \tag{5}$$

for which $l = m$. We must first find s.

$$2\,ydy = \frac{8}{9m}(x - m)^2 dx,$$

$$dy = \frac{4}{9m}\frac{(x - m)^2}{y} dx,$$

$$ds^2 = \frac{2x + m}{3m} dx^2,$$

$$s = \frac{1}{\sqrt{3m}} \int_m^x (2x + m)^{\frac{1}{2}} dx = \frac{1}{3\sqrt{3m}}(2x + m)^{\frac{3}{2}} - m,$$

$$\rho' = s + m = \frac{1}{3\sqrt{3m}}(2x + m)^{\frac{3}{2}},$$

$$x = x' + \frac{2x + m}{3},$$

$$y = y' + \frac{4}{27m}\frac{(2x + m)(x - m)^2}{y},$$

$$x' = \frac{x - m}{3},$$

$$y' = -\frac{4}{9y}(x - m)^2 = -\frac{4x'^2}{y},$$

$$x = 3x' + m,$$

$$y = -\frac{4x'^2}{y'}.$$

Substituting in (5) the values of x and y just obtained, we have

$$y'^2 = 2mx'$$

as the equations of the required involute.

EXAMPLE.

Find an involute of $ay^2 = x^3$.

113. An involute of the cycloid is easily found. Take equations I. Art. 100 (C).

$$\left.\begin{aligned} x &= a\theta + a\sin\theta \\ y &= -a + a\cos\theta \end{aligned}\right\}.$$

Let
$$\rho' = s,$$

$$dx = a(1+\cos\theta)d\theta \qquad = 2a\cos^2\frac{\theta}{2}\,d\theta,$$

$$dy = -a\sin\theta . d\theta \qquad = -2a\sin\frac{\theta}{2}\cos\frac{\theta}{2}\,d\theta,$$

$$ds^2 = 2a^2 d\theta^2(1+\cos\theta) = 4a^2 d\theta^2 \cos^2\frac{\theta}{2},$$

$$s = 2a\int_0^\theta \cos\frac{\theta}{2}\,d\theta \qquad = 4a\sin\frac{\theta}{2},$$

$$x = x' + 4a\sin\frac{\theta}{2}\cos\frac{\theta}{2} = x' + 2a\sin\theta,$$

$$y = y' - 4a\sin^2\frac{\theta}{2} \qquad = y' - 2a(1-\cos\theta),$$

$$\left.\begin{aligned} x' &= a\theta - a\sin\theta \\ y' &= a - a\cos\theta \end{aligned}\right\}$$

a cycloid with its cusp at the summit of the given cycloid.

EXAMPLE.

From the equations of a circle

$$\left.\begin{aligned} x &= a\cos\phi \\ y &= a\sin\phi \end{aligned}\right\}$$

obtain the equations of the involute of the circle. Let $l = 0$.

Ans. $\left.\begin{aligned} x' &= a(\cos\phi + \phi\sin\phi) \\ y' &= a(\sin\phi - \phi\cos\phi) \end{aligned}\right\}.$

Intrinsic Equation of a Curve.

114. An equation connecting the length of the arc, measured from a fixed point of any curve to a variable point, with the angle between the tangent at the fixed point and the tangent at the variable point, is the *intrinsic equation* of the curve. If the fixed point is the origin and the fixed tangent the axis of X, the variables in the *intrinsic* equation are s and τ.

We have already such an equation for the catenary

$$s = a \tan \tau, \qquad \text{Art. 101 (3)}, \qquad [1]$$

the origin being the lowest point of the curve.

The intrinsic equation of a circle is obviously

$$s = a\tau, \qquad [2]$$

whatever origin we may take.

The intrinsic equation of the tractrix is easily obtained. We have

$$y = -a \sin \tau, \qquad \text{Art. 102 (1)},$$

and

$$s = a \log \frac{a}{y}; \qquad \text{Art. 102, Ex. 1.}$$

hence

$$s = a \log(-\csc \tau)$$

where τ is measured from the axis of X, and s is measured from the point where the curve crosses the axis of Y. As the curve is tangent to the axis of Y, we must replace τ by $\tau - 90°$, and we get

$$s = a \log \sec \tau \qquad [3]$$

as the intrinsic equation of the tractrix.

EXAMPLE.

Show that the intrinsic equation of an inverted cycloid, when the vertex is origin, is

$$s = 4a \sin \tau; \qquad (1)$$

when the cusp is origin, is

$$s = 4a(1 - \cos \tau). \qquad (2)$$

115. To find the intrinsic equation of the epicycloid we can use the results obtained in Art. 106.

$$dx=(a+b)\left(\sin\frac{a+b}{b}\theta-\sin\theta\right)d\theta=2(a+b)\cos\frac{a+2b}{2b}\theta\sin\frac{a}{2b}\theta.d\theta,$$

$$dy=(a+b)\left(\cos\theta-\cos\frac{a+b}{b}\theta\right)d\theta=2(a+b)\sin\frac{a+2b}{2b}\theta\sin\frac{a}{2b}\theta.d\theta,$$

by the formulas of Trigonometry

$$\sin\alpha-\sin\beta=2\cos\tfrac{1}{2}(\alpha+\beta)\sin\tfrac{1}{2}(\alpha-\beta),$$

$$\cos\beta-\cos\alpha=2\sin\tfrac{1}{2}(\alpha+\beta)\sin\tfrac{1}{2}(\alpha-\beta);$$

$$\tan\tau=\frac{dy}{dx}=\tan\frac{a+2b}{2b}\theta,$$

hence $$\tau=\frac{a+2b}{2b}\theta;$$

$$s=\frac{4b(a+b)}{a}\left(1-\cos\frac{a}{2b}\theta\right)\text{ by Art. 106[1];}$$

therefore $$s=\frac{4b(a+b)}{a}\left(1-\cos\frac{a}{a+2b}\tau\right) \qquad [1]$$

is the intrinsic equation of the epicycloid, with the *cusp* as origin.

If we take the origin at a vertex instead of at a cusp

$$s=\frac{4b(a+b)}{a}+s',$$

$$\tau=\frac{\pi(a+2b)}{2a}+\tau';$$

hence $$s'=\frac{4b(a+b)}{a}\sin\frac{a}{a+2b}\tau';$$

or $$s=\frac{4b(a+b)}{a}\sin\frac{a}{a+2b}\tau \qquad [2$$

is the intrinsic equation of an epicycloid referred to a vertex.

EXAMPLE.

Obtain the intrinsic equation of the hypocycloid in the forms

$$s=\frac{4b(a-b)}{a}\left(1-\cos\frac{a}{a-2b}\tau\right), \qquad (1)$$

$$s=\frac{4b(a-b)}{a}\sin\frac{a}{a-2b}\tau. \qquad (2)$$

116. The intrinsic equation of the Logarithmic Spiral is found without difficulty.

We have $r=be^{\frac{\phi}{a}}$, (Art. 109),

and $s=\sqrt{1+a^2}\,(r_1-r_0)$. (Art. 110).

If we measure the arc from the point where the spiral crosses the initial line, $r_0=b$, and we have

$$s=b\sqrt{1+a^2}\,(e^{\frac{\phi}{a}}-1).$$

In polar coördinates $\tau=\phi+\epsilon$, and in this case $\epsilon=\tan^{-1}a$; if we measure our angle from the tangent at the beginning of the arc we must subtract ϵ from the value just given, and we have

$$s=b(\sqrt{1+a^2})\,(e^{\frac{\tau}{a}}-1);$$

or, more briefly, $s=k(c^\tau-1)$, k and c being constants.

117. If we wish to get the intrinsic equation of a curve directly from the equation in rectangular coördinates, the following method will serve:

Let the axis of X be tangent to the curve at the point we take as origin.

$$\tan\tau=\frac{dy}{dx}; \qquad (1)$$

and as the equation of the curve enables us to express y in terms of x, (1) will give us x in terms of τ, say $x=F\tau$;

then $\qquad dx = F'\tau . d\tau,$ $\qquad$ divide by ds;

$$\frac{dx}{ds} = F'\tau\frac{d\tau}{ds}, \text{ but } \frac{dx}{ds} = \cos\tau;$$

hence $\qquad ds = \sec\tau F'\tau . d\tau. \qquad (2)$

Integrating both members we shall have the required intrinsic equation.

For example, let us take $x^2 = 2my$, which is tangent to the axis of X at the origin.

$$2xdx = 2mdy,$$

$$\frac{dy}{dx} = \tan\tau = \frac{x}{m},$$

$$dx = m\sec^2\tau . d\tau,$$

$$\frac{dx}{ds} = \cos\tau = m\sec^2\tau\frac{d\tau}{ds},$$

$$ds = m\sec^3\tau . d\tau, \qquad (1)$$

$$s = m\int\frac{d\tau}{\cos^3\tau} = \frac{m}{2}\left[\frac{\sin\tau}{\cos^2\tau} + \log\tan\left(\frac{\pi}{4} + \frac{\tau}{2}\right)\right] + C,$$

$$s = 0 \text{ when } \tau = 0; \qquad \therefore C = 0;$$

$$s = \frac{m}{2}\left[\frac{\sin\tau}{\cos^2\tau} + \log\tan\left(\frac{\pi}{4} + \frac{\tau}{2}\right)\right]. \qquad (2)$$

EXAMPLES.

(1) Devise a method when the curve is tangent to the axis of Y, and apply it to $y^2 = 2mx$.

(2) Obtain the intrinsic equation of $y^2 = \frac{8}{27m}(x - m)^3$.

(3) Obtain the intrinsic equation of the involute of a circle. (Art. 113, Ex.)

118. The *evolute* or the *involute* of a curve is easily found from its intrinsic equation.

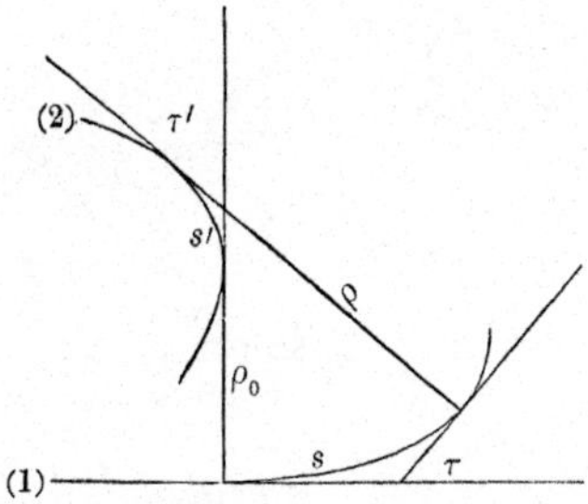

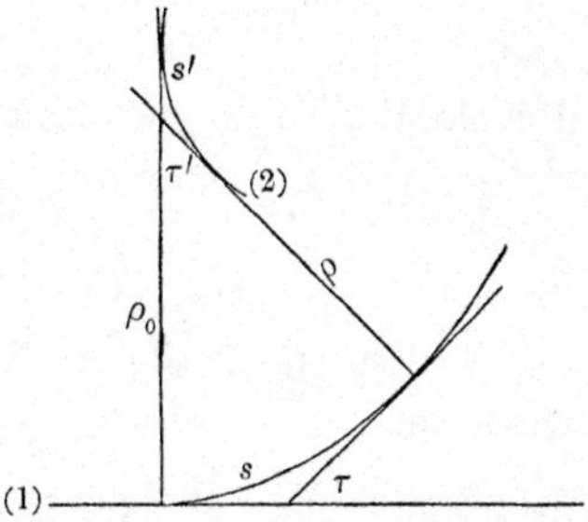

If the curvature of the given curve decreases as we pass along the curve, ρ increases, and

$$s' = \rho - \rho_0. \qquad \text{(I. Art. 96).}$$

If the curvature increases, ρ decreases, and

$$s' = \rho_0 - \rho.$$

Hence always $$s' = \pm(\rho - \rho_0); \qquad [1]$$

$$\rho = \frac{ds}{d\tau}, \qquad \text{(I. Arts. 86 and 90).}$$

We see from the figure that $\tau' = \tau$.

Hence $$s' = \pm\left[\left(\frac{ds}{d\tau}\right)_{\tau=\tau'} - \left(\frac{ds}{d\tau}\right)_{\tau=0}\right],$$

or, as we shall write it for brevity,

$$s = \pm \frac{ds}{d\tau}\Bigg|_0^\tau. \qquad [2]$$

119. The evolute of the *tractrix* $s = a \log\sec\tau$ is

$$s = a\frac{d\log\sec\tau}{d\tau}\Bigg|_0^\tau = a\tan\tau, \quad \text{the catenary.}$$

The evolute of the circle $s = a\tau$ is

$$s = a\frac{d\tau}{d\tau}\Bigg|_0^\tau = 0, \quad \text{a point.}$$

The evolute of the cycloid $s = 4a(1 - \cos\tau)$ is

$$s = 4a\left.\frac{d(1-\cos\tau)}{d\tau}\right|_0^\tau = 4a\sin\tau,$$

an equal cycloid, with its vertex at the origin.

EXAMPLES.

(1) Prove that the evolute of the logarithmic spiral is an equal logarithmic spiral.

(2) Find the evolute of a parabola.

(3) Find the evolute of the catenary.

120. The evolute of an epicycloid is a similar epicycloid, with each vertex at a cusp of the given curve.

Take the equation

$$s = \frac{4b(a+b)}{a}\left(1 - \cos\frac{a}{a+2b}\tau\right). \quad \text{Art. 115 [1].}$$

For the evolute,

$$s = \frac{4b(a+b)}{a}\cdot\left.\frac{d\left(1-\cos\frac{a}{a+2b}\tau\right)}{d\tau}\right|_0^\tau,$$

$$s = \frac{4b(a+b)}{a+2b}\sin\frac{a}{a+2b}\tau. \qquad [1]$$

The form of [1] is that of an epicycloid referred to a vertex as origin; let us find a' and b', the radii of the fixed and rolling circles.

$$s = \frac{4b'(a'+b')}{a'}\sin\frac{a'}{a'+2b'}\tau, \qquad \text{by Art. 115 [2];}$$

hence,
$$\frac{4b'(a'+b')}{a'} = \frac{4b(a+b)}{a+2b},$$

$$\frac{a'}{a'+2b'} = \frac{a}{a+2b}.$$

Solving these equations, we get

$$a' = \frac{a^2}{a+2b},$$

$$b' = \frac{ab}{a+2b},$$

$$\frac{a'}{b'} = \frac{a}{b},$$

and the radii of the fixed and rolling circles have the same ratio in the evolute as in the original epicycloid; therefore the two curves are similar.

EXAMPLE.

Show that the evolute of a hypocycloid is a similar hypocycloid.

121. We have seen that in *involute* and *evolute* τ has the same value; that is, $\tau = \tau'$.

If s' and τ' refer to the *evolute*, and s and τ to the *involute*, we have found that

$$s' = \frac{ds}{d\tau}\bigg|_0^{\tau'},$$

or

$$s' = \frac{ds}{d\tau'} - l, \quad l \text{ being a constant,}$$

the length of the radius of curvature at the origin.

$$(s' + l)\,d\tau' = ds,$$

$$s = \int_0^{\tau} (s' + l)\,d\tau'$$

is the equation of the *involute*.

The involute of the catenary $s = a \tan\tau$ is, when $l = 0$,

$$s = a\int_0^{\tau} \tan\tau . d\tau = a \log\sec\tau, \quad \textit{the tractrix.}$$

The involute of the cycloid $s = 4a\sin\tau$ when $l = 0$ is

$$s = 4a\int_0^\tau \sin\tau . d\tau = 4a(1 - \cos\tau),$$

an equal cycloid referred to its cusp as origin.

The involute of a cycloid referred to its cusp $s = 4a(1 - \cos\tau)$ when $l = 0$ is

$$s = 4a\int_0^\tau (1 - \cos\tau)\,d\tau = 4a(\tau + \sin\tau),$$

a curve we have not studied.

The involute of a circle $s = a\tau$ when $l = 0$ is

$$s = a\int_0^\tau \tau . d\tau = \frac{a\tau^2}{2}.$$

122. While any given curve has but one evolute, it has an infinite number of involutes, since the equation of the involute

$$s' = \int_0^\tau (s + l)\,d\tau$$

contains an arbitrary constant l; and the nature of the involute will in general be different for different values of l.

If we form the involute of a given curve, taking a particular value for l, and form the involute of this involute, taking the same value of l, and so on indefinitely, the curves obtained will continually approach the logarithmic spiral.

Let $$s = f\tau \qquad (1)$$

be the given curve.

$$s = \int_0^\tau (l + f\tau)\,d\tau = l\tau + \int_0^\tau f\tau . d\tau$$

is the first involute;

$$s = \int_0^\tau \left(l + l\tau + \int_0^\tau f\tau . d\tau\right) d\tau = l\tau + \frac{l\tau^2}{2} + \int_0^\tau \int_0^\tau f\tau . d\tau^2$$

is the second involute;

$$s = l\tau + \frac{l\tau^2}{2} + \frac{l\tau^3}{3\,!} + \cdots\cdots + \frac{l\tau^n}{n\,!} + \int_0^{\tau_n} f\tau . d\tau^n \qquad (2)$$

is the nth involute.

By Maclaurin's Theorem,

$$f\tau = fo + \tau f'o + \frac{\tau^2}{2\,!}f''o + \frac{\tau^3}{3\,!}f'''o + \cdots\cdots.$$

But $s = 0$ when $\tau = 0$; hence $fo = 0$, and

$$f\tau = A_1\tau \;+ \frac{A_2}{2\,!}\tau^2 + \frac{A_3}{3\,!}\tau^3 + \cdots\cdots,$$

$$\int_0^\tau f\tau.d\tau \;= \frac{A_1}{2}\tau^2 + \frac{A_2}{3\,!}\tau^3 + \frac{A_3}{4\,!}\tau^4 + \cdots\cdots,$$

$$\int_0^{\tau^2} f\tau.d\tau^2 = \frac{A_1}{3\,!}\tau^3 + \frac{A_2}{4\,!}\tau^4 + \frac{A_3}{5\,!}\tau^5 + \cdots\cdots,$$

$$\int_0^{\tau^n} f\tau.d\tau^n = \frac{A_1\tau^{n+1}}{(n+1)\,!} + \frac{A_2\tau^{n+2}}{(n+2)\,!} + \frac{A_3\tau^{n+3}}{(n+3)\,!} + \cdots\cdots; \qquad (3)$$

as n increases indefinitely all the terms of (3) approach zero (I. Art. 133), and the limiting form of (2) is

$$s = l\tau + \frac{l\tau^2}{2\,!} + \frac{l\tau^3}{3\,!} + \cdots\cdots$$

$$= l\left(1 + \frac{\tau}{1} + \frac{\tau^2}{2\,!} + \frac{\tau^3}{3\,!} + \cdots\cdots - 1\right),$$

$$s = l(e^\tau - 1) \qquad \text{by I. Art. 133 [2]},$$

which is a logarithmic spiral.

123. The equation of a curve in rectangular coördinates is readily obtained from the intrinsic equation.

Given $$s = f\tau,$$

we know that $$\sin\tau = \frac{dy}{ds},$$

and $$\cos\tau = \frac{dx}{ds};$$

hence $$dx = \cos\tau ds = \cos\tau f'\tau.d\tau,$$

$$dy = \sin\tau ds = \sin\tau f'\tau.d\tau,$$

$$\left.\begin{aligned} x &= \int_0^\tau \cos\tau f'\tau.d\tau \\ y &= \int_0^\tau \sin\tau f'\tau.d\tau \end{aligned}\right\}.$$

The elimination of τ between these equations will give us the equation of the curve in terms of x and y. Let us apply this method to the catenary.

$$s = a \tan\tau,$$

$$ds = a \sec^2\tau . d\tau,$$

$$x = a\int_0^\tau \sec\tau . d\tau = a \log\sqrt{\frac{1+\sin\tau}{1-\sin\tau}},$$

$$y = a\int_0^\tau \sec\tau \tan\tau . d\tau = a(\sec\tau - 1),$$

$$e^{\frac{2x}{a}} = \frac{1+\sin\tau}{1-\sin\tau},$$

$$\sin\tau = \frac{e^{\frac{2x}{a}}-1}{e^{\frac{2x}{a}}+1} = \frac{e^{\frac{x}{a}} - e^{-\frac{x}{a}}}{e^{\frac{x}{a}} + e^{-\frac{x}{a}}},$$

$$\sec\tau = \tfrac{1}{2}(e^{\frac{x}{a}} + e^{-\frac{x}{a}}),$$

$$y = \frac{a}{2}(e^{\frac{x}{a}} + e^{-\frac{x}{a}}) - a,$$

the equation of the catenary referred to its lowest point as origin.

Curves in Space.

124. The length of the arc of a curve of double curvature is the limit of the sum of the chords of smaller arcs into which the given arc may be broken up, as the number of these smaller arcs is indefinitely increased. Let (x, y, z), $(x + dx,\ y + \Delta y,\ z + \Delta z)$ be the coördinates of the extremities of any one of the small arcs in question; $dx, \Delta y, \Delta z$ are infinitesimal; $\sqrt{dx^2 + \Delta y^2 + \Delta z^2}$ is the length of the chord of the arc. In dealing with the limit of the sum of these chords, any one may be replaced by a quantity differing from it by infinitesimals of higher order than the first. $\sqrt{dx^2 + dy^2 + dz^2}$ is such a value;

hence $$s = \int_{x=x_0}^{x=x_1} \sqrt{dx^2 + dy^2 + dz^2}.$$

Let us rectify the helix.

$$\left.\begin{aligned} x &= a\cos\theta \\ y &= a\sin\theta \\ z &= k\theta \end{aligned}\right\}. \qquad \text{(I. Art. 214.)}$$

$$dx = -a\sin\theta . d\theta,$$

$$dy = a\cos\theta . d\theta,$$

$$dz = kd\theta,$$

$$ds^2 = (a^2 + k^2)d\theta^2,$$

$$s = (a^2 + k^2)^{\frac{1}{2}}\int_{\theta_0}^{\theta_1} d\theta = \sqrt{a^2 + k^2}(\theta_1 - \theta_0).$$

EXAMPLES.

(1) Find the length of the curve $\left(y = \frac{x^2}{2a}, z = \frac{x^3}{6a^2}\right)$.

Ans. $s = x + z + l.$

(2) $y = 2\sqrt{ax} - x,\ z = x - \frac{2}{3}\sqrt{\frac{x^3}{a}}.$ *Ans.* $s = x + y - z + l.$

CHAPTER X.

AREAS.

125. We have found and used a formula for the area bounded by a given curve, the axis of X, and a pair of ordinates.

$$A = \int y dx.$$

We can readily get this formula as a definite integral. The area in the figure is the sum of the slices into which it is divided by the ordinates; if Δx, the base of each slice, is indefinitely decreased, the slice is infinitesimal. The area of any slice differs from $y\Delta x$ by less than $\Delta y \Delta x$, which is of the second order if Δx is the principal infinitesimal. We have then

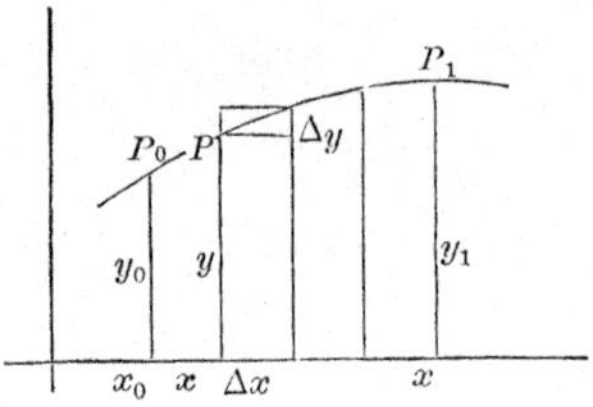

$$A = \underset{\Delta x \doteq 0}{\text{limit}} \sum_{x=x_0}^{x=x_1} y\Delta x \qquad \text{by I. Art. 161.}$$

Hence

$$A = \int_{x_0}^{x_1} y dx. \qquad [1]$$

If the curve in question lies above the axis of X, and x_0 is less than x_1, each ordinate is positive, each Δx is positive, each term of the sum whose limit is required is positive, the sum is positive, and the limit of the sum or the area sought is positive. If, however, the curve lies below the axis of X, and x_0 is less than x_1, each ordinate is negative, each Δx is positive, each term of the sum is negative, the sum is negative, and the limit of the sum or the area sought is negative. If, then, the curve happens to cross the axis between x_0 and x_1, formula [1] gives us the difference between the portion of the area above the axis of X and the portion below the axis of X, but throws no light upon the magnitudes of the separate portions. Consequently, in any actual geometrical problem it is usually necessary to find the portion of the required area above the axis of X and the portion below the axis of X separately; and for this purpose it is essential to know at what points the curve crosses the axis. Indeed, if the problem is in the least complicated, it is necessary to begin by carefully tracing the given curve from its equation, and then to keep its form and position in mind during the whole process of solution.

EXAMPLES.

(1) Show that $\int_{y_0}^{y_1} xdy$ is the area bounded by a curve, the axis of Y, and perpendiculars let fall from the ends of the bounding arc upon the axis of Y.

(2) If the axes are inclined at the angle ω, show that these formulas become

$$A = \sin\omega \int_{x_0}^{x_1} ydx = \sin\omega \int_{y_0}^{y_1} xdy.$$

(3) Find the area bounded by the axis of X, the curve $x^2 + 4y = 0$, and the ordinate of the point corresponding to the abscissa 4. *Ans.* $5\frac{1}{3}$.

(4) Find the area bounded by the axis of X, the curve $y = x^3$, and the ordinates corresponding to the abscissæ -2 and 2. *Ans.* 8.

(5). Find the area bounded by the axis of X, the axis of Y, the curve $y = \cos x$, and the ordinate corresponding to the abscissa 3π. *Ans.* 6.

126. In polar coördinates we can regard the area between two radii vectores and the curve as the limit of the sum of sectors.

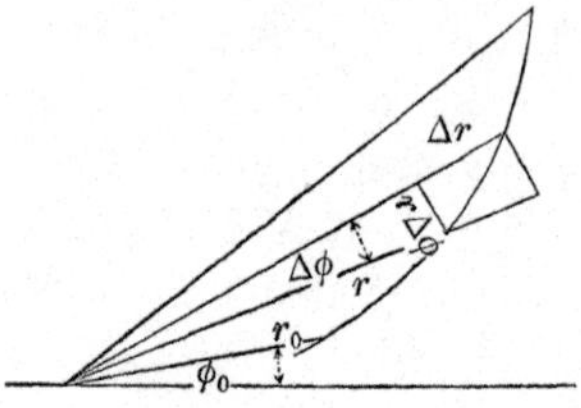

The area in question is the sum of the smaller sectorial areas, any one of which differs from $\frac{1}{2}r^2\Delta\phi$ by less than the difference between the two circular sectors $\frac{1}{2}(r+\Delta r)^2\Delta\phi$ and $\frac{1}{2}r^2\Delta\phi$; that is, by less than $r\Delta r\Delta\phi + \frac{(\Delta r)^2\Delta\phi}{2}$, which is of the second order if $\Delta\phi$ is the principal infinitesimal.

Hence
$$A = \underset{\Delta\phi \doteq 0}{\text{limit}}\left[\tfrac{1}{2}\sum_{\phi=\phi_0}^{\phi=\phi_1} r^2\Delta\phi\right],$$

$$A = \tfrac{1}{2}\int_{\phi_0}^{\phi_1} r^2 d\phi.$$

127. Let us find the area between the catenary, the axis of X, the axis of Y, and any ordinate.

$$A = \int_0^x y dx = \frac{a}{2}\int_0^x (e^{\frac{x}{a}} + e^{-\frac{x}{a}})dx,$$

$$A = \frac{a^2}{2}(e^{\frac{x}{a}} - e^{-\frac{x}{a}}),$$

but
$$\frac{a}{2}(e^{\frac{x}{a}} - e^{-\frac{x}{a}}) = s,$$ by Art. 101.

Hence
$$A = as,$$

and the area in question is the length of the arc multiplied by the distance of the lowest point of the curve from the origin.

128. Let us find the area between the tractrix and the axis of X.

We have
$$dx = -\frac{dy}{y}\sqrt{a^2 - y^2}.$$ (Art. 102.)

$$A = \int y dx = -\int dy\sqrt{a^2 - y^2}.$$

The area in question is

$$A = -\int_a^0 dy\sqrt{a^2 - y^2} = \frac{\pi a^2}{4},$$

which is the area of the quadrant of a circle with a as radius.

EXAMPLE.

Give, by the aid of infinitesimals, a geometric proof of the result just obtained for the tractrix.

129. In the last section we found the area between a curve and its asymptote, and obtained a finite result. Of course this means that, as our second bounding ordinate recedes from the origin, the area in question, instead of increasing indefinitely, approaches a finite limit, which is the area obtained. Whether the area between a curve and its asymptote is finite or infinite will depend upon the nature of the curve.

Let us find the area between an hyperbola and its asymptote.

The equation of the hyperbola referred to its asymptotes as axes is

$$xy = \frac{a^2 + b^2}{4}.$$

Let ω be the angle between the asymptotes; then

$$A = \sin\omega \int_0^\infty y dx = \frac{a^2 + b^2}{4} \sin\omega \int_0^\infty \frac{dx}{x} = \infty.$$

Take the curve $$y^2 x = 4a^2(2a - x),$$

or $$y^2 = 4a^2 . \frac{2a - x}{x};$$

any value of x will give two values of y equal with opposite signs; therefore the axis of x is an axis of symmetry of the curve.

When $x = 2a$, $y = 0$; as x decreases, y increases; and when $x = 0$, $y = \infty$. If x is negative, or greater than $2a$, y is imaginary. The shape of the curve is something like that in the figure, the axis of Y being an asymptote. The area between the curve and the asymptote is then either

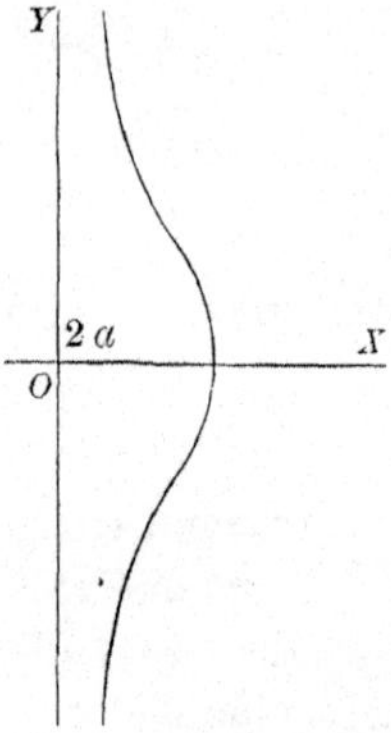

$$A = 2\int_0^{2a} y\,dx \quad \text{or} \quad A = 2\int_0^{\infty} x\,dy;$$

by the first formula,

$$A = 4a\int_0^{2a}\sqrt{\frac{2a-x}{x}}\cdot dx = 4a^2\pi;$$

by the second,

$$A = 16a^3\int_0^{\infty}\frac{dy}{y^2+4a^2} = 4a^2\pi.$$

EXAMPLES.

(1) Find the area between the curve $y^2(x^2+a^2) = a^2x^2$ and its asymptote $y = a$. *Ans.* $A = 2a^2$.

(2) Find the area between $y^2(2a-x) = x^3$ and its asymptote $x = 2a$. *Ans.* $A = 3\pi a^2$.

(3) Find the area bounded by the curve $y^2 = \dfrac{x^2(a+x)}{a-x}$ and its asymptote $x = a$.

Ans. $A = 2a^2\left(1+\dfrac{\pi}{4}\right)$.

130. If the coördinates of the points of a curve are expressed in terms of an auxiliary variable, no new difficulty is presented.

Take the case of the circle $x^2+y^2 = a^2$, which may be written

$$\left.\begin{aligned} x &= a\cos\phi \\ y &= a\sin\phi \end{aligned}\right\};$$

$$dy = a\cos\phi\,d\phi.$$

The whole area $A = a^2\displaystyle\int_0^{2\pi}\cos^2\phi\,d\phi = \pi a^2.$

EXAMPLES.

(1) The whole area of an ellipse $\left.\begin{array}{l} x = a\cos\phi \\ y = b\sin\phi \end{array}\right\}$ is πab.

(2) The area of an arch of the cycloid is $3\pi a^2$.

(3) The area of an arch of the companion to the cycloid $x = a\theta$, $y = a(1-\cos\theta)$ is $2\pi a^2$.

131. If we wish to find the area between two curves, or the area bounded by a closed curve, the altitude of our elementary rectangle is the difference between the two values of y, which correspond to a single value of x. If the area between two curves is required, we must find the abscissas of their points of intersection, and they will be our limits of integration; if the whole area bounded by a closed curve is required, we must find the values of x belonging to the points of contact of tangents parallel to the axis of Y.

Let us find the whole area of the curve

$$a^4 y^2 + b^2 x^4 = a^2 b^2 x^2,$$

or

$$a^4 y^2 = b^2 x^2 (a^2 - x^2).$$

The curve is symmetrical with reference to the axis of X, and passes through the origin. It consists of two loops whose areas must be found separately. Let us find where the tangents are parallel to the axis of Y.

$$y = \frac{b}{a^2} x\sqrt{a^2 - x^2},$$

$$\frac{dy}{dx} = \frac{b}{a^2}\cdot\frac{a^2 - 2x^2}{\sqrt{a^2 - x^2}} = \tan\tau.$$

$\tau = \frac{\pi}{2}$ when $\tan\tau = \infty$, that is, when $x = \pm a$.

$$A = 2\frac{b}{a^2}\int_{-a}^{0} x\sqrt{a^2 - x^2}.dx + 2\frac{b}{a^2}\int_0^a x\sqrt{a^2 - x^2}.dx = \tfrac{4}{3}ab.$$

Again; find the whole area of $(y-x)^2 = a^2 - x^2$.

$$y = x \pm \sqrt{a^2 - x^2},$$

$$A = \int (y' - y'')\,dx = \int 2\sqrt{a^2 - x^2}\,.\,dx.$$

To find the limits of integration, we must see where $\tau = \frac{\pi}{2}$.

$$\frac{dy}{dx} = \frac{\sqrt{a^2 - x^2} \mp x}{\sqrt{a^2 - x^2}} = \infty \text{ when } x = \pm a.$$

$$A = 2\int_{-a}^{a} \sqrt{a^2 - x^2}\,.\,dx = \pi a^2.$$

EXAMPLES.

(1) Find the area of the loop of the curve $y^2 = \frac{x^2(a+x)}{a-x}$.

Ans. $2a^2\left(1 - \frac{\pi}{4}\right)$.

(2) Find the area between the curves $y^2 - 4ax = 0$ and $x^2 - 4ay = 0$.

Ans. $\frac{16a^2}{3}$.

(3) Find the whole area of the curve $x^{\frac{2}{3}} + y^{\frac{2}{3}} = a^{\frac{2}{3}}$. *Ans.* $\frac{3}{8}\pi a^2$.

(4) Find the area of a loop of $a^2y^4 = x^4(a^2 - x^2)$. *Ans.* $\frac{4a^2}{5}$.

(5) Find the whole area of the curve

$$2y^2(a^2 + x^2) - 4ay(a^2 - x^2) + (a^2 - x^2)^2 = 0.$$

Ans. $a^2\pi\left(4 - \frac{5\sqrt{2}}{2}\right)$.

132. We have seen that in polar coördinates

$$A = \tfrac{1}{2}\int_{\phi_0}^{\phi_1} r^2\,d\phi.$$

Let us try one or two examples.

(*a*) To find the whole area of a circle.

The polar equation is $r = a$.

$$A = \tfrac{1}{2}\int_0^{2\pi} a^2\,d\phi = \pi a^2.$$

(b) To find the area of the cardioide $r = 2a(1 - \cos\phi)$.

$$A = \tfrac{1}{2}\int_0^{2\pi} 4a^2(1 - \cos\phi)^2 d\phi = 2a^2\int_0^{2\pi}(1 - 2\cos\phi + \cos^2 d\phi)d\phi,$$

$$A = 6a^2\pi.$$

(c) To find the area between an arch of the epicycloid and the circumference of the fixed circle.

$$\left.\begin{aligned} x &= (a+b)\cos\theta - b\cos\frac{a+b}{b}\theta \\ y &= (a+b)\sin\theta - b\sin\frac{a+b}{b}\theta \end{aligned}\right\}.$$

We can get the area bounded by two radii vectores and the arch in question, and subtract the area of the corresponding sector of the fixed circle.

Changing to polar coördinates,

$$x = r\cos\phi,$$
$$y = r\sin\phi.$$

We want $\frac{1}{2}\int r^2 d\phi$.

$$\tan\phi = \frac{y}{x},$$

$$\sec^2\phi d\phi = \frac{xdy - ydx}{x^2};$$

but, since $\qquad x = r\cos\phi, \quad \sec\phi = \frac{r}{x};$

hence $\qquad \frac{r^2 d\phi}{x^2} = \frac{xdy - ydx}{x^2},$

and $\qquad r^2 d\phi = xdy - ydx;$

$$dx = (a+b)\left(-\sin\theta + \sin\frac{a+b}{b}\theta\right)d\theta,$$

$$dy = (a+b)\left(\cos\theta - \cos\frac{a+b}{b}\theta\right)d\theta.$$

$$xdy - ydx = (a+b)(a+2b)\left(1 - \cos\frac{a}{b}\theta\right)d\theta = r^2 d\phi.$$

Our limits of integration are obviously 0 and $\frac{2b\pi}{a}$.

Hence $$A = \tfrac{1}{2}(a+b)(a+2b)\int_0^{\frac{2b\pi}{a}}\left(1-\cos\frac{a}{b}\theta\right)d\theta,$$

$$A = \frac{b\pi}{a}(a+b)(a+2b),$$

is the area of the sector of the epicycloid. Subtract the area of the circular sector πab, and we get

$$A = \frac{b^2(3a+2b)}{a}\pi$$

as the area in question.

(*d*) To find the area of a loop of the curve $r^2 = a^2\cos 2\phi$.

For any value of ϕ the values of r are equal with opposite signs. Hence the origin is a centre.

When $\phi = 0$, $r = \pm a$; as ϕ increases, r decreases in length till $\phi = \frac{\pi}{4}$, when $r = 0$; as soon as $\phi > \frac{\pi}{4}$, r is imaginary. If ϕ decreases from 0, r decreases in length until $\phi = -\frac{\pi}{4}$, when $r = 0$; and when $\phi < \frac{\pi}{4}$, r is imaginary. To get the area of a loop, then, we must integrate from $\phi = -\frac{\pi}{4}$ to $\phi = \frac{\pi}{4}$.

$$A = \tfrac{1}{2}\int_{-\frac{\pi}{4}}^{\frac{\pi}{4}} r^2 d\phi = \tfrac{1}{2}a^2\int_{-\frac{\pi}{4}}^{\frac{\pi}{4}}\cos 2\phi . d\phi = \frac{a^2}{2}.$$

EXAMPLES.

(1) Find the area of a sector of the parabola $r = \dfrac{m}{1+\cos\phi}$.

(2) Find the area of a loop of the curve $r^2\cos\phi = a^2\sin 3\phi$.

Ans. $\dfrac{3a^2}{4} - \dfrac{a^2}{2}\log 2$.

(3) Find the whole area of the curve $r = a(\cos 2\phi + \sin 2\phi)$.

Ans. πa^2.

(4) Find the area of a loop of the curve $r\cos\phi = a\cos 2\phi$.

Ans. $\left(2 - \dfrac{\pi}{2}\right)a^2$.

(5) Find the area between $r = a(\sec\phi + \tan\phi)$ and its asymptote $r\cos\phi = 2a$.

Ans. $\left(\dfrac{\pi}{2} + 2\right)a^2$.

133. When the equation of a curve is given in rectangular coördinates, we can often simplify the problem of finding its area by transforming to polar coördinates.

For example, let us find the area of

$$(x^2+y^2)^2 = 4a^2x^2 + 4b^2y^2.$$

Transform to polar coördinates.

$$r^4 = 4r^2(a^2\cos^2\phi + b^2\sin^2\phi),$$

$$r^2 = 4\,(a^2\cos^2\phi + b^2\sin^2\phi),$$

$$A = 2\int_0^{2\pi}(a^2\cos^2\phi + b^2\sin^2\phi)\,d\phi = 2\pi(a^2+b^2).$$

EXAMPLES.

(1) Find the area of a loop of the curve $(x^2+y^2)^3 = 4a^2x^2y^2$.

Ans. $\frac{\pi a^2}{8}$.

(2) Find the whole area of the curve $\frac{x^2}{a^4}+\frac{y^2}{b^4}=\frac{1}{c^2}\left(\frac{x^2}{a^2}+\frac{y^2}{b^2}\right)^2$.

Ans. $\frac{\pi c^2}{2ab}(a^2+b^2)$.

(3) Find the area of a loop of the curve $y^3 - 3axy + x^3 = 0$.

Ans. $\frac{3a^2}{2}$.

134. The area between a curve and its evolute can easily be found from the intrinsic equation of the curve.

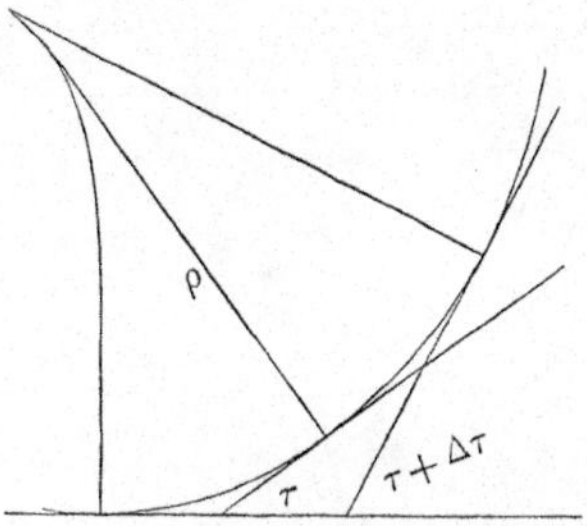

It is easily seen that the area bounded by the radii of curvature at two points infinitely near, by the curve and by the evolute, differs from $\frac{1}{2}\rho^2 d\tau$ by an infinitesimal of higher order. The area bounded by two given radii vectores, the curve and the evolute, is then

$$A = \frac{1}{2}\int_{\tau_0}^{\tau_1}\rho^2\,d\tau.$$

$$\rho = \frac{ds}{d\tau}.$$

Hence
$$A = \tfrac{1}{2}\int_{\tau_0}^{\tau_1}\left(\frac{ds}{d\tau}\right)^2 d\tau.$$

For example, the area between a cycloid and its evolute is

$$A = \tfrac{1}{2}\int_{\tau_0}^{\tau_1}\left(\frac{d(4a\sin\tau)}{d\tau}\right)^2 d\tau$$

$$= 8a^2\int_{\tau_0}^{\tau_1}\cos^2\tau d\tau.$$

Let
$$\tau_0 = 0 \quad\text{and}\quad \tau_1 = \frac{\pi}{2};$$

$$A = 8a^2\int_0^{\frac{\pi}{2}}\cos^2\tau d\tau = 2\pi a^2.$$

EXAMPLES.

(1) Find the area between a circle and its evolute.

(2) Find the area between the circle and its involute.

Holditch's Theorem.

135. If a line of fixed length move with its ends on any closed curve which is always concave toward it, the area between the curve and the locus of a given point of the moving line is equal to the area of an ellipse, of which the segments into which the line is divided by the given point are the semi-axes.

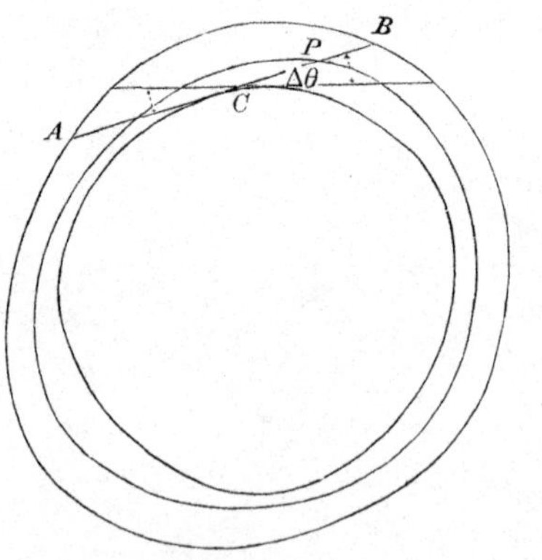

Let the figure represent the given curve, the locus of P, and the envelope of the moving line.

Let $AP = a$ and $PB = b$, and let $CB = \rho$, C being the point of contact of the moving line with its envelope. Let $AB = a + b = c$.

The area between the first curve and the second is the area between the first curve and the envelope, minus the area between the second curve and the envelope.

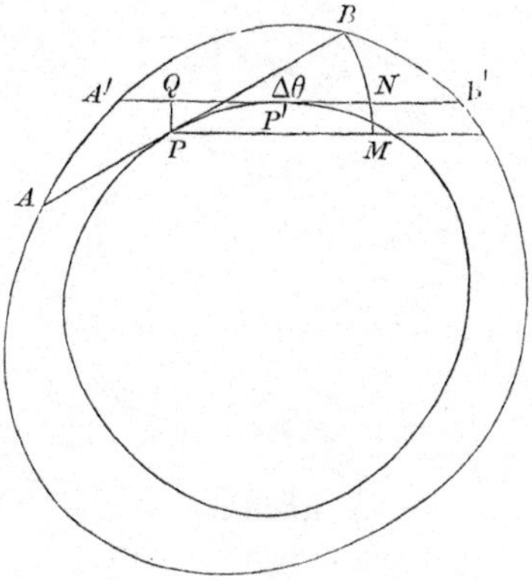

Let θ be the angle which the moving line makes at any instant with some fixed direction. Let the figure represent two near positions of the moving line; $\Delta\theta$, the angle between these positions, being the principal infinitesimal.

$PB = \rho, \;\; P'B' = \rho + \Delta\rho.$

The area $PBB'P'P$ differs from $\frac{1}{2}\rho^2 d\theta$ by an infinitesimal of higher order than the first.

$\frac{1}{2}\rho^2 d\theta$ is the area of $PBMP$, and differs from $PP'NB$ by less than the rectangle on PM and PQ, which is of higher order than the first, by I. Art. 153. But $PP'NB$ differs from $PP'B'B$ by less than the rectangle on BN and NB', which is of higher order than the first, since NB', which is less than $PP' + \Delta\rho$, is infinitesimal and $\Delta\theta$ is infinitesimal.

The area between the first curve and the envelope is then $\frac{1}{2}\int_0^{2\pi}\rho^2 d\theta$; or, since we can take $PP'A'A$ just as well for our elementary area, $\frac{1}{2}\int_0^{2\pi}(c-\rho)^2 d\theta.$

Hence $$\frac{1}{2}\int_0^{2\pi}\rho^2 d\theta = \frac{1}{2}\int_0^{2\pi}(c-\rho)^2 d\theta;$$

whence $$2c\int_0^{2\pi}\rho d\theta = 2c^2\pi,$$

or $$\int_0^{2\pi}\rho d\theta = \pi c. \tag{1}$$

The area between the second curve and the envelope is $\frac{1}{2}\int_0^{2\pi}(\rho - b)^2 d\theta.$

The area between the first curve and the second is then

$$A = \tfrac{1}{2}\int_0^{2\pi}\rho^2 d\theta - \tfrac{1}{2}\int_0^{2\pi}(\rho - b)^2 d\theta$$

$$= b\int_0^{2\pi}\rho d\theta - b^2\pi$$

$$= \pi bc - b^2\pi \qquad \text{by (1)},$$

$$= \pi b(a+b) - b^2\pi,$$

$$A = \pi ab, \qquad (2)$$

which is the area of an ellipse of which a and b are semi-axes.

Q. E. D.

EXAMPLES.

(1) If a line of fixed length move with its extremities on two lines at right angles with each other, the area of the locus of a given point of the line is that of an ellipse on the segments of the line as semi-axes.

(2) The result of (1) holds even when the fixed lines are not perpendicular.

Areas by Double Integration.

136. If we take x and y as the coördinates of any point P within our area, x and y will be independent variables, and we can find the area bounded by two given curves, $y = fx$ and $y = Fx$, by a double integration. Suppose the area in question divided into slices by lines drawn parallel to the axis of Y, and these slices subdivided into parallelograms by lines drawn parallel to the axis of X. The area of any one of the small parallelograms is $\Delta y\Delta x$. If we keep x constant, and take the sum of these rectangles from $y = fx$ to $y = Fx$, we shall get a result differing from the area of the corresponding slice by less than

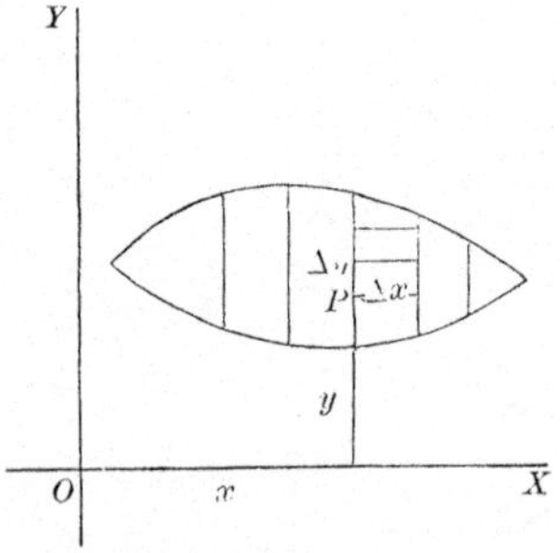

$2\,\Delta x \Delta y$, which is infinitesimal of the second order if Δx and Δy are of the first order.

Hence $$\int_{fx}^{Fx} \Delta x . dy = \Delta x \int_{fx}^{Fx} dy$$

is the area of the slice in question. If now we take the limit of the sum of all these slices, choosing our initial and final values of x, so that we shall include the whole area, we shall get the area required.

Hence $$A = \int_{x_0}^{x_1} \left(\int_{fx}^{Fx} dy \right) dx.$$

In writing a double integral, the parentheses are usually omitted for the sake of conciseness, and this formula is given as

$$A = \int_{x_0}^{x_1} \int_{fx}^{Fx} dy dx,$$

the order in which the integrations are to be performed being the same as if the parentheses were actually written.

If we begin by keeping y constant, and integrating with respect to x, we shall get the area of a slice formed by lines parallel to the axis of X, and we shall have to take the limit of the sum of these slices varying y in such a way as to include the whole area desired. In that case we should use the formula

$$A = \int_{y_0}^{y_1} \int_{f^{-1}y}^{F^{-1}y} dx dy.$$

137. For example, let us find the area bounded by the parabolas $y^2 = 4\,ax$ and $x^2 = 4\,ay$.

The parabolas intersect at the origin and at the point $(4\,a, 4\,a)$.

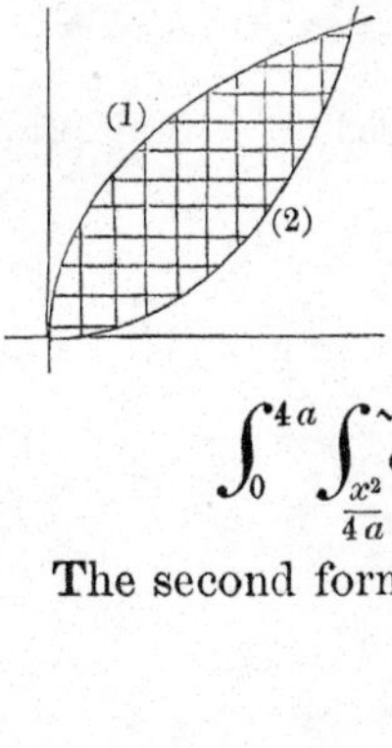

$$A = \int_0^{4a} \int_{\frac{x^2}{4a}}^{\sqrt{4ax}} dy dx, \quad \text{or} \quad A = \int_0^{4a} \int_{\frac{y^2}{4a}}^{\sqrt{4ay}} dx dy;$$

$$\int_{\frac{x^2}{4a}}^{\sqrt{4ax}} dy = \sqrt{4\,ax} - \frac{x^2}{4\,a};$$

$$\int_0^{4a} \int_{\frac{x^2}{4a}}^{\sqrt{4ax}} dy dx = \int_0^{4a} \left(\sqrt{4\,ax} - \frac{x^2}{4\,a} \right) dx = \frac{16}{3} a^2.$$

The second formula gives the same result.

EXAMPLES.

(1) Find the area of a rectangle by double integration; of a parallelogram; of a triangle.

(2) Find the area between the parabola $y^2 = ax$ and the circle $y^2 = 2ax - x^2$.

Ans. $2\left(\frac{\pi a^2}{4} - \frac{2a^2}{3}\right)$.

(3) Find the whole area of the curve $(y - mx - c)^2 = a^2 - x^2$.

Ans. πa^2.

138. If we use polar coördinates we can still find our areas by double integration.

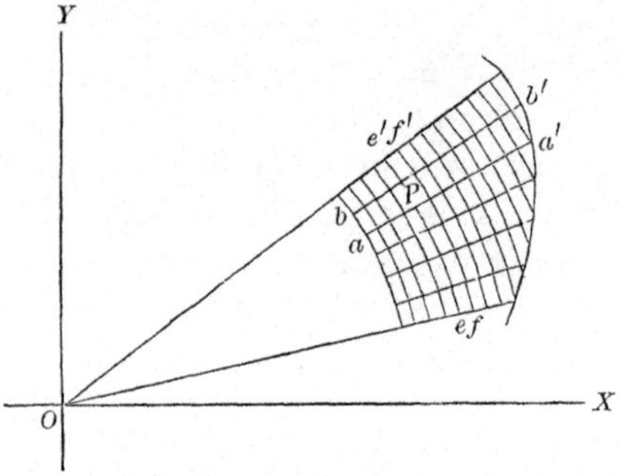

Let $r = f\phi$ and $r = F\phi$ be two curves. Divide the area between them into slices by drawing radii vectores; then subdivide these slices by drawing arcs of circles, with the origin as centre.

Let P, with coördinates r and ϕ, be any point within the space whose area is sought. The curvilinear rectangle at P has the base $r\Delta\phi$ and the altitude Δr; its area differs from $r\Delta\phi\Delta r$ by an infinitesimal of higher order than $r\Delta\phi\Delta r$.

The area of any slice as $aba'b'$ is $\int_{f\phi}^{F\phi} r\Delta\phi dr$, ϕ and $\Delta\phi$ being constant, that is $\Delta\phi\int_{f\phi}^{F\phi} rdr$. The whole area, the limit of the sum of such slices is $A = \int_{\phi_0}^{\phi_1}\int_{f\phi}^{F\phi} rdrd\phi$. (1)

Or we may first sum our rectangles, keeping r unchanged, and we get as the area of $efe'f'$

$$r\Delta r\int_{f^{-1}\phi}^{F^{-1}\phi} d\phi, \text{ and } A = \int_{r_0}^{r_1}\int_{f^{-1}\phi}^{F^{-1}\phi} rd\phi dr. \qquad (2)$$

It must be kept in mind that r in (1) and (2) is the radius vector of any point within the area sought, and not of a point on the boundary.

For example, the area between two concentric circles, $r = a$ and $r = b$, is

$$A = \int_b^a \int_0^{2\pi} r\,d\phi\,dr = \int_0^{2\pi} \int_b^a r\,dr\,d\phi = \pi(a^2 - b^2).$$

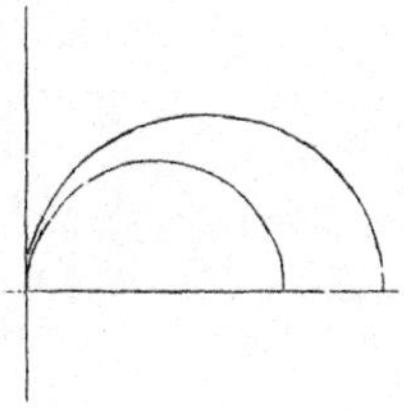

Again, let us find the area between two tangent circles and a diameter through the point of contact.

Let a and b be the two radii,

$$r = 2a\cos\phi \qquad (1)$$

and

$$r = 2b\cos\phi \qquad (2)$$

are the equations of the two circles.

$$A = \int_0^{\frac{\pi}{2}} \int_{2b\cos\phi}^{2a\cos\phi} r\,dr\,d\phi = 2(a^2 - b^2)\int_0^{\frac{\pi}{2}} \cos^2\phi\,d\phi = \frac{\pi}{2}(a^2 - b^2).$$

If we wish to reverse the order of our integrations we must break our area into two parts by an arc described from the origin as a centre, and with $2b$ as a radius; then we have

$$A = \int_0^{2b} \int_{\cos^{-1}\frac{r}{2b}}^{\cos^{-1}\frac{r}{2a}} r\,d\phi\,dr + \int_{2b}^{2a} \int_0^{\cos^{-1}\frac{r}{2a}} r\,d\phi\,dr$$

$$= \int_0^{2b} r\left(\cos^{-1}\frac{r}{2a} - \cos^{-1}\frac{r}{2b}\right)dr + \int_{2b}^{2a} r\cos^{-1}\frac{r}{2a}\,dr$$

$$= \frac{\pi}{2}(a^2 - b^2).$$

EXAMPLE.

Find the area between the axis of X and two coils of the spiral $r = a\phi$.

CHAPTER XI.

AREAS OF SURFACES.

Surfaces of Revolution.

139. If a plane curve $y = fx$ revolves about the axis of X, the area of the surface generated is the limit of the sum of the areas generated by the chords of the infinitesimal arcs into which the whole arc may be broken up. Each of these chords will generate the surface of the frustum of a cone of revolution if it revolves completely around the axis; and the area of the surface of a frustum of a cone of revolution is, by elementary Geometry, one-half the sum of the circumferences of the bases multiplied by the slant height. The frustum generated by the chord in the figure will have an area differing by infinitesimals of higher order from $\pi(y + y + \Delta y)\Delta s$ or from $2\pi y ds$. The area generated by any given arc is then

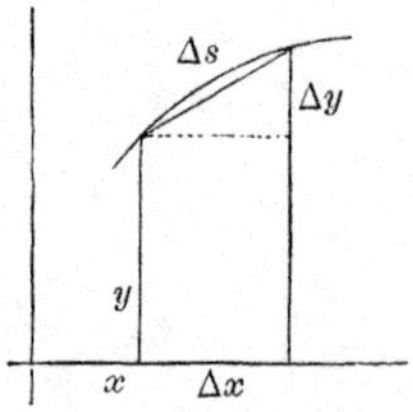

$$S = 2\pi \int_{y_0}^{y_1} y ds. \qquad [1]$$

If the arc revolves through an angle θ instead of making a complete revolution, the surface generated is

$$S = \theta \int_{y_0}^{y_1} y ds. \qquad [2]$$

It must be noted that [1] and [2] will give a positive value for S if the generating curve lies wholly above the axis of X at the start, and a negative value for S if it lies wholly below the axis of X at the start. If the curve happens to cross the axis of X between the points whose ordinates are y_0 and y_1, [1] and [2] give not the area of the surface generated by the curve in question, but the difference between the areas generated by the portion originally above the axis, and the portion originally below the axis.

EXAMPLE.

Show that if the arc revolves about the axis of Y, $S = 2\pi \int_{x_0}^{x_1} x\,ds$.

140. To find the area of a cylinder of revolution.

Take the axis of the cylinder as the axis of X. Let a be the altitude and b the radius of the base of the cylinder. The equation of the revolving line is

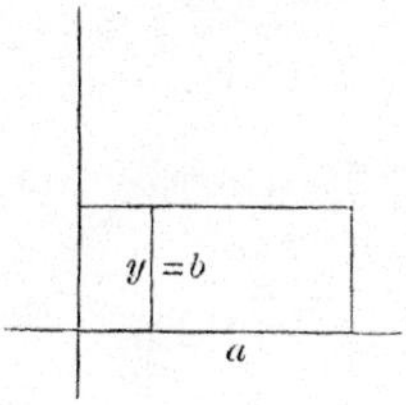

$$y = b;$$

$$dy = 0,$$

$$ds = \sqrt{dx^2 + dy^2} = dx;$$

$$S = 2\pi \int_b^a y\,dx = 2\pi ab,$$

or the product of the altitude by the circumference of the base.

Again, let us find the surface of a zone.

The equation of the generating circle is

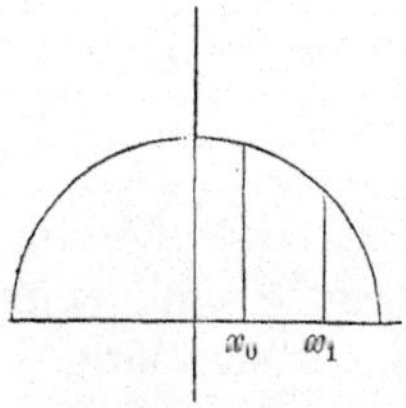

$$x^2 + y^2 = a^2;$$

$$ds = \frac{a\,dx}{y};$$

$$S = 2\pi \int_{x_0}^{x_1} a\,dx = 2a\pi(x_1 - x_0).$$

If $x_0 = -a$ and $x_1 = a$, $S = 4a^2\pi$.

Hence the surface of a zone is the altitude of the zone multiplied by the circumference of a great circle, and the surface of a sphere is equal to the areas of four great circles.

Again, take the surface generated by the revolution of a cycloid about its base.

$$\left.\begin{aligned} x &= a\theta - a\sin\theta \\ y &= a - a\cos\theta \end{aligned}\right\};$$

$$ds = a\,d\theta\sqrt{2(1 - \cos\theta)}, \qquad \text{by Art. 105;}$$

$$S = 2\pi \int_0^{2\pi} a^2\sqrt{2}\,.(1 - \cos\theta)^{\frac{3}{2}}\,d\theta = \tfrac{64}{3}\pi a^2.$$

EXAMPLES.

(1) The area of the surface generated by the revolution of the ellipse
$$\frac{x^2}{a^2}+\frac{y^2}{b^2}=1$$
about the axis of X is $2\pi ab\left(\sqrt{1-e^2}+\frac{\sin^{-1}e}{e}\right)$;

about the axis of Y is $2\pi a^2\left(1+\frac{1-e^2}{2e}\log\frac{1+e}{1-e}\right)$,

where $e^2=\frac{a^2-b^2}{a^2}$.

(2) Find the area of the surface generated by the revolution of the catenary about the axis of X; about the axis of Y.

(3) The whole surface generated by the revolution of the tractrix about its asymptote is $4\pi a^2$.

(4) The area generated by the revolution of a cycloid about its vertical axis is $8\pi a^2(\pi-\frac{4}{3})$.

(5) The area generated by the revolution of a cycloid about the tangent at its vertex is $\frac{32}{3}\pi a^2$.

(6) The area generated by the revolution of the curve $x^{\frac{2}{3}}+y^{\frac{2}{3}}=a^{\frac{2}{3}}$ about its axis is $\frac{12}{5}\pi a^2$.

141. If we know the area generated by the revolution of a curve about any axis, we can get the area generated by the revolution about any parallel axis by an easy transformation of coördinates.

Given the surface generated by the arc from s_0 to s_1 about OX, to find the area generated by the same arc when it revolves about $O'X'$.

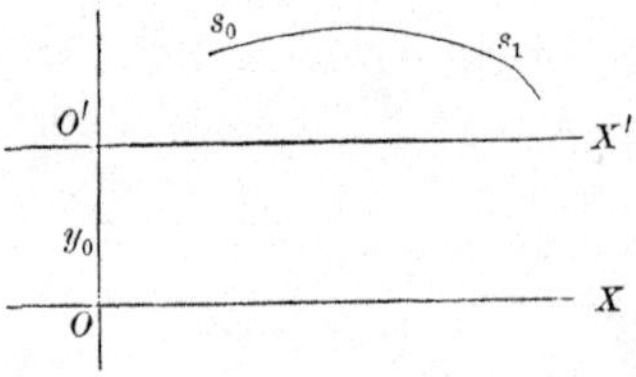

Let S be the surface about OX, and S' about $O'X'$.

We have

$$S=2\pi\int_{s_0}^{s_1}y\,ds,\quad S'=2\pi\int_{s_0}^{s_1}y'ds'.$$

By Anal. Geom.,
$$x = x',$$
$$y = y_0 + y'.$$

Hence $$dx = dx', \quad dy = dy', \quad ds = ds',$$

and $$S = 2\pi\int_{s_0}^{s_1}(y_0 + y')ds = 2\pi y_0(s_1 - s_0) + 2\pi\int_{s_0}^{s_1}y'ds,$$
$$= 2\pi y_0(s_1 - s_0) + S'.$$

Therefore $$S' = S - 2\pi y_0(s_1 - s_0). \qquad [1]$$

$s_1 - s_0$ is the length of the revolving curve; $2\pi y_0$ is the circumference of a circle of which y_0 is the radius. Hence the new area is equal to the old area minus the area of a cylinder whose length is the length of the given arc and whose base is a circle of which the distance between the two lines is radius.

In using this principle careful attention must be paid to the sign of y_0, and it must be noted that the original formula $S = 2\pi\int_{s_0}^{s_1}yds$ will always give a negative value for the area of the surface generated, if the revolving arc starts from below the axis; and hence, that the surface generated by the revolution of any curve about an axis of symmetry will come out zero.

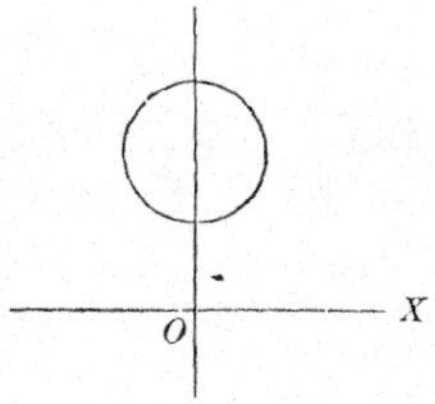

As an example of the use of the principle, let us find the surface of a ring.

Let a be the distance of the centre of the circle from the axis, and b the radius of the circle. Since the area generated by the revolution of the circle about a diameter is zero, the required area is
$$2\pi b . 2\pi a = 4\pi^2 ab.$$

EXAMPLE.

Find the area of the ring generated by the revolution of a cycloid about any axis parallel to its base.

Ans. $S = 4ab\pi\left(\pi + \dfrac{16a + 12b}{3b}\right).$

142. If we use polar coördinates,

$$S = 2\pi \int_{s_0}^{s_1} y ds$$

becomes
$$S = 2\pi \int_{s_0}^{s_1} r \sin\phi . ds.$$

where
$$ds = \sqrt{dr^2 + r^2 d\phi^2}.$$

For example; let us find the area of the surface generated by the revolution of the upper half of a cardioide about the horizontal axis.

$$r = 2a(1 - \cos\phi);$$

$$dr = 2a \sin\phi . d\phi,$$

$$ds^2 = 8a^2(1 - \cos\phi) d\phi^2,$$

$$S = 2\pi \int_0^{\pi} 4\sqrt{2} a^2 (1 - \cos\phi)^{\frac{3}{2}} \sin\phi . d\phi.$$

$$S = \tfrac{128}{5} \pi a^2.$$

EXAMPLES.

(1) Find the surface of a sphere from the polar equation.

(2) Find the surface of a paraboloid of revolution from the polar equation of the parabola

$$r = \frac{m}{1 - \cos\phi}.$$

Cylindrical Surfaces.

143. If a cylindrical surface is generated by a line which is always parallel to the axis of Z, the area of the portion bounded by two positions of the generating line, the plane of XY, and any curve whose projection on the plane of XZ is given, is easily found.

Let $ABCD$ be the cylindrical area required.

Let
$$y = fx \qquad (1)$$
be the equation of AB, the line of intersection of the surface with the plane XY; and let
$$z = Fx \qquad (2)$$
be the equation of C_1D_1, the projection of CD on the plane of XZ.

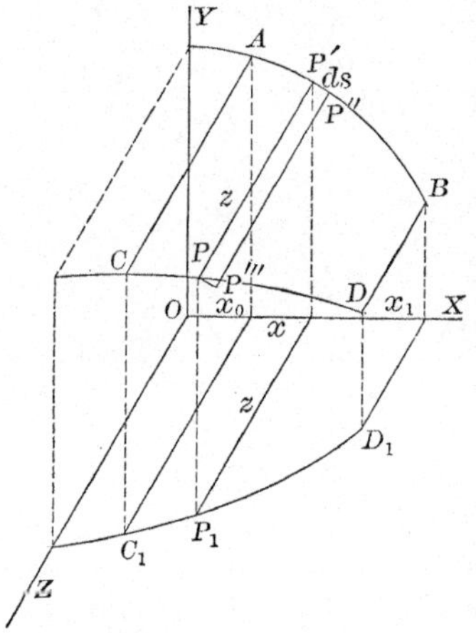

If x, y, z are the coördinates of any point P of CD, the required area is evidently the limit of the sum of rectangles, of which $PP'P''P'''$ is any one. The area of $PP'P''P'''$ differs by an infinitesimal of higher order than ds from zds, and therefore the required area $S = \int_{x_0}^{x_1} zds$.

x, z are the coördinates of P_1, and satisfy (2), and $ds = \sqrt{dx^2 + dy^2}$ where x, y are the coördinates of P' and satisfy (1).

We have, then,
$$S = \int_{x_0}^{x_1} z\sqrt{dx^2 + dy^2}. \qquad [3]$$

For example, let AB be the quadrant of a circle, and let the projection of the required area on the plane of XZ be the quadrant of an equal circle, so that the surface required is one-eighth of the surface of a *groin*.

Here
$$x^2 + y^2 = a^2, \qquad (4)$$
and
$$x^2 + z^2 = a^2; \qquad (5)$$
$$ds = \sqrt{dx^2 + dy^2} = \frac{a}{y}dx = \frac{adx}{\sqrt{a^2 - x^2}},$$
and
$$z = \sqrt{a^2 - x^2}.$$

Therefore $S = \int_0^a \sqrt{a^2 - x^2} \cdot \frac{a\,dx}{\sqrt{a^2 - x^2}} = a \int_0^a dx = a^2.$

Again, let us find the area of the curved surface of the portion of a cylinder of revolution included within a spherical surface, whose centre lies on the surface of the cylinder, and whose radius is equal to the diameter of the cylinder.

If the centre of the sphere is taken as the origin, and a diametral plane of the cylinder as the plane of XZ, the surface required is four times that indicated in the figure.

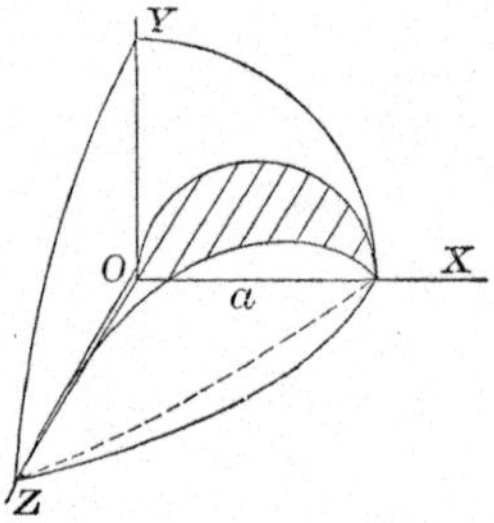

The equation of the cylinder is

$$x^2 - ax + y^2 = 0, \qquad (6)$$

and of the sphere

$$x^2 + y^2 + z^2 - a^2 = 0. \qquad (7)$$

Subtract (6) from (7), and we get

$$z^2 + ax - a^2 = 0 \qquad (8)$$

as the equation of a cylindrical surface perpendicular to the plane XZ, and passing through all the points of intersection of (6) and (7). (8) is, then, the equation of the projection on the plane of XZ of the line of intersection of the given spherical surface and the given cylindrical surface.

From (6), $ds = \sqrt{dx^2 + dy^2} = \frac{a}{2y} dx = \frac{a\,dx}{2\sqrt{ax - x^2}}.$

From (8), $z = \sqrt{a^2 - ax}.$

Hence $S = \int_0^a \sqrt{a^2 - ax} \cdot \frac{a\,dx}{2\sqrt{ax - x^2}} = \frac{a\sqrt{a}}{2} \int_0^a \frac{\sqrt{a - x}\,.\,dx}{\sqrt{x}\sqrt{a - x}}$

$$= \frac{a\sqrt{a}}{2} \int_0^a \frac{dx}{\sqrt{x}} = a^2;$$

and the whole area required,

$$4S = 4a^2.$$

EXAMPLES.

(1) Find the area cut from the cylindrical surface whose base in the plane XY is a quadrant of the curve $x^{\frac{2}{3}} + y^{\frac{2}{3}} = a^{\frac{2}{3}}$ by the plane $x = z$.
Ans. $\frac{3}{5}a^2$.

(2) Find the area of that portion of a cylindrical surface whose base in the plane of XY is a quadrant of the ellipse $\frac{x^2}{a^2} + \frac{y^2}{b^2} = 1$, and whose projection on the plane of XZ is bounded by the curve $a^2z^2 = b^2x^2(a^2 - x^2)$.
Ans. $S = \frac{ab(a^2+ab+b^2)}{3(a+b)}$.

(3) Let the base of the cylindrical surface be a tractrix, whose vertex lies at a distance a to the left of the origin, and whose asymptote is the axis of Y, while its projection on the plane of XZ is bounded by the parabola $z^2 = -2mx$.
Ans. $S = 2a\sqrt{2ma}$.

(4) Let the base of the cylindrical surface be the upper half of a cycloid, having its vertex at the origin and its base parallel to the axis of Y, and at a distance $2a$ from the origin, while its projection on the plane of XZ is bounded by the parabola $z^2 = 2mx$.
Ans. $S = 4a\sqrt{am}$.

Any Surface.

144. Let x, y, z be the coördinates of any point P of the surface, and $x + \Delta x$, $y + \Delta y$, $z + \Delta z$ the coördinates of a second point Q infinitely near the first. Draw planes through P and Q parallel to the planes of XY and YZ. These planes will intercept a curved quadrilateral PQ on the surface; its projection pq, a rectangle, on the plane of XZ; and a parallelogram $p'q'$ not shown in the figure, on the tangent plane at P, of which pq is the projection. PQ will differ from $p'q'$ by an infinitesimal of higher order, and therefore our required surface will be the limit of the sum of the parallelograms of which $p'q'$ is any one.

If β is the angle the tangent plane at P makes with XZ, $p'q'\cos\beta = pq$ or $p'q' = pq\sec\beta = \Delta x\Delta z\sec\beta$, and σ, our surface required, is equal to the double integral

$$\sigma = \int\int \sec\beta dxdz$$

taken between limits so chosen as to embrace the whole surface.

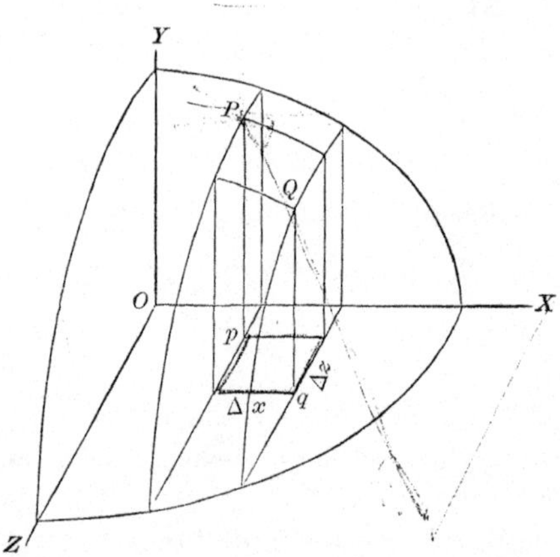

The limit of the sum of the parallelograms, of which $p'q'$ is a type, will be the required surface if the limit of the sum of the rectangles, of which pq is a type, is the projection of the surface in question on the plane of XZ; so that the values of x and z between which we integrate in $\sigma = \int\int \sec\beta dxdz$ are precisely those we should use if we were finding the area of the projection of σ by the double integration $\int\int dxdz$. (v. Art. 136.)

The equation of the tangent plane at P is

$(x - x_0)D_{x_0}f + (y - y_0)D_{y_0}f + (z - z_0)D_{z_0}f = 0$, by I. Art. 217, (x_0, y_0, z_0) standing for the coördinates of the point of contact, and $f(x, y, z) = 0$ being the equation of the surface.

The direction cosines of the perpendicular from the origin upon the plane are

$$\cos\alpha = \frac{D_{x_0}f}{\sqrt{(D_{x_0}f)^2 + (D_{y_0}f)^2 + (D_{z_0}f)^2}},$$

$$\cos\beta = \frac{D_{y_0}f}{\sqrt{(D_{x_0}f)^2 + (D_{y_0}f)^2 + (D_{z_0}f)^2}},$$

$$\cos\gamma = \frac{D_{z_0}f}{\sqrt{(D_{x_0}f)^2 + (D_{y_0}f)^2 + (D_{z_0}f)^2}}$$

by Anal. Geom. of Three Dimensions.

Hence, dropping the accents,

$$\sigma = \int\int \frac{\sqrt{(D_x f)^2 + (D_y f)^2 + (D_z f)^2}}{D_y f} dxdz. \qquad [1]$$

By considering the projections upon the other coördinate planes we shall find

$$\sigma = \int\int \frac{\sqrt{(D_x f)^2 + (D_y f)^2 + (D_z f)^2}}{D_x f} dydz; \qquad [2]$$

$$\sigma = \int\int \frac{\sqrt{(D_x f)^2 + (D_y f)^2 + (D_z f)^2}}{D_z f} dxdy. \qquad [3]$$

In each of the formulas the derivatives are partial derivatives.

Let us find the area of the portion of the surface of the sphere

$$x^2 + y^2 + z^2 = a^2$$

intercepted by the three coördinate planes.

$$D_x f = 2x,$$

$$D_y f = 2y,$$

$$D_z f = 2z,$$

$$\sqrt{(D_x f)^2 + (D_y f)^2 + (D_z f)^2} = 2a.$$

$$\sigma = \int_0^a \int_0^{\sqrt{a^2 - z^2}} \frac{a}{x} dydz; \qquad (1)$$

or

$$\sigma = \int_0^a \int_0^{\sqrt{a^2 - x^2}} \frac{a}{y} dzdx; \qquad (2)$$

or

$$\sigma = \int_0^a \int_0^{\sqrt{a^2 - y^2}} \frac{a}{z} dxdy. \qquad (3)$$

For, in the second one, which agrees best with the figure, we must take our limits so that the limit of the sum of the projections may be the quadrant in which the sphere is cut by the plane XZ; and the equation of this section is obtained by letting $y=0$ in the equation of the sphere, and is

$$x^2+z^2=a^2,$$

whence $$z=\sqrt{a^2-x^2}.$$

If we take as our limits in the integral $\int \frac{a}{y}\,dz$ zero and $\sqrt{a^2-x^2}$ we shall get the area whose projection is a strip running from the axis of X to the curve; then, taking $\int\left(\int \frac{a}{y}\,dz\right)dx$ from 0 to a, we shall get the area whose projection is the sum of all these strips, and that is our required surface.

$$y=\sqrt{a^2-x^2-z^2},$$

$$\sigma=a\int_0^a\int_0^{\sqrt{a^2-x^2}}\frac{dzdx}{\sqrt{a^2-x^2-z^2}};$$

$$\int\frac{dz}{\sqrt{a^2-x^2-z^2}}=\sin^{-1}\frac{z}{\sqrt{a^2-x^2}}$$

if we regard x as constant;

$$\int_0^{\sqrt{a^2-x^2}}\frac{dz}{\sqrt{a^2-x^2-z^2}}=\frac{\pi}{2};$$

$$\sigma=a\int_0^a\frac{\pi}{2}\,dx=\frac{\pi a^2}{2},$$

the required area. Formulas (1) and (3) give the same result.

145. Suppose two cylinders of revolution drawn tangent to each other, and perpendicular to the plane of a great circle of a sphere, each having the radius of the great circle as a diameter; required the surface of the sphere not included by the cylinders.

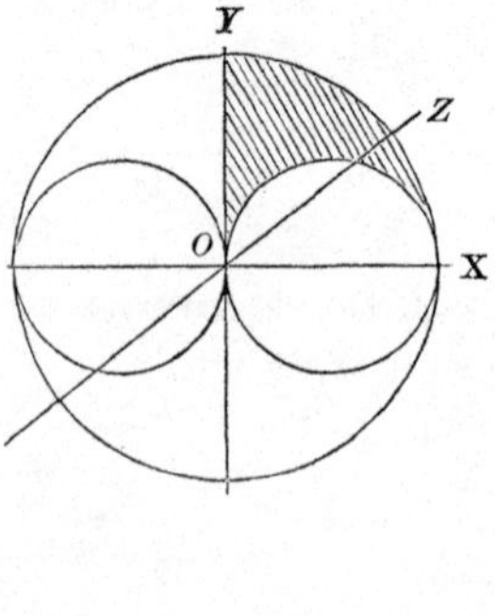

The surface required is eight times the surface of which the shaded portion of the figure is the projection.

If we take the plane of the great circle as the plane of XY,

$$x^2 - ax + y^2 = 0 \qquad (1)$$

is the equation of the cylinder, and

$$x^2 + y^2 + z^2 = a^2 \qquad (2)$$

of the sphere.

We have $\sigma = \int\int \frac{\sqrt{(D_xf)^2 + (D_yf)^2 + (D_zf)^2}}{D_zf} dydx.$

From (2)

$$D_xf = 2x,$$

$$D_yf = 2y,$$

$$D_zf = 2z;$$

$$(D_xf)^2 + (D_yf)^2 + (D_zf)^2 = 4a^2.$$

Hence $\sigma = \int\int \frac{a}{z} dydx = a\int\int \frac{dydx}{\sqrt{a^2 - x^2 - y^2}}.$

Our limits of integration for y are $\sqrt{ax - x^2}$ and $\sqrt{a^2 - x^2}$; for x are 0 and a.

$$\sigma = a\int_0^a \int_{\sqrt{ax - x^2}}^{\sqrt{a^2 - x^2}} \frac{dydx}{\sqrt{a^2 - x^2 - y^2}}.$$

$$\int_{\sqrt{ax - x^2}}^{\sqrt{a^2 - x^2}} \frac{dy}{\sqrt{a^2 - x^2 - y^2}} = \sin^{-1} \frac{y}{\sqrt{a^2 - x^2}}\Bigg]_{\sqrt{ax - x^2}}^{\sqrt{a^2 - x^2}} = \frac{\pi}{2} - \sin^{-1}\sqrt{\frac{x}{a + x}}.$$

To find $\int_0^a \sin^{-1}\sqrt{\frac{x}{a+x}}.dx$ we must integrate by parts.

Let $$u = \sin^{-1}\sqrt{\frac{x}{a + x}},$$

and $$dv = dx;$$

$$v = x,$$

$$du = \frac{1}{2(a + x)}\sqrt{\frac{a}{x}}.dx;$$

$$\int \sin^{-1}\sqrt{\frac{x}{a + x}}.dx = x \sin^{-1}\sqrt{\frac{x}{a + x}} - \frac{\sqrt{a}}{2}\int \frac{\sqrt{x}}{a + x}.dx.$$

Let $$w = \sqrt{x}\,; \quad 2\,w dw = dx$$

and $$\int \frac{\sqrt{x}.dx}{a+x} = 2\int \frac{w^2\,dw}{a+w^2} = 2\int\left(1 - \frac{a}{a+w^2}\right)dw.$$

$$\int \frac{\sqrt{x}.dx}{a+x} = 2\left(w - \sqrt{a}\tan^{-1}\frac{w}{\sqrt{a}}\right),$$

$$\int_0^a \sin^{-1}\sqrt{\frac{x}{a+x}}.dx$$
$$= a\sin^{-1}\frac{\sqrt{2}}{2} + a\tan^{-1}1 - a = \frac{a\pi}{4} + \frac{a\pi}{4} - a = \frac{a\pi}{2} - a,$$

$$\sigma = a\left(\frac{a\pi}{2} - \frac{a\pi}{2} + a\right) = a^2.$$

$8\sigma = 8a^2$ is the whole surface in question.

146. Let us find the area of the curved surface of a right cone whose base is the curve $x^{\frac{2}{3}} + y^{\frac{2}{3}} = a^{\frac{2}{3}}$, and whose altitude is c.

If we take the vertex of the cone as the origin of coördinates, and its axis as the axis of Z, the equation of its curved surface is

$$x^{\frac{2}{3}} + y^{\frac{2}{3}} = \left(\frac{az}{c}\right)^{\frac{2}{3}}, \tag{1}$$

and the projection of the surface on the plane of XY is bounded by the curve

$$x^{\frac{2}{3}} + y^{\frac{2}{3}} = a^{\frac{2}{3}}. \tag{2}$$

From (1) we get

$$\frac{\sqrt{(D_x f)^2 + (D_y f)^2 + (D_z f)^2}}{D_z f} = \sqrt{1 + \frac{c^2}{a^2}\,\frac{(x^{\frac{2}{3}} + y^{\frac{2}{3}})^2}{x^{\frac{2}{3}}y^{\frac{2}{3}}}},$$

where x, y are the coördinates of any point within the projection of the base of the cone.

Since the four faces of the cone are equal, the required surface

$$\sigma = \frac{4}{a}\int_0^a \int_0^{(a^{\frac{2}{3}} - x^{\frac{2}{3}})^{\frac{3}{2}}} x^{-\frac{1}{3}}y^{-\frac{1}{3}}\sqrt{a^2x^{\frac{2}{3}}y^{\frac{2}{3}} + c^2(x^{\frac{2}{3}} + y^{\frac{2}{3}})^2}\,.\,dy\,dx. \tag{3}$$

Let us substitute $v^3 = x$ and $w^3 = y$, whence $dx = 3v^2 dv$ and $dy = 3w^2 dw$, and we have

$$\sigma = \frac{36}{a}\int_0^{a^{\frac{1}{3}}}\int_0^{(a^{\frac{2}{3}}-v^2)^{\frac{1}{2}}} vw\sqrt{a^2v^2w^2 + c^2(v^2+w^2)^2}\,.\,dw\,dv\,;$$

or, since in a definite integral it makes no difference what letters we use for the variables,

$$\sigma = \frac{36}{a}\int_0^{a^{\frac{1}{3}}}\int_0^{(a^{\frac{2}{3}}-x^2)^{\frac{1}{2}}} xy\sqrt{a^2x^2y^2 + c^2(x^2+y^2)^2}\,.\,dy\,dx. \qquad (4)$$

The x and y in (4), however, must not be confounded with the x and y in (3).

The integral in (4) is precisely that which we should have to find if we sought the area of a surface of such a nature that its projection on the plane of XY was a quadrant of the circle $x^2 + y^2 = a^{\frac{2}{3}}$, and the secant of the angle made by the tangent plane at any point (x, y, z) of the surface with the plane of XY was $xy\sqrt{a^2x^2y^2 + c^2(x^2+y^2)^2}$.

In the latter problem there is nothing to prevent our replacing x and y in $xy\sqrt{a^2x^2y^2 + c^2(x^2+y^2)^2}$ by their values in terms of r and ϕ, the polar coördinates of any point of the projection $x^2 + y^2 = a^{\frac{2}{3}}$, and dividing this projection into polar elements instead of rectangular elements, and then integrating between the limits which we should use if we were finding the area of the projection by the formula $A = \int\int r d\phi dr$.

We have, then,

$$\sigma = \frac{36}{a}\int_0^{\frac{\pi}{2}}\int_0^{a^{\frac{1}{3}}} r^2 \sin\phi\cos\phi\sqrt{a^2r^4\sin^2\phi\cos^2\phi + c^2r^4}\,.\,rdrd\phi,$$

or

$$\sigma = \frac{36}{a}\int_0^{\frac{\pi}{2}}\int_0^{a^{\frac{1}{3}}} r^5 \sin\phi\cos\phi\sqrt{a^2\sin^2\phi\cos^2\phi + c^2}\,.\,dr\,d\phi,$$

$$\sigma = 6a\int_0^{\frac{\pi}{2}} \sin\phi\cos\phi\sqrt{a^2\sin^2\phi\cos^2\phi + c^2}\,.\,d\phi.$$

Substitute $u = \sin^2\phi$, and

$$\sigma = 3a\int_0^1 \sqrt{a^2u(1-u)+c^2}\,.\,du,$$

$$\sigma = \tfrac{3}{4}\left[2ac + (a^2+4c^2)\tan^{-1}\frac{a}{2c}\right].$$

EXAMPLES.

(1) Find the area included by the cylinders described in Art. 145 by direct integration.

(2) A square hole is cut through a sphere, the axis of the hole coinciding with a diameter of the sphere; find the area of the surface removed.

(3) A cylinder is constructed on a single loop of the curve $r = a\cos n\phi$, having its generating lines perpendicular to the plane of this curve; determine the area of the portion of the surface of the sphere $x^2+y^2+z^2=a^2$ which the cylinder intercepts.

Ans. $\frac{4a^2}{n}\left(\frac{\pi}{2}-1\right)$.

(4) Find the area of the portion of the surface of the cone described in Art. 146 included by the cylinder $x^2+y^2=b^2$.

Ans. $\frac{2b^2}{a}\left[2\sqrt{a^2+3c^2}\tan^{-1}\left(\frac{\sqrt{a^2+3c^2}}{c}\right) - a\tan^{-1}\frac{a}{2c}\right]$.

(5) Find the area of the portion of the surface of the sphere $x^2+y^2+z^2=2ay$ cut out by one nappe of the cone $Ax^2+Bz^2=y^2$.

Ans. $\frac{4\pi a^2}{\sqrt{(1+A)(1+B)}}$.

(6) Find the area of the portion of the surface of the sphere $x^2+y^2+z^2=2ay$ lying within the paraboloid $y=Ax^2+Bz^2$.

Ans. $\frac{2\pi a}{\sqrt{AB}}$.

(7) The centre of a regular hexagon moves along a diameter of a given circle (radius $= a$), the plane of the hexagon being perpendicular to this diameter, and its magnitude varying in such a manner that one of its diagonals always coincides with a chord of the circle; find the surface generated.

Ans. $a^2(2\pi+3\sqrt{3})$.

CHAPTER XII.

VOLUMES.

Single Integration.

147. If sections of a solid are made by parallel planes, and a set of cylinders drawn, each having for its base one of the sections, and for its altitude the distance between two adjacent cutting planes, the limit of the sum of the volumes of these cylinders, as the distance between the sections is indefinitely decreased, is the volume of the solid.

We shall take as established by Geometry the fact that the volume of a cylinder or prism is the product of the area of its base by its altitude.

It follows from what has just been said, that if, in a given solid, all of a set of parallel sections are equal, the volume of the solid is its base by its altitude, no matter how irregular its form.

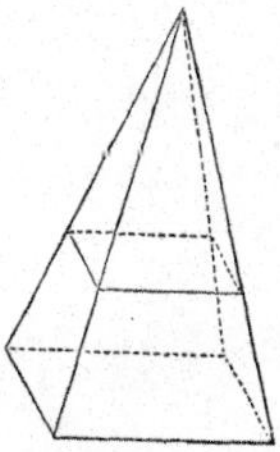

Let us find the volume of a pyramid having b for the area of its base, and a for its altitude.

Divide the pyramid by planes parallel to the base, and let z be the area of a section at the distance x from the vertex.

We know from Geometry that $\frac{z}{b} = \frac{x^2}{a^2}$.

Hence
$$z = \frac{b}{a^2} x^2.$$

Let the distance between two adjacent sections be dx; then the volume of the cylinder on z is

$$\frac{b}{a^2} x^2 dx,$$

and V, the required volume of the pyramid, is

$$V = \frac{b}{a^2} \int_0^a x^2 dx = \frac{ab}{3}.$$

Precisely the same reasoning applies to any cone, which will therefore have for its volume one-third the product of its base by its altitude.

EXAMPLE.

Find the volume of the frustum of a pyramid or of a cone.

148. If a line move keeping always parallel to a given plane, and touching a plane curve and a straight line parallel to the plane of the curve, the surface generated is called a *conoid*. Let us find the volume of a *conoid* when the director line and curve are perpendicular to the given plane.

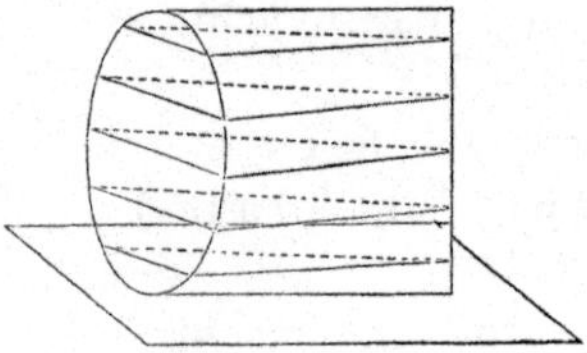

Divide the conoid into laminae by planes parallel to the fixed plane.

Let Δy be the distance between two adjacent sections, and let x be the length of the line in which any section cuts the base of the conoid; let a be the altitude and b the area of the base of the figure. Any one of our elementary cylinders will have for its volume $\frac{1}{2}ax\Delta y$, since the area of its triangular base is $\frac{1}{2}ax$, and we have $V = \frac{1}{2}a\int xdy$, the limits of integration being so taken as to embrace the whole solid. $\int xdy$ between the limits in question is the area of the base of the conoid; hence its volume,

$$V = \tfrac{1}{2}ab.$$

EXAMPLES.

(1) Find the volume of a conoid when the director line and curve are not perpendicular to the given plane.

(2) A woodman fells a tree 2 feet in diameter, cutting halfway through from each side. The lower face of each cut is horizontal, and the upper face makes an angle of 45° with the horizontal. How much wood does he cut out?

149. To find the volume of an ellipsoid.

$$\frac{x^2}{a^2}+\frac{y^2}{b^2}+\frac{z^2}{c^2}=1.$$

Take the cutting planes parallel to the plane of XY. A section at the distance z from the origin will have

$$\frac{x^2}{a^2}+\frac{y^2}{b^2}=1-\frac{z^2}{c^2}=\frac{c^2-z^2}{c^2}$$

for its equation, and $\frac{a}{c}\sqrt{c^2-z^2}$ and $\frac{b}{c}\sqrt{c^2-z^2}$ for its semi-axes; hence its area will be $\frac{\pi ab}{c^2}(c^2-z^2)$.

Any of the elementary cylinders will have for its volume $\frac{\pi ab}{c^2}(c^2-z^2)\Delta z$, and we shall have for the whole solid

$$V=\frac{\pi ab}{c^2}\int_{-c}^{c}(c^2-z^2)\,dz.$$

$$V=\tfrac{4}{3}\pi abc.$$

If a, b, and c are equal, the ellipsoid is a sphere, and

$$V=\tfrac{4}{3}\pi a^3.$$

EXAMPLES.

(1) Find the volume included between an hyperboloid of one sheet

$$\frac{x^2}{a^2}+\frac{y^2}{b^2}-\frac{z^2}{c^2}=1,$$

and its asymptotic cone

$$\frac{x^2}{a^2}+\frac{y^2}{b^2}-\frac{z^2}{c^2}=0.$$

Ans. It is equal to a cylinder of the same altitude as the solid in question, and having for a base the section made by the plane of XY.

(2) Find the whole volume of the solid bounded by the surface

$$\frac{x^2}{a^2}+\frac{y^2}{b^2}+\frac{z^4}{c^4}=1. \qquad \textit{Ans.}\ \frac{8\pi abc}{5}.$$

(3) Find the volume cut from the surface

$$\frac{z^2}{c}+\frac{y^2}{b}=2x$$

by a plane parallel to the plane of (YZ) at a distance a from it.

Ans. $\pi a^2\sqrt{(bc)}$.

(4) The centre of a regular hexagon moves along a diameter of a given circle (radius $=a$), the plane of the hexagon being perpendicular to this diameter, and its magnitude varying in such a manner that one of its diagonals always coincides with a chord of the circle; find the volume generated.

Ans. $2\sqrt{3}.a^3$.

(5) A circle (radius $=a$) moves with its centre on the circumference of an equal circle, and keeps parallel to a given plane which is perpendicular to the plane of the given circle; find the volume of the solid it will generate.

Ans. $\frac{2a^3}{3}(3\pi+8)$.

Solids of Revolution. Single Integration.

150. If a solid is generated by the revolution of a plane curve $y=fx$ about the axis of x, sections made by planes perpendicular to the axis are circles. The area of any such circle is πy^2, the volume of the elementary cylinder is $\pi y^2\Delta x$, and

$$V=\pi\int_{x_0}^{x}y^2dx$$

is the volume of the solid generated.

For example; let us find the volume of the solid generated by the revolution of one branch of the tractrix about the axis of X. Here we must integrate from $x=0$ to $x=\infty$.

$$V=\pi\int_0^{\infty}y^2\,dx.$$

We have $$dx=-\frac{(a^2-y^2)^{\frac{1}{2}}}{y}dy \qquad \text{(Art. 102 [2].)}$$

in the case of the tractrix;

hence
$$V = -\pi \int_{x=0}^{x=\infty} y(a^2 - y^2)^{\frac{1}{2}}\,dy.$$

When $x = 0$, $y = a$, and when $x = \infty$, $y = 0$.

Therefore
$$V = -\pi \int_a^0 y(a^2 - y^2)^{\frac{1}{2}}\,dy = \frac{\pi a^3}{3}.$$

EXAMPLES.

(1) If the plane curve revolves about the axis of Y,

$$V = \pi \int_{y_0}^{y_1} x^2\,dy.$$

(2) The volume of a sphere is $\frac{4}{3}\pi a^3$.

(3) The volume of the solid formed by the revolution of a cycloid about its base is $5\pi^2 a^3$.

(4) The curve $y^2(2a - x) = x^3$ revolves about its asymptote; show that the volume generated is $2\pi^2 a^3$.

(5) The curve $x^{\frac{2}{3}} + y^{\frac{2}{3}} = a^{\frac{2}{3}}$ revolves about the axis of X; show that the volume generated is $\frac{96}{315}\pi a^3$.

Solids of Revolution. Double Integration.

151. If we suppose the area of the revolving curve broken up into infinitesimal rectangles as in Art. 137, the element $\Delta x \Delta y$ at any point P, whose coördinates are x and y, will generate a ring the volume of which will differ from $2\pi y \Delta x \Delta y$ by an amount which will be an infinitesimal of higher order than the second if we regard Δx and Δy as of the first order. For the ring in question is obviously greater than a prism having the same cross-section $\Delta x \Delta y$, and having an altitude equal to the inner circumference $2\pi y$ of the ring, and is less than a prism having $\Delta x \Delta y$ for its base and $2\pi(y + \Delta y)$, the outer circumference of the ring, for its altitude; but these two prisms differ by $2\pi \Delta x (\Delta y)^2$, which is of the third order.

$\Delta x \int_0^y 2\pi y dy$, where the upper limit of integration is the ordinate of the point of the curve immediately above P, and must be expressed in terms of x by the aid of the equation of the revolving curve, will give us the elementary cylinder used in Art. 150.

The whole volume required will be the limit of the sum of these cylinders; that is,

$$V = 2\pi \int_{x_0}^{x_1} \int_0^y y dy dx. \qquad [1]$$

If the figure revolved is bounded by two curves, the required volume can be found by the formula just obtained, if the limits of integration are suitably chosen.

Let us consider the following example:

A paraboloid of revolution has its axis coincident with the diameter of a sphere, and its vertex in the surface of the sphere; required the volume between the two surfaces.

Let $$y^2 = 2mx \qquad (1)$$

be the parabola, and $$x^2 + y^2 - 2ax = 0 \qquad (2)$$

be the circle, which form the paraboloid and the sphere by their revolution. The abscissas of their points of intersection are 0 and $2(a - m)$.

We have $$V = 2\pi \int \int y dy dx,$$

and, in performing our first integration, our limits must be the values of y obtained from equations (1) and (2).

We get $$V = \pi \int [2(a - m)x - x^2] dx,$$

and here our limits of integration are 0 and $2(a - m)$.

Hence $$V = \tfrac{4}{3}\pi(a - m)^3 = \frac{\pi h^3}{6},$$

if h is the altitude of the solid in question.

EXAMPLES.

(1) A cone of revolution and a paraboloid of revolution have the same vertex and the same base; required the volume between them. *Ans.* $\frac{\pi m h^2}{3}$, where h is the altitude of the cone.

(2) Find the volume included between a right cone, whose vertical angle is 30°, and a sphere of given radius touching it along a circle.

Ans. $\frac{\pi r^3}{6}$.

Solids of Revolution. Polar Formula.

152. If we use polar coördinates, and suppose the revolving area broken up, as in Art. 138, into elements of which $rd\phi dr$ is the one at any point P whose coördinates are r and ϕ, the element $rd\phi dr$ will generate a ring whose volume will differ from $2\pi r^2 \sin\phi d\phi dr$ by an infinitesimal of higher order than the second, if we regard $d\phi$ and dr as of the first order; for it will be less than a prism having for its base $rd\phi dr$, and for its altitude $2\pi(r+dr)\sin(\phi+d\phi)$, and greater than a prism having the same base and the altitude $2\pi r\sin\phi$; and these prisms differ by an amount which is infinitesimal of higher order than the second.

We shall have then

$$V = 2\pi\int\int r^2 \sin\phi dr d\phi, \qquad [1]$$

the limits being so taken as to bring in the whole of the generating area.

For example; let us find the volume generated by the revolution of a cardioide about its axis.

$$r = 2a(1-\cos\phi)$$

is the equation of the cardioide;

$$V = 2\pi\int\int r^2 \sin\phi dr d\phi.$$

Our first integral must be taken between the limits $r=0$ and $r = 2a(1-\cos\phi)$, and is

$$\frac{8a^3}{3}(1-\cos\phi)^3 \sin\phi d\phi.$$

$$V = \frac{16}{3}a^3\pi\int_0^\pi(1-\cos\phi)^3\sin\phi d\phi,$$

$$V = \frac{64}{3}\pi a^3.$$

EXAMPLE.

A right cone has its vertex on the surface of a sphere, and its axis coincident with the diameter of the sphere passing through that point; find the volume common to the cone and the sphere.

Volume of any Solid. Triple Integration.

153. If we suppose our solid divided into parallelopipeds by planes parallel to the three coördinate planes, the elementary

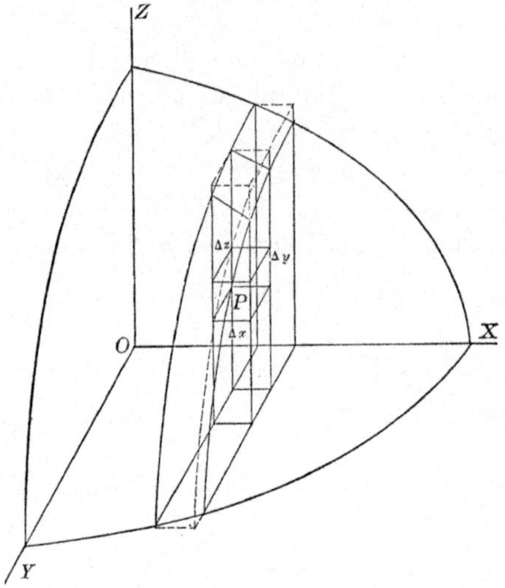

parallelopiped at any point (x,y,z) within the solid will have for its volume $\Delta x\Delta y\Delta z$, or, if we regard x, y, and z as independent, $dxdydz$; and the whole volume

$$V=\int\int\int dxdydz, \qquad [1]$$

the limits being so chosen as to embrace the whole solid.

The integrations are independent, and may be performed in any order if the limits are suitably chosen.

As it is important to have a perfectly clear conception of the geometrical interpretation of each step in the process of finding a volume by a triple integration, we will consider one case in detail.

Let the integrations be performed in the order indicated by the formula

$$V = \int\int\int dz dy dx.$$

If the limits are correctly chosen, our first integration gives us the volume of a prism one of whose lateral edges passes through any chosen point $P, (x, y, z)$ within the solid, is parallel to the axis of Z, and reaches directly across the solid from surface to surface, while the base of the prism is the rectangle $dydx$; our second integration gives the volume of a right cylinder whose base is a plane section of the solid, passes through the point P, and is parallel to the plane YZ, and whose altitude is dx; and our third integration gives the volume of the whole solid.

The limits in our first integration are, then, the values of z belonging to the point in the lower bounding surface and the point in the upper bounding surface which have the coördinates x and y; the limits in the second integration are the values of y belonging to the two points in the perimeter of the projection of the solid in the plane of XY which have the coördinate x; and the limits in the third integration are the least value and the greatest value of x belonging to points on the perimeter of the projection of the solid on the plane of XY.

It is easily seen from what has just been said that the limits in the second and third integrations are precisely those we should use if we were finding the area of the projection of the solid by the formula

$$A = \int\int dy dx.$$

Of course, it is necessary to have a clear idea of the form of the solid whose volume is required.

For example, let us find the volume of the portion of the ellipsoid

$$\frac{x^2}{a^2} + \frac{y^2}{b^2} + \frac{z^2}{c^2} = 1$$

cut off by the coördinate planes.

$$V = \int\int\int dz\,dy\,dx,$$

and our limits are, for z, 0 and $c\sqrt{1-\frac{x^2}{a^2}-\frac{y^2}{b^2}}$; for y, 0 and $b\sqrt{1-\frac{x^2}{a^2}}$; and for x, 0 and a. For, starting at any point (x,y,z) and integrating on the hypothesis that z alone varies, we get a column of our elementary parallelopipeds having $dxdy$ as a base and passing through the point (x,y,z). To make this column reach from the plane XY to the surface, z must increase from the value zero to the value belonging to the point on the surface of the ellipsoid which has the coördinates x and y; that is, to the value $c\sqrt{1-\frac{x^2}{a^2}-\frac{y^2}{b^2}}$. Then, integrating on the hypothesis that y alone varies, we shall sum these columns and shall get a slice of the solid passing through (x,y,z) and having the thickness dx. To make this slice reach completely across the solid, we must let y increase from the value zero to the greatest value it can have in the slice in question; that is, to the value which is the ordinate of that point of the section of the ellipsoid by the plane XY which has the abscissa x. The section in question has the equation

$$\frac{x^2}{a^2}+\frac{y^2}{b^2}=1;$$

therefore the required value of y is $b\sqrt{1-\frac{x^2}{a^2}}$.

Last, in integrating on the hypothesis that x alone varies, we must choose our limits so as to include all the slices just described, and must increase x from zero to a.

$$\int dz = z = c\sqrt{1-\frac{x^2}{a^2}-\frac{y^2}{b^2}}$$

between the limits 0 and $c\sqrt{1-\frac{x^2}{a^2}-\frac{y^2}{b^2}}$

$$c\int\sqrt{1-\frac{x^2}{a^2}-\frac{y^2}{b^2}}\,.\,dy$$

$$=\frac{c}{b}\int\sqrt{b^2\left(1-\frac{x^2}{a^2}\right)-y^2}\,.\,dy$$

$$=\frac{1}{2}\frac{c}{b}\left[y\sqrt{b^2\left(1-\frac{x^2}{a^2}\right)-y^2}+b^2\left(1-\frac{x^2}{a^2}\right)\sin^{-1}\frac{y}{b\sqrt{1-\frac{x^2}{a^2}}}\right]$$

$$=\frac{\pi bc}{4}\left(1-\frac{x^2}{a^2}\right)$$

between the limits 0 and $b\sqrt{1-\frac{x^2}{a^2}}$.

$$\frac{\pi bc}{4}\int_0^a\left(1-\frac{x^2}{a^2}\right)dx=\frac{\pi abc}{6},$$

the volume required.

EXAMPLES.

(1) Find the volume obtained in the present article, performing the integrations in the order indicated by the formula,

$$V=\int\int\int dxdydz.$$

(2) Find the volume cut off from the surface

$$\frac{z^2}{c}+\frac{y^2}{b}=2x$$

by a plane parallel to that of YZ, at a distance a from it.

Ans. $\pi a^2\sqrt{(bc)}$.

(3) Find the volume enclosed by the surfaces,

$$x^2+y^2=cz,\quad x^2+y^2=ax,\quad z=0.\qquad \textit{Ans.}\ \frac{3\pi a^4}{32c}.$$

(4) Obtain the volume bounded by the surface

$$z=a-\sqrt{x^2+y^2}$$

and the planes $x=z$ and $x=0$. *Ans.* $\frac{2a^3}{9}$.

(5) Find the volume of the conoid bounded by the surface $z^2 + \frac{a^2 y^2}{x^2} = c^2$ and the planes $x = 0$ and $x = a$. *Ans.* $\frac{\pi c^2 a}{2}$.

154. If we use polar coördinates we can take as our element of volume

$$r^2 \sin\phi dr d\phi d\theta,$$

an expression easily obtained from the element $2\pi r^2 \sin\phi dr d\phi$ used in Art. 152.

Then $$V = \int\int\int r^2 \sin\phi dr d\phi d\theta,$$

where the order of the integrations is usually immaterial if the limits are properly chosen.

EXAMPLES.

(1) Find the volume of a sphere by polar coördinates.

(2) Find the whole volume of the solid bounded by

$$(x^2 + y^2 + z^2)^3 = 27 a^3 xyz.$$

Suggestion: Transform to polar coördinates. *Ans.* $\frac{9}{2} a^3$.

CHAPTER XIII.

CENTRES OF GRAVITY.

155. The *moment of a force about an axis* perpendicular to its line of direction is the product of the magnitude of the force by the perpendicular distance of its line of direction from the axis, and measures the tendency of the force to produce rotation about the axis.

The *force exerted by gravity* on any material body is proportional to the mass of the body, and may be measured by the mass of the body.

The *Centre of Gravity* of a body is a point so situated that the force of gravity produces no tendency in the body to rotate about any axis passing through this point.

The subject of centres of gravity belongs to Mechanics, and we shall accept the definitions and principles just stated as data for mathematical work, without investigating the mechanical grounds on which they rest.

156. Suppose the points of a body referred to a set of three rectangular axes fixed in the body, and let $x, y, \bar{z}$ be the coördinates of the centre of gravity. Place the body with the axes of X and Z horizontal, and consider the tendency of the particles of the body to produce rotation about an axis through $(\bar{x}, \bar{y}, \bar{z})$ parallel to OZ, under the influence of gravity. Represent the mass of an elementary parallelopiped at any point (x,y,z) by dm. The force exerted by gravity on dm is measured by dm, and its line of direction is vertical. If the mass of dm were concentrated at P, the moment of the force exerted on dm about the

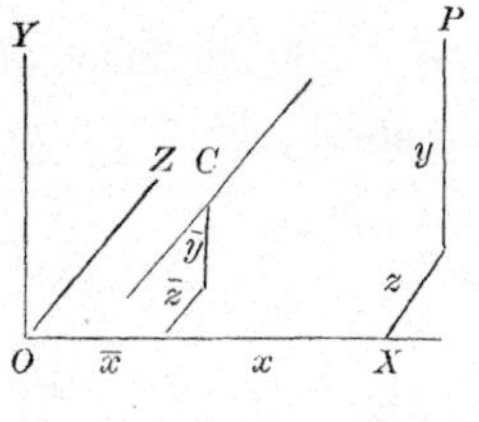

axis through C would be $(x-\bar{x})dm$, and this moment would represent the tendency of dm to rotate about the axis in question; the tendency of the whole body to rotate about this axis would be $\Sigma(x-\bar{x})dm$. If now we decrease dm indefinitely, the error committed in assuming that the mass of dm is concentrated at P decreases indefinitely, and we shall have as the true expression for the tendency of the whole body to rotate about the axis through C, $\int(x-\bar{x})dm$; but this must be zero.

Hence
$$\int(x-\bar{x})dm = 0,$$
$$\int xdm - \bar{x}\int dm = 0,$$
$$\bar{x} = \frac{\int xdm}{\int dm}. \qquad [1]$$

If we place the body so that the axes of Y and X are horizontal, the same reasoning will give us

$$\bar{y} = \frac{\int ydm}{\int dm}; \qquad [2]$$

and in like manner we cán get

$$\bar{z} = \frac{\int zdm}{\int dm}. \qquad [3]$$

Since $\int dm$ is the mass of the whole body, if we represent it by M we shall have

$$\bar{x} = \frac{\int xdm}{M},$$

$$\bar{y} = \frac{\int ydm}{M},$$

$$\bar{z} = \frac{\int zdm}{M}.$$

EXAMPLE.

Show that the effect of gravity in making a body tend to rotate about any given axis is precisely the same as if the mass of the body were concentrated at its centre of gravity.

157. The mass of any homogeneous body is the product of its volume by its density. If the body is not homogeneous, the density at any point will be a function of the position of that point. Let us represent it by κ. Then we may regard dm as equal to κdv if dv is the element of volume, and we shall have

$$\bar{x} = \frac{\int x\kappa dv}{\int \kappa dv} \qquad [1]$$

and corresponding formulas for $\bar{y}$ and $\bar{z}$.

If the body considered is homogeneous, κ is constant, and we shall have

$$\bar{x} = \frac{\int x dv}{\int dv} = \frac{\int x dv}{V}, \qquad [2]$$

$$\bar{y} = \frac{\int y dv}{\int dv} = \frac{\int y dv}{V}, \qquad [3]$$

$$\bar{z} = \frac{\int z dv}{\int dv} = \frac{\int z dv}{V}. \qquad [4]$$

In any particular problem we have only to express dv in terms of the coördinates.

Plane Area.

158. If we use rectangular coördinates, and are dealing with a plane area, where the weight is uniformly distributed, we have

$$dv = dA = dxdy. \qquad \text{(Art. 136).}$$

Hence, by 157, [2] and [3],

$$\left.\begin{aligned}\bar{x}&=\frac{\int\int xdxdy}{\int\int dxdy}\\ \bar{y}&=\frac{\int\int ydxdy}{\int\int dxdy}\end{aligned}\right\}. \qquad [1]$$

If we use polar coördinates,

$$dv = dA = rd\phi dr,$$

and

$$\left.\begin{aligned}\bar{x}&=\frac{\int\int r^2\cos\phi\, d\phi dr}{\int\int rd\phi dr}\\ \bar{y}&=\frac{\int\int r^2\sin\phi\, d\phi dr}{\int\int rd\phi dr}\end{aligned}\right\}. \qquad [2]$$

For example; let us find the *centre of gravity* of the area between the *cissoid* and its asymptote. From the equation of the cissoid

$$y^2 = \frac{x^3}{a-x},$$

we see that the curve is symmetrical with respect to the axis of X, passes through the origin, and has the line $x = a$ as an asymptote. From the symmetry of the area in question, $\bar{y} = 0$, and we need only find $\bar{x}$.

$$\bar{x} = \frac{\int_0^a\int_{-y}^{y} xdydx}{\int_0^a\int_{-y}^{y} dydx} = \frac{\int_0^a xydx}{\int_0^a ydx},$$

$$\bar{x} = \frac{\int_0^a \frac{x^{\frac{5}{2}}}{(a-x)^{\frac{1}{2}}}dx}{\int_0^a \frac{x^{\frac{3}{2}}}{(a-x)^{\frac{1}{2}}}dx} = \frac{\frac{5}{6}a\int_0^a \frac{x^{\frac{3}{2}}}{(a-x)^{\frac{1}{2}}}dx}{\int_0^a \frac{x^{\frac{3}{2}}}{(a-x)^{\frac{1}{2}}}dx}; \text{ by Art. 64 [4].}$$

$$\bar{x} = \tfrac{5}{6}a.$$

As an example of the use of the polar formulas [2], let us find the *centre of gravity* of the cardioide

$$r = 2a(1 - \cos\phi).$$

Here, from the fact that the axis of X is an axis of symmetry, we know that $\bar{y} = 0$.

$$\bar{x} = \frac{\int_0^{2\pi}\int_0^r r^2\cos\phi\, dr d\phi}{\int_0^{2\pi}\int_0^r r dr d\phi}$$

$$= \frac{\frac{1}{3}\int_0^{2\pi} r^3\cos\phi\, d\phi}{\frac{1}{2}\int_0^{2\pi} r^2 d\phi} = \frac{\frac{8a^3}{3}\int_0^{2\pi}(1-\cos\phi)^3\cos\phi\, d\phi}{2a^2\int_0^{2\pi}(1-\cos\phi)^2 d\phi},$$

$$\int_0^{2\pi}(\cos\phi - 3\cos^2\phi + 3\cos^3\phi - \cos^4\phi)\,d\phi = -\tfrac{15}{4}\pi;$$

and $$\int_0^{2\pi}(1 - 2\cos\phi + \cos^2\phi)\,d\phi = 3\pi.$$

Hence $$\bar{x} = -\tfrac{5}{3}a.$$

EXAMPLES.

1. Show that formulas [1] hold even when we use *oblique coördinates*.

2. Find the centre of gravity of a segment of a parabola cut off by any chord.

Ans. $\bar{x} = \frac{3}{5}a$, $\bar{y} = 0$. If the axes are the tangent parallel to the chord and the diameter bisecting the chord.

3. Find the centre of gravity of the area bounded by the semi-cubical parabola $ay^2 = x^3$ and a double ordinate. *Ans.* $\bar{x} = \frac{5}{7}x$.

4. Find the centre of gravity of a semi-ellipse, the bisecting line being any diameter.

Ans. If the bisecting diameter is taken as the axis of Y, and the conjugate diameter as the axis of X, $\bar{x} = \frac{4a}{3\pi}$, $\bar{y} = 0$.

5. Find the centre of gravity of the curve $y^2 = b^2\frac{a-x}{x}$.

Ans. $\bar{x} = \frac{1}{4}a$.

6. Find the centre of gravity of the cycloid.

Ans. $\bar{x} = a\pi$, $\bar{y} = \frac{5}{6}a$.

7. Find the centre of gravity of the lemniscate $r^2 = a^2 \cos 2\phi$.

Ans. $\bar{x} = \frac{\pi\sqrt{2}}{8}a$.

8. Find the centre of gravity of a circular sector.

Ans. If we take the radius bisecting the sector as the axis of X, and represent the angle of the sector by 2α, $\bar{x} = \frac{2}{3}\frac{a \sin\alpha}{\alpha}$.

9. Find the centre of gravity of the segment of an ellipse cut off by a quadrantal chord. *Ans.* $\bar{x} = \frac{2}{3}\frac{a}{\pi - 2}$, $\bar{y} = \frac{2}{3}\frac{b}{\pi - 2}$.

10. Find the centre of gravity of a quadrant of the area of the curve $x^{\frac{2}{3}} + y^{\frac{2}{3}} = a^{\frac{2}{3}}$. *Ans.* $\bar{x} = \bar{y} = \frac{256}{315}\frac{a}{\pi}$.

159. If we are dealing with a homogeneous solid formed by the revolution of a plane curve about the axis of X, we have

$$dv = 2\pi y dy dx. \qquad \text{(Art. 151 [1])}$$

Hence, by Art. 157 [2],

$$\bar{x} = \frac{\int\int xydxdy}{\int\int ydxdy}. \qquad [1]$$

If we use polar coördinates,

$$dv = 2\pi r^2 \sin\phi\, dr\, d\phi. \qquad \text{(Art. 152 [1].)}$$

Hence
$$\bar{x} = \frac{\int\int r^3 \sin\phi \cos\phi\, dr\, d\phi}{\int\int r^2 \sin\phi\, dr\, d\phi}. \qquad [2]$$

For example; let us find the centre of gravity of a hemisphere. The equation of the revolving curve is $x^2 + y^2 = a^2$.

$$\bar{x} = \frac{\int_0^a \int_0^{\sqrt{a^2-x^2}} xy\,dy\,dx}{\int_0^a \int_0^{\sqrt{a^2-x^2}} y\,dy\,dx} = \frac{\frac{1}{8}a^4}{\frac{1}{3}a^3} = \frac{3}{8}a.$$

If we use polar coördinates the equation of the revolving curve is $r = a$.

Here
$$\bar{x} = \frac{\int_0^a \int_0^{\frac{\pi}{2}} r^3 \sin\phi \cos\phi\, d\phi\, dr}{\int_0^a \int_0^{\frac{\pi}{2}} r^2 \sin\phi\, d\phi\, dr} = \frac{\frac{1}{8}a^4}{\frac{1}{3}a^3} = \frac{3}{8}a.$$

EXAMPLES.

1. Find the centre of gravity of the solid formed by the revolution of the sector of a circle about one of its extreme radii.

Ans. $\bar{x} = \frac{3}{4}a\cos^2\frac{1}{2}\beta$, where β is the angle of the sector.

2. Find the centre of gravity of the segment of a paraboloid of revolution cut off by a plane perpendicular to the axis.

Ans. $\bar{x} = \frac{2}{3}a$, where $x = a$ is the plane.

3. Find the centre of gravity of the solid formed by scooping out a cone from a given paraboloid of revolution, the bases of the two volumes being coincident as well as their vertices.

Ans. The centre of gravity bisects the axis.

4. A cardioide is made to revolve about its axis; find the centre of gravity of the solid generated. *Ans.* $\bar{x} = -\frac{8}{5}a.$

5. Obtain formulas for the centre of gravity of any homogeneous solid.

6. Find the centre of gravity of the solid bounded by the surface $z^2 = xy$ and the five planes $x=0$, $y=0$, $z=0$, $x=a$, $y=b$.

Ans. $\bar{x} = \frac{3}{5}a$, $\bar{y} = \frac{3}{5}b$, $\bar{z} = \frac{9}{32}a^{\frac{1}{2}}b^{\frac{1}{2}}.$

160. If we are dealing with the arc of a plane curve, the formulas of Art. 157 reduce to

$$\bar{x} = \frac{\int x ds}{\int ds}, \qquad [1]$$

$$\bar{y} = \frac{\int y ds}{\int ds}. \qquad [2]$$

EXAMPLES.

1. Find the centre of gravity of an arc of a circle, taking the diameter bisecting the arc as the axis of X and the centre as the origin. *Ans.* $\bar{x} = \frac{ac}{s}$, where c is the chord of the arc.

2. Find the centre of gravity of the arc of the curve $x^{\frac{2}{3}} + y^{\frac{2}{3}} = a^{\frac{2}{3}}$ between two successive cusps. *Ans.* $\bar{x} = \bar{y} = \frac{2}{5}a.$

3. Find the centre of gravity of the arc of a semi-cycloid.

Ans. $\bar{x} = (\pi - \frac{4}{3})a$, $\bar{y} = -\frac{2}{3}a.$

4. Find the centre of gravity of the arc of a catenary cut off by any horizontal chord.

Ans. $\bar{x} = 0$, $\bar{y} = \frac{ax + sy}{2s}$, where $2s$ is the length of the arc.

5. Obtain formulas for the centre of gravity of a surface of revolution, the weight being uniformly distributed over the surface.

6. Find the centre of gravity of any zone of a sphere.

Ans. The centre of gravity bisects the line joining the centres of the bases of the zone.

7. A cardioide revolves about its axis; find the centre of gravity of the surface generated. *Ans.* $\bar{x} = -\frac{100}{63}a.$

8. Find the centre of gravity of the surface of a hemisphere when the density at each point of the surface varies as its perpendicular distance from the base of the hemisphere.

Ans. $\bar{x} = \frac{2}{3}a.$

9. Find the centre of gravity of a quadrant of a circle, the density at any point of which varies as the *nth* power of its distance from the centre. *Ans.* $\bar{x} = \bar{y} = \frac{n+2}{n+3} \cdot \frac{2a}{\pi}.$

10. Find the centre of gravity of a hemisphere, the density of which varies as the distance from the centre of the sphere.

Ans. $\bar{x} = \frac{2}{5}a.$

Properties of Guldin.

161. I. If a plane area revolve about an axis external to itself through any assigned angle, the volume of the solid generated will be equal to a prism whose base is the revolving area and whose altitude is the length of the path described by the centre of gravity of the area.

II. If the arc of a plane curve revolve about an external axis in its own plane through any assigned angle, the area of the surface generated will be equal to that of a rectangle, one side of which is the length of the revolving curve, and the other the length of the path described by its centre of gravity.

First; let the area in question revolve about the axis of X through an angle Θ. The ordinate of the centre of gravity of the area in question is

$$\bar{y} = \frac{\int\int y\,dx\,dy}{\int\int dx\,dy}, \qquad \text{by Art. 158 [1].}$$

The length of the path described by the centre of gravity

$$\bar{y}\Theta = \frac{\Theta\int\int y\,dxdy}{\int\int dxdy}. \qquad (1)$$

The volume generated is

$$V = \Theta\int\int y\,dxdy, \qquad \text{by Art. 151.}$$

Hence

$$V = \bar{y}\Theta\int\int dxdy.$$

But $\int\int dxdy$ is the revolving area, and the first theorem is established.

We leave the proof of the second theorem to the student.

EXAMPLES.

1. Find the surface and volume of a sphere, regarding it as generated by the revolution of a semicircle.

2. Find the surface and volume of the solid generated by the revolution of a cycloid about its base.

3. Find the volume and the surface of the ring generated by the revolution of a circle about an external axis.

Ans. $V = 2\pi^2a^2b$, $S = 4\pi^2ab$, where b is the distance of the centre of the circle from the axis.

4. Find the volume of the ring generated by the revolution of an ellipse about an external axis.

Ans. $V = 2\pi^2abc$, where c is the distance of the centre of the ellipse from the axis.

CHAPTER XIV.

LINE, SURFACE, AND SPACE INTEGRALS.

162. Any variable which depends for its value solely upon the position of a point, as, for example, any function of the rectangular or polar coördinates of the point, may be called a *point-function.*

A point-function is said to be continuous along a given line if its value changes continuously as the point, on whose position the function depends for its value, moves along the line; it is said to be continuous over a given surface if its value changes continuously as the point is made to move at pleasure over the surface; and it is said to be continuous throughout a given space if its value changes continuously as the point is made to move about at pleasure within the space.

163. If a given line is divided in any way into infinitesimal elements, and the length of each element is multiplied by the value a given point-function, which is continuous along the line, has at some point within the element, the limit approached by the sum of these products as each element is indefinitely decreased, is called the *line integral* of the given function along the line in question.

If a given surface is divided in any way into infinitesimal elements such that the distance between the two most widely separated points within each element is infinitesimal, and the area of each element is multiplied by the value a given point-function, which is continuous over the surface, has at some point within the element, the limit approached by the sum of these products as each element is indefinitely decreased, is called the *surface integral* of the given function over the surface in question.

If a given space is divided in any way into infinitesimal elements such that the distance between the two most widely separated points within each element is infinitesimal, and the volume of each element is multiplied by the value a given point-function, which is continuous throughout the space, has at some point within the element, the limit approached by the sum of these products as each element is indefinitely decreased, is called the *space integral* of the given function throughout the space in question.

It is easily seen that the line integral of unity along a given line is the length of the line; that the surface integral of unity over a given surface is the area of the surface; and that the space integral of unity throughout a given space is the volume of the space.

In the chapter on Centres of Gravity we have had numerous simple examples of line, surface, and space integrals.

164. That the value of a line, surface, or space integral is independent of the position in each element of the point at which the value of the given function is taken can be proved as follows: The distance apart of any two points in the same infinitesimal element is infinitesimal (Art. 163), therefore the values of a continuous function taken at any two points in the same element will differ in general by an infinitesimal; the products obtained by multiplying these two values by the magnitude of the element will, then, differ by an infinitesimal of higher order than that of the element; therefore, in forming the integral either of these products may be used in place of the other without changing the result. (I. Art. 161.)

165. The line integral of a function along a given line is absolutely independent of the manner in which the line is broken up into infinitesimal elements, and is equal to the length of the line multiplied by the *mean value* of the function along the line; the *mean value* of the function being defined as follows: Suppose a set of points uniformly distributed along the

line, that is, so distributed that the number of points in any portion of the line is proportional to the length of the portion; take the value of the function at each of these points; divide the sum of these values by the number of the points; and the limit approached by this quotient as the number of the points is indefinitely increased is the *mean value* of the given function along the line; and this mean value is in general finite and determinate.

To prove our proposition, we have only to consider in detail the method of finding the mean value in question. Let the number of points in a unit of length of the line be k. Then, no matter how the line is broken up into infinitesimal elements, the number of points in each element is k times the length of the element. Since any two values of the function corresponding to points in the same element differ by an infinitesimal, in finding our limit we may replace all values corresponding to points in the same element by any one; hence the sum of the values corresponding to points in the same element may be replaced by one value multiplied by the number of points taken in that element, that is, this sum may be replaced by k times the product of one value by the length of the element; and the sum of the values corresponding to all the points taken in the line may be replaced by k times the sum of the terms obtained by multiplying the length of each element by the value of the function at some point within the element. When we divide this sum by the whole number of points considered, that is, by k times the length of the line, the k's cancel out, and the required mean value reduces to the limit of the numerator divided by the length of the line, and the limit of the numerator is the line integral of the function along the line. Therefore the line integral is the mean value of the function multiplied by the length of the line.

The same proof may be given for a surface integral or for a space integral. The former is the product of the area of the surface by the mean value of the function over the surface; the latter is the volume of the space multiplied by the mean value of the function throughout the space; and both are independent of the way in which the surface or space may be divided into infinitesimal elements.

166. If the line along which the integral is taken is a plane curve, it is easy to get a geometrical representation of the integral. For, if at every point of the line a perpendicular to the plane of the line is erected whose length is equal to the value of the function at the point, the line integral required clearly represents the area of the cylindrical surface containing the perpendiculars if the values are all of the same sign, and represents the difference of the areas of the portions of the cylindrical surface which lie on opposite sides of the line if the values of the function are not all of the same sign.

A similar construction shows that a surface integral over a plane surface may be represented by a volume or by the differences of volumes. Consequently, in each case if the function is finite and continuous, the integral is finite and determinate.

167. As examples of line, surface, and space integrals, we will calculate a few *moments of inertia.*

The *moment of inertia* of a body about a given axis may be defined as the space integral of the product of the density at any point of the body by the square of the distance of the point from the axis; the integral being taken throughout the space occupied by the body.

If the body considered is a material surface or a material line, the integral reduces to a surface integral or to a line integral.

In the examples taken below the body is supposed to be homogeneous.

(*a*) The moment of inertia of a circumference about a given diameter.

Using polar coördinates and taking the diameter as our axis,

$$I=\int_0^{2\pi} a^2\sin^2\phi\cdot ka\,d\phi = ka^3\pi$$

$$=\tfrac{1}{2}Ma^2, \qquad [1]$$

if I is the moment of inertia, and a the radius, k the density, and M the mass of the circumference in question.

(*b*) The moment of inertia of the perimeter of a square about an axis passing through the centre of the square and parallel to a side.

$$I = 2\int_{-a}^{a} y^2 k dy + 2\int_{-a}^{a} a^2 k dx$$

$$= \tfrac{4}{3} ka^3 + 4 ka^3 = \tfrac{16}{3} ka^3$$

$$= \tfrac{2}{3} Ma^2, \qquad [2]$$

if $2a$ is the length of a side.

(*c*) The moment of inertia of a circle about a diameter.

$$I = \int_0^a \int_0^{2\pi} r^2 \sin^2\phi \,.\, krd\phi dr = \tfrac{1}{4} k\pi a^4$$

$$= \tfrac{1}{4} Ma^2. \qquad [3]$$

(*d*) The moment of inertia of a square about an axis through the centre of the square and parallel to a side.

$$I = \int_{-a}^{a} \int_{-a}^{a} y^2 kdxdy = \tfrac{4}{3} ka^4$$

$$= \tfrac{1}{3} Ma^2. \qquad [4]$$

(*e*) The moment of inertia of the surface of a sphere about a diameter.

$$I = \int_0^{2\pi} \int_0^{\pi} a^2 \sin^2\phi \,.\, ka^2 \sin\phi d\phi d\theta = \tfrac{8}{3} k\pi a^4$$

$$= \tfrac{2}{3} Ma^2. \qquad [5]$$

(*f*) The moment of inertia of the surface of a cube about an axis parallel to an edge and passing through the centre.

$$I = 4\int_{-a}^{a} \int_{-a}^{a} (a^2 + z^2) kdxdz + 2\int_{-a}^{a} \int_{-a}^{a} (y^2 + z^2) kdydz$$

$$= \tfrac{64}{3} ka^4 + \tfrac{16}{3} ka^4$$

$$= \tfrac{10}{9} Ma^2. \qquad [6]$$

(*g*) The moment of inertia of a sphere about a diameter.

$$I = \int_0^{2\pi}\int_0^{\pi}\int_0^{a} r^2 \sin^2\phi \,.\, kr^2 \sin\phi \, dr d\phi d\theta = \tfrac{8}{15} k\pi a^5$$

$$= \tfrac{2}{5} Ma^2. \qquad [7]$$

(*h*) The moment of inertia of a cube about an axis through the centre and parallel to an edge.

$$I = \int_{-a}^{a}\int_{-a}^{a}\int_{-a}^{a} (y^2 + z^2) k dx dy dz = \tfrac{16}{3} ka^5$$

$$= \tfrac{2}{3} Ma^2. \qquad [8]$$

EXAMPLES.

Find the moments of inertia of the following bodies:

(1) Of a straight line about a perpendicular through an extremity; about a perpendicular through its middle point.
Ans. $\frac{1}{3}Ml^2$; $\frac{1}{12}Ml^2$.

(2) Of the circumference of a circle about an axis through its centre perpendicular to its plane. *Ans.* Ma^2.

(3) Of a circle about an axis through its centre perpendicular to its plane. *Ans.* $\frac{1}{2}Ma^2$.

(4) Of a rectangle whose sides are $2a$, $2b$, about an axis through its centre perpendicular to its plane; about an axis through its centre parallel to the side $2b$.
Ans. $\frac{1}{3}M(a^2+b^2)$; $\frac{1}{3}Ma^2$.

(5) Of an ellipse about its major axis; about its minor axis; about an axis through the centre perpendicular to the plane of the ellipse. *Ans.* $\frac{1}{4}Mb^2$; $\frac{1}{4}Ma^2$; $\frac{1}{4}M(a^2+b^2)$.

(6) Of an ellipsoid about the axis a. *Ans.* $\frac{1}{5}M(b^2+c^2)$.

(7) Of a rectangular parallelopiped about an axis through the centre parallel to the edge $2a$. *Ans.* $\frac{1}{3}M(b^2+c^2)$.

(8) Of a segment of a parabola about the principal axis.
Ans. $\frac{1}{5}Mb^2$, where $2b$ is the breadth of the segment.

168. *If* u, $D_x u$, *and* $D_y u$ *are finite, continuous, and single-valued for all points in a given plane surface bounded by a closed curve* T, *the surface integral of* $D_x u$ *taken over the surface is equal to the line integral of* $u \cos \alpha$ *taken around the whole bounding curve*, where α is the angle made with the axis of X by the external normal at any point of the boundary.

This may be formulated thus:

$$\int\int D_x u dx dy = \int u \cos \alpha \, . \, ds. \qquad [1]$$

Let the axes be chosen so that the surface in question lies in the first quadrant, and divide the projection of T on the axis of Y into infinitesimal elements of which any one is dy.

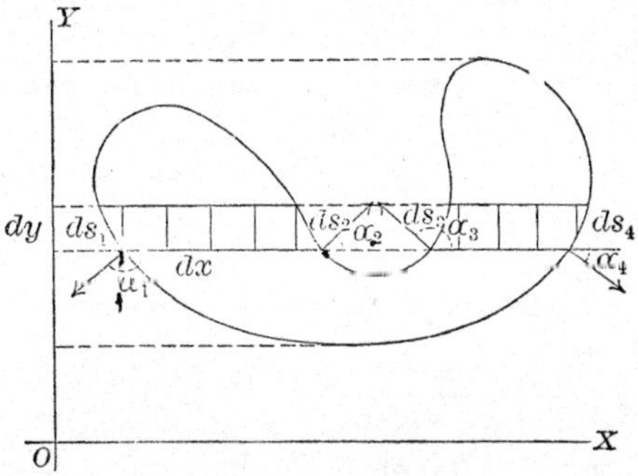

On each of these elements as a base erect a rectangle; and since T is a closed curve, each of these rectangles will cut it an even number of times.

Let us call the values of u at the points where the lower side of any one of these rectangles cuts T, u_1, u_2, u_3, u_4, etc., respectively; the angles which this side makes with the exterior normals at these points, α_1, α_2, α_3, α_4, etc.; and the elements which the rectangle cuts from T, ds_1, ds_2, ds_3, ds_4, etc.

It is evident that whenever a line parallel to the axis of X cuts into the surface bounded by T, the corresponding value of α is obtuse and its cosine negative; that whenever it cuts out,

α is acute and its cosine positive; and that any value of α is the angle which the contour T itself makes at the point in question with the axis of Y if we suppose the contour traced by a point moving so as to keep the bounded surface always on the left hand.

We have then approximately,

$$dy = -ds_1 \cdot \cos\alpha_1 = ds_2 \cdot \cos\alpha_2 = -ds_3 \cdot \cos\alpha_3 = ds_4 \cdot \cos\alpha_4 = \cdots. \quad [2]$$

If, now, in $\int\int D_x u dx dy$ we perform the integration with respect to x, and introduce the proper limits, we shall have

$$\int\int D_x u dx dy = \int dy(-u_1 + u_2 - u_3 + u_4 \cdots); \quad [3]$$

and the second member indicates that we are to form a quantity corresponding to that in parenthesis for every rectangle which cuts T, to multiply it by the base of the rectangle, and then to take the limit of the sum of the results as all the bases are indefinitely decreased.

By [2],

$$dy(-u_1 + u_2 - u_3 + u_4 \cdots)$$

$$= u_1 \cos\alpha_1 ds_1 + u_2 \cos\alpha_2 ds_2 + u_3 \cos\alpha_3 ds_3 + u_4 \cos\alpha_4 ds_4 + \cdots; \quad [4]$$

and the limit of the sum of the values any one of which is represented by the second member of [4] is clearly $\int u \cos\alpha ds$ taken around the whole of T.

EXAMPLE.

Prove that under the conditions stated in the last article

$$\int\int D_y u dx dy = \int u \cos\beta . ds,$$

where β is the angle made with the axis of Y by the exterior normal.

169. As an illustration of the last proposition, let us find the centre of gravity of a semicircle.

We have
$$\bar{y} = \frac{k}{M}\int\int y\,dx\,dy. \qquad (1)$$

But we may write $y = D_x(xy)$. Hence, by Art. 168,

$$\bar{y} = \frac{k}{M}\int\int y\,dx\,dy = \frac{k}{M}\int xy\cos\alpha\,ds$$

$$= \frac{k}{M}\left(\int_0^\pi a\cos\phi\, a\sin\phi\cos\phi\, a\,d\phi + \int_{-a}^{a} x.0.\cos\frac{\pi}{2}\cdot dx\right)$$

$$= \frac{k}{k\frac{\pi a^2}{2}}\cdot\frac{2}{3}a^3 + 0 = \frac{4a}{3\pi},$$

which agrees with the result of Ex. 8, Art. 158.

As a second example, we shall find the moment of inertia of a circle about a diameter.

We have

$$I = k\int\int y^2\,dx\,dy = k\int xy^2\cos\phi\,.\,ds$$

$$= k\int_0^{2\pi} a\cos\phi\, a^2\sin^2\phi\cos\phi\, a\,d\phi$$

$$= ka^4\int_0^{2\pi}\sin^2\phi\cos^2\phi\,d\phi = \frac{k}{4}\pi a^4 = \frac{1}{4}Ma^2,$$

which agrees with the result of (*c*), Art. 167.

EXAMPLES.

(1) Find the centre of gravity of a semicircle, using the theorem $\int\int D_y u\,dx\,dy = \int u\cos\beta\,.\,ds$.

(2) Find the moment of inertia of a circle about an axis through its centre perpendicular to its plane, using the principles $\int\int D_x u\,dx\,dy = \int u\cos\alpha\,.\,ds$ and $\int\int D_y u\,dx\,dy = \int u\cos\beta\,.\,ds$.

170. Since, as we have seen in Art. 168, α is the angle which the curve T makes with the axis of Y; if we trace the curve so as to keep the bounded space on our left, it follows that $\cos\alpha . ds = dy$.

Hence
$$\int\int D_x u dx dy = \int u dy; \qquad [1]$$

and in like manner,
$$\int\int D_y u dx dy = -\int u dx; \qquad [2]$$

the first integral in [1] and [2] being taken over the bounded surface, and the second around the bounding curve.

For example, the moment of inertia of a square about an axis through the centre and parallel to a side is

$$I = k\int\int y^2 dx dy. \qquad ((d) \text{ Art. } 167.)$$

By [1],
$$\int\int y^2 dx dy = \int x y^2 dy,$$

and the last integral is to be taken around the perimeter.

Hence
$$I = k\left[\int_{-a}^{a} a y^2 dy + \int_{a}^{-a} (-a y^2 dy)\right] = 2ka\int_{-a}^{a} y^2 dy = \tfrac{4}{3}ka^4$$
$$= \tfrac{1}{3}Ma^2.$$

EXAMPLE.

Work Ex. 8, Art. 167, by the aid of (2).

171. *If* U, D_xU, D_yU, *and* D_zU *are finite, continuous, single-valued functions throughout the space bounded by a given closed surface* T, *the space integral of* D_xU *taken throughout the space in question is equal to the surface integral, taken over the bounding surface, of* U $\cos\alpha$, where α is the angle made with the axis of X by the exterior normal at any point of the surface.

This may be formulated thus:

$$\int\int\int D_x U dx dy dz = \int U \cos\alpha . dS. \qquad [1]$$

The proof is almost identical with that given in Art. 168, except that for elementary rectangle we use elementary prism. We shall merely indicate the steps.

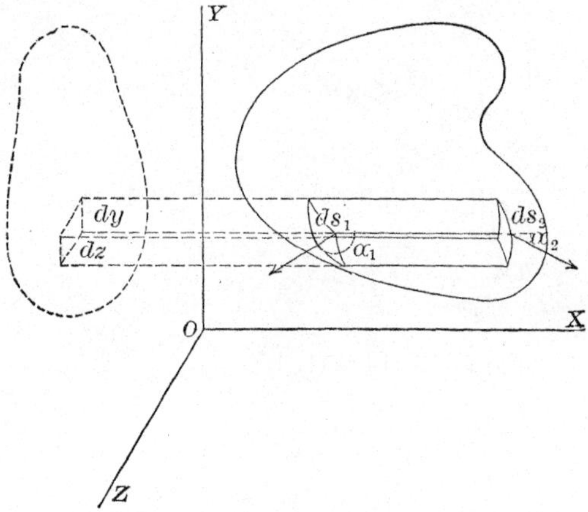

$$dydz = -dS_1\cos\alpha_1 = dS_2\cos\alpha_2 = -dS_3\cos\alpha_3 = \cdots$$

$$\int\int\int D_xU\,dxdydz = \int\int dydz\,[-U_1 + U_2 - U_3 \cdots]$$

= the limit of the sum of terms of the form

$$U_1\cos\alpha_1 . dS_1 + U_2\cos\alpha_2 . dS_2 + U_3\cos\alpha_3 . dS_3 + \cdots$$

$$= \int U\cos\alpha . dS.$$

EXAMPLE.

Prove that under the conditions of the last article

$$\int\int\int D_yU\,dxdydz = \int U\cos\beta . dS,$$

and
$$\int\int\int D_zU\,dxdydz = \int U\cos\gamma . dS,$$

where β and γ are the angles made with the axes of Y and Z respectively by the exterior normal to the bounding surface.

172. As an illustration, let us find the centre of gravity of a hemisphere.

We have

$$\bar{x} = \frac{k}{M}\int\int\int x\,dx\,dy\,dz = \frac{k}{M}\int\frac{x^2}{2}\cos\alpha\,.\,dS$$

$$= \frac{k}{2M}\int_0^{2\pi}\int_0^{\frac{\pi}{2}} a^2\cos^2\phi\cos\phi\, a^2\sin\phi\, d\phi\, d\theta$$

$$= \frac{a^4 k}{2M}\int_0^{2\pi}\int_0^{\frac{\pi}{2}}\cos^3\phi\sin\phi\, d\phi\, d\theta$$

$$= \frac{a^4 k}{\frac{4}{3}\pi a^3 k}\cdot\frac{\pi}{2} = \frac{3}{8}a\,;$$

which agrees with the result of Art. 159.

EXAMPLE.

Find the moment of inertia of a sphere about a diameter; of a cube about an axis through the centre parallel to an edge. Make your work depend upon finding the value of a surface integral.

CHAPTER XV.

MEAN VALUE AND PROBABILITY.

173. The application of the Integral Calculus to questions in Mean Value and Probability is a matter of decided interest; but lack of space will prevent our doing more than solving a few problems in illustration of some of the simplest of the methods and devices ordinarily employed. A full and admirable treatment of the subject is given in "Williamson's Integral Calculus" (London: Longmans, Green, & Co.); and numerous interesting problems are published with their solutions in "The Mathematical Visitor" and "The Annals of Mathematics."

174. The *mean* of n quantities is their sum divided by their number. If the number of quantities considered is supposed to increase indefinitely according to some given law, the problem of finding the limiting value approached by their mean usually calls for the Integral Calculus. The *mean value* of a continuous function of one, two, or three independent variables has been carefully defined in Art. 165, and has been proved to depend upon a line, surface, or space integral.

(*a*) Let us find the mean distance of all the points on the circumference of a circle from a given point on the circumference.

If we take the given point as origin, the distances whose mean is required are the radii vectores of points uniformly distributed along the circumference of the circle.

The required mean is, therefore, by Art. 165, equal to

the quotient obtained by dividing the line integral of r taken around the circumference by the length of the circumference; that is,

$$M = \frac{\int r ds}{2\pi a}.$$

The polar equation of the circle is

$$r = 2a\cos\phi;$$

$$ds = 2a d\phi,$$

$$M = \frac{1}{2\pi a}\int_{-\frac{\pi}{2}}^{\frac{\pi}{2}} 4a^2\cos\phi\, d\phi = \frac{4a}{\pi},$$

the required *mean value*.

(*b*) Let us find the mean distance of points on the surface of a circle from a fixed point on the circumference.

Here, by Art. 165, the required mean is the surface integral of r taken over the circle, divided by the area of the circle; that is,

$$M = \frac{1}{\pi a^2}\int_{-\frac{\pi}{2}}^{\frac{\pi}{2}}\int_0^{2a\cos\phi} r^2 dr d\phi = \frac{32a}{9\pi}.$$

(*c*) The problem of finding the mean distance of points on the surface of a square from a corner of the square can be simplified slightly by considering merely one of the halves into which the square is divided by a diagonal.

Here

$$M = \frac{2}{a^2}\int_0^{\frac{\pi}{4}}\int_0^{a\sec\phi} r \,.\, r dr d\phi$$

$$= \frac{a}{3}\left(\sqrt{2} + \log\tan\frac{3\pi}{8}\right).$$

(*d*) As an example of a device often employed, we shall now solve the problem, To find the mean distance between two points within a given circle.

If M be the required mean, the sum of the whole number of cases can be represented by $(\pi r^2)^2 M$, r being the radius of the circle; since for each position of the first point the number of positions of the second point is proportional to the area of the circle, and may be measured by that area; and as the number of possible positions of the first point may also be measured by the area of the circle, the whole number of cases to be considered is represented by the square of the area; and the sum of all the distances to be considered must be the product of the mean distance by the number.

Let us see what change will be produced in this sum by increasing r by the infinitesimal dr; that is, let us find $d(\pi^2 r^4 M)$.

If the first point is anywhere on the annulus $2\pi r.dr$, which we have just added, its mean distance from the other points of the circle is $\dfrac{32r}{9\pi}$, by (*b*).

Therefore, the sum of the new distances to be considered, if the first point is on the annulus, is $\dfrac{32r}{9\pi}.\pi r^2.2\pi r dr$; but the second point may be on the annulus, instead of the first; so that to get the sum of all the new cases brought in by increasing r by dr, we must double the value just obtained.

Hence
$$d(\pi^2 r^4 M) = \tfrac{128}{9}\pi r^4 dr,$$
$$\pi^2 a^4 M = \tfrac{128}{9}\pi \int_0^a r^4 dr = \tfrac{128}{45}\pi a^5,$$
$$M = \frac{128a}{45\pi}.$$

175. In solving questions in *Probability*, we shall assume that the student is familiar with the elements of the theory as given in "Todhunter's Algebra."

(*a*) A man starts from the bank of a straight river, and walks till noon in a random direction; he then turns and walks in another random direction; what is the probability that he will reach the river by night?

Let θ be the angle his first course makes with the river. If the angle through which he turns at noon is less than $\pi - 2\theta$, he will reach the river by night. For any given value of θ, then, the required probability is $\frac{\pi - 2\theta}{2\pi}$. The probability that θ shall lie between any given value θ_0 and $\theta_0 + d\theta$ is $\frac{d\theta}{\frac{1}{2}\pi}$.

The chance that his first course shall make an angle with the river between θ_0 and $\theta_0 + d\theta$, and that he shall get back, is

$$\frac{\pi - 2\theta}{2\pi} \cdot \frac{d\theta}{\frac{1}{2}\pi} = \frac{(\pi - 2\theta)\,d\theta}{\pi^2}.$$

As θ is equally likely to have any value between 0 and $\frac{\pi}{2}$, the required probability,

$$p = \int_0^{\frac{1}{2}\pi} \frac{(\pi - 2\theta)\,d\theta}{\pi^2} = \tfrac{1}{4}.$$

(*b*) A floor is ruled with equidistant straight lines; a rod, shorter than the distance between the lines, is thrown at random on the floor; to find the chance of its falling on one of the lines.

Let x be the distance of the centre of the rod from the nearest line; θ the inclination of the rod to a perpendicular to the parallels passing through the centre of the rod; $2a$ the common distance of the parallels; $2c$ the length of the rod.

In order that the rod may cross a line, we must have $c\cos\theta > x$; the chance of this for any given value x_0 of x is $\frac{1}{\frac{1}{2}\pi}\cos^{-1}\frac{x_0}{c}$.

The probability that x will have the value x_0 is $\frac{dx}{a}$. The probability required is

$$p = \frac{2}{\pi a}\int_0^c \cos^{-1}\frac{x}{c}\,dx = \frac{2c}{\pi a}.$$

This problem may be solved by another method which possesses considerable interest.

Since all values of x from 0 to a, and all values of θ from $-\frac{\pi}{2}$ to $\frac{\pi}{2}$ are equally probable, the whole number of cases that can arise may be represented by

$$\int_{-\frac{1}{2}\pi}^{\frac{1}{2}\pi}\int_0^a dx d\theta = \pi a.$$

The number of favorable cases will be represented by

$$\int_{-\frac{1}{2}\pi}^{\frac{1}{2}\pi}\int_0^{c\cos\theta} dx d\theta = 2c.$$

Hence $$p = \frac{2c}{\pi a}.$$

(*c*) To find the probability that the distance of two stars, taken at random in the northern hemisphere, shall exceed 90°.

Let α be the latitude of the first star. With the star as a pole, describe an arc of a great circle, dividing the hemisphere into two lunes; the probability that the distance of the second star from the first will exceed 90° is the ratio of the lune not containing the first star to the hemisphere, and is equal to $\frac{(\frac{1}{2}\pi - \alpha)}{\pi}$. The probability that the latitude of the first star will be between α and $\alpha + d\alpha$ is the ratio of the area of the zone, whose bounding circles have the latitudes α and $\alpha + d\alpha$ respectively, to the area of the hemisphere, and is

$$\frac{2\pi a^2 \cos\alpha\, d\alpha}{2\pi a^2} = \cos\alpha\, d\alpha.$$

Hence $$p = \int_0^{\frac{\pi}{2}} \frac{(\frac{1}{2}\pi - \alpha)}{\pi}\cos\alpha\, d\alpha = \frac{1}{\pi}.$$

(*d*) A random straight line meets a closed convex curve; what is the probability that it will meet a second closed convex curve within the first?

If an infinite number of random lines be drawn in a plane, all directions are equally probable; and lines having any given direction will be disposed with equal frequency all over the plane. If we determine a line by its distance p from the origin, and by the angle α which p makes with the axis of X, we can get all the lines to be considered by making p and α vary between suitable limits by equal infinitesimal increments.

In our problem, the whole number of lines meeting the external curve can be represented by $\int\int dp d\alpha$. If the origin is within the curve, the limits for p must be zero, and the perpendicular distance from the origin to a tangent to the curve; and for α must be zero and 2π. If we call this number N, we shall have

$$N = \int_0^{2\pi} p d\alpha,$$

p being now the perpendicular from the origin to the tangent.

If we regard the distance from a given point of any closed convex curve along the curve to the point of contact of a tangent, and then along the tangent to the foot of the perpendicular let fall upon it from the origin, as a function of the α used above, its differential is easily seen to be $p d\alpha$. If we sum these differentials from $\alpha = 0$ to $\alpha = 2\pi$, we shall get the perimeter of the given curve.

Hence $$N = \int_0^{2\pi} p d\alpha = L,$$

where L is the perimeter of the curve in question. By the same reasoning, we can see that n, the number of the random lines which meet the inner curve, is equal to l, its perimeter. For p, the required probability, we shall have

$$p = \frac{l}{L}.$$

EXAMPLES.

(1) A number n is divided at random into two parts; find the mean value of their product. *Ans.* $\frac{n^2}{6}$.

(2) Find the mean value of the ordinates of a semicircle, supposing the series of ordinates taken equidistant. *Ans.* $\frac{\pi}{4}a$.

(3) Find the mean value of the ordinates of a semicircle, supposing the ordinates drawn through equidistant points on the circumference. *Ans.* $\frac{2a}{\pi}$.

(4) Find the mean values of the roots of the quadratic $x^2 - ax + b = 0$, the roots being known to be real, but b being unknown but positive. *Ans.* $\frac{5a}{6}$ and $\frac{a}{6}$.

(5) Prove that the mean of the radii vectores of an ellipse, the focus being the origin, is equal to half the minor axis when they are drawn at equal angular intervals, and is equal to half the major axis when they are drawn so that the abscissas of their extremities increase uniformly.

(6) Suppose a straight line divided at random into three parts; find the mean value of their product. *Ans.* $\frac{a^3}{60}$.

(7) Find the mean square of the distance of a point within a given square (side $= 2a$) from the centre of the square. *Ans.* $\frac{2}{3}a^2$.

(8) A chord is drawn joining two points taken at random on a circumference; find the mean area of the less of the two segments into which it divides the circle. *Ans.* $\frac{\pi a^2}{4} - \frac{a^2}{\pi}$.

(9) Find the mean latitude of all places north of the equator. *Ans.* 32°.7.

(10) Find the mean distance of points within a sphere from a given point of the surface. *Ans.* $\frac{6}{5}a$.

(11) Find the mean distance of two points taken at random within a sphere. *Ans.* $\frac{36}{35}a$.

(12) Two points are taken at random in a given line a; find the chance that their distance shall exceed a given value c. *Ans.* $\left(\frac{a-c}{a}\right)^2$.

(13) Find the chance that the distance of two points within a square shall not exceed a side of the square. *Ans.* $\pi - \frac{13}{6}$.

(14) A line crosses a circle at random; find the chances that a point, taken at random within the circle, shall be distant from the line by less than the radius of the circle. *Ans.* $1 - \dfrac{2}{3\pi}$.

(15) A random straight line crosses a circle; find the chance that two points, taken at random in the circle, shall lie on opposite sides of the line. *Ans.* $\dfrac{128}{45\pi^2}$.

(16) A random straight line is drawn across a square; find the chance that it intersects two opposite sides. *Ans.* $\frac{1}{2} - \dfrac{\log 2}{\pi}$.

(17) Two arrows are sticking in a circular target; find the chance that their distance apart is greater than the radius.

Ans. $\dfrac{3\sqrt{3}}{4\pi}$.

(18) From a point in the circumference of a circular field a projectile is thrown at random with a given velocity which is such that the diameter of the field is equal to the greatest range of the projectile: find the chance of its falling within the field.

Ans. $\frac{1}{2} - \frac{2}{\pi}(\sqrt{2} - 1)$.

(19) On a table a series of equidistant parallel lines is drawn, and a cube is thrown at random on the table. Supposing that the diagonal of the cube is less than the distance between consecutive straight lines, find the chance that the cube will rest without covering any part of the lines.

Ans. $1 - \dfrac{4a}{\pi c}$, where a is the edge of the cube and c the distance between consecutive lines.

(20) A plane area is ruled with equidistant parallel straight lines, the distance between consecutive lines being c. A closed curve, having no singular points, whose greatest diameter is less

than c, is thrown down on the area. Find the chance that the curve falls on one of the lines.

Ans. $\frac{l}{\pi c}$, where l is the perimeter of the curve.

(21) During a heavy rain-storm, a circular pond is formed in a circular field. If a man undertakes to cross the field in the dark, what is the chance that he will walk into the pond?

CHAPTER XVI.

ELLIPTIC INTEGRALS.

176. In attempting to solve completely the problem of the motion of a simple pendulum by the methods of I. Chapter VIII. we encounter an integral of great importance which we have not yet considered. The problem is closely analogous to that of the Cycloidal pendulum (I. Art. 119).

For the sake of simplicity we shall suppose the pendulum bob to start from the lowest point of its circular path with the initial velocity that would be acquired by a particle falling freely in a vacuum through the distance y_0; and this by I. Art. 114 [1] is $\sqrt{2gy_0}$.

Forming our differential equation of motion as in I. Art. 118, but taking the positive direction of the axis of Y upward, we have

$$\frac{d^2s}{dt^2} = -g\frac{dy}{ds}. \qquad (1)$$

Multiplying by $2\dfrac{ds}{dt}$ and integrating,

$$v^2 = \left(\frac{ds}{dt}\right)^2 = -2gy + C,$$

or, determining C,

$$v^2 = \left(\frac{ds}{dt}\right)^2 = 2g(y_0 - y). \qquad (2)$$

If the starting-point is taken as the origin, the equation of the circular path is $x^2 + y^2 - 2ay = 0$, whence

$$\left(\frac{ds}{dt}\right)^2 = \frac{a^2}{2ay - y^2}\left(\frac{dy}{dt}\right)^2,$$

and we have

$$\frac{a}{\sqrt{2ay - y^2}}\frac{dy}{dt} = \sqrt{2g(y_0 - y)},$$

or
$$dt = \frac{a\,dy}{\sqrt{2g}\,.\,\sqrt{(y_0 - y)(2ay - y^2)}}.$$

Integrating, and determining the arbitrary constant, we get

$$t = \frac{a}{\sqrt{2g}}\int_0^y \frac{dy}{\sqrt{(y_0 - y)(2ay - y^2)}} \qquad (3)$$

as the time required to reach that point of the path which has the ordinate y.

The substitution of $x^2 = \frac{y}{y_0}$ reduces (3) to the form

$$t = \sqrt{\frac{a}{g}}\int_0^x \frac{dx}{\sqrt{(1 - x^2)\left(1 - \frac{y_0}{2a}x^2\right)}}, \qquad (4)$$

where the integral is of the form

$$\int_0^x \frac{dx}{\sqrt{(1 - x^2)(1 - k^2x^2)}}, \qquad (5)$$

k^2 being positive and less than unity if y_0 is less than $2a$. An examination of equation (2) will show that if this is true, the pendulum will oscillate between the two points of the arc which have the ordinate y_0.

If y_0 is greater than $2a$, the pendulum will make complete revolutions. For this case the substitution of $x^2 = \frac{y}{2a}$ in (3) will reduce it to

$$t = a\sqrt{\frac{2}{gy_0}}\int_0^x \frac{dx}{\sqrt{(1 - x^2)\left(1 - \frac{2a}{y_0}x^2\right)}}, \qquad (6)$$

where the integral is of the form (5), k^2 being positive and less than unity.

The time required for the pendulum to reach its greatest height — that is, in the first case, the time of a half-vibration, and in the second case, the time of a half-revolution — will depend upon

$$\int_0^1 \frac{dx}{\sqrt{(1 - x^2)(1 - k^2x^2)}}. \qquad (7)$$

177. The length of an arc of an Ellipse, measured from the extremity of the minor axis, has been found to be (Art. 107)

$$s = \int_0^x \sqrt{\frac{a^2 - e^2 x^2}{a^2 - x^2}} \cdot dx. \tag{1}$$

If we replace $\frac{x}{a}$ by x, (1) becomes

$$s = a\int_0^x \sqrt{\frac{1 - e^2 x^2}{1 - x^2}} \cdot dx, \tag{2}$$

and the integral is of the form

$$\int_0^x \sqrt{\frac{1 - k^2 x^2}{1 - x^2}} \cdot dx, \tag{3}$$

where k^2 is positive and less than unity.

The length of an Elliptic quadrant depends upon the integral

$$\int_0^1 \sqrt{\frac{1 - k^2 x^2}{1 - x^2}} \cdot dx. \tag{4}$$

178. It can be shown by an elaborate investigation, for which we have not room, that the integral of any algebraic function, which is irrational through containing under the square root sign an algebraic polynomial of the third or fourth degree, can by suitable transformations be made to depend upon one or more of the three integrals

$$F(k, x) = \int_0^x \frac{dx}{\sqrt{(1 - x^2)(1 - k^2 x^2)}}, \qquad [1]$$

$$E(k, x) = \int_0^x \sqrt{\frac{1 - k^2 x^2}{1 - x^2}} \cdot dx, \qquad [2]$$

$$\Pi(n, k, x) = \int_0^x \frac{dx}{(1 + nx^2)\sqrt{(1 - x^2)(1 - k^2 x^2)}}, \qquad [3]$$

which are known as the Elliptic Integrals of the first, second, and third class respectively.

k, which may always be taken positive and less than 1, is called the *modulus;* and n, which may be taken real, is called the *parameter* of the integral.

$$K = F(k, 1) = \int_0^1 \frac{dx}{\sqrt{(1-x^2)(1-k^2x^2)}}, \qquad [4]$$

and

$$E = E(k, 1) = \int_0^1 \sqrt{\frac{1-k^2x^2}{1-x^2}} \cdot dx, \qquad [5]$$

are known as the Complete Elliptic Integrals of the first and second classes.

179. The substitution of $x = \sin\phi$ in the Elliptic Integrals reduces them to the following simpler forms.

$$F(k, \phi) = \int_0^\phi \frac{d\phi}{\sqrt{1-k^2\sin^2\phi}} = \int_0^\phi \frac{d\phi}{\Delta\phi}. \qquad [1]$$

$$E(k, \phi) = \int_0^\phi \sqrt{1-k^2\sin^2\phi} \,.\, d\phi = \int_0^\phi \Delta\phi \,.\, d\phi. \qquad [2]$$

$$\Pi(n, k, \phi) = \int_0^\phi \frac{d\phi}{(1+n\sin^2\phi)\sqrt{1-k^2\sin^2\phi}} = \int_0^\phi \frac{d\phi}{(1+n\sin^2\phi)\,\Delta\phi}. \qquad [3]$$

$$K = \int_0^{\frac{\pi}{2}} \frac{d\phi}{\sqrt{1-k^2\sin^2\phi}} = \int_0^{\frac{\pi}{2}} \frac{d\phi}{\Delta\phi}. \qquad [4]$$

$$E = \int_0^{\frac{\pi}{2}} \sqrt{1-k^2\sin^2\phi} \,.\, d\phi = \int_0^{\frac{\pi}{2}} \Delta\phi \,.\, d\phi. \qquad [5]$$

ϕ is called the *amplitude* of the Elliptic Integral, and $\Delta\phi = \sqrt{1-k^2\sin^2\phi}$ is called the *delta* of ϕ, or more simply, delta ϕ, and is regarded as a new trigonometric function: it is always taken with the positive sign, and has an analogy with $\cos\phi$.

For a given value of k, $\Delta\phi$ is easily seen to be a periodic function of ϕ having the period π. It has its maximum value 1 when $\phi = 0$ and when $\phi = \pi$, and its minimum value $\sqrt{1-k^2}$, which is usually represented by k' and called the *complementary modulus*, when $\phi = \frac{\pi}{2}$; and $\Delta\left(\frac{\pi}{2}+a\right) = \Delta\left(\frac{\pi}{2}-a\right)$.

Landen's Transformation.

180. The approximate numerical value of an Elliptic Integral of the first class, when k and ϕ are given, is easily computed by the aid of two valuable reduction formulas due to *Landen*.

If in
$$F(k, \phi) = \int_0^\phi \frac{d\phi}{\sqrt{1-k^2\sin^2\phi}}$$

we replace ϕ by ϕ_1, ϕ_1 and ϕ being connected by the relation

$$\tan\phi = \frac{\sin 2\phi_1}{k+\cos 2\phi_1}, \tag{1}$$

which is easily transformable into either of the following:

$$k\sin\phi = \sin(2\phi_1 - \phi), \tag{2}$$

$$\tan(\phi - \phi_1) = \frac{1-k}{1+k}\tan\phi_1, \tag{3}$$

$$\int_0^\phi \frac{d\phi}{\sqrt{1-k^2\sin^2\phi}} \quad \text{reduces to} \quad \frac{2}{1+k}\int_0^{\phi_1} \frac{d\phi_1}{\sqrt{1-\frac{4k}{(1+k)^2}\sin^2\phi_1}},$$

which is also an Elliptic Integral of the first class, but has a different modulus and a different amplitude from those of the given integral.

The steps of the process are as follows:

From (1) we easily find

$$\sin^2\phi = \frac{\sin^2 2\phi_1}{1+k^2+2k\cos 2\phi_1},$$

whence
$$\sqrt{1-k^2\sin^2\phi} = \frac{1+k\cos 2\phi_1}{\sqrt{1+k^2+2k\cos 2\phi_1}}.$$

Differentiating (1), we get

$$\sec^2\phi\,d\phi = \frac{2(1+k\cos 2\phi_1)}{(k+\cos 2\phi_1)^2}\,d\phi_1;$$

but from (1),
$$\sec^2\phi = \frac{1+k^2+2k\cos 2\phi_1}{(k+\cos 2\phi_1)^2};$$

hence
$$d\phi = \frac{2(1+k\cos 2\phi_1)}{1+k^2+2k\cos 2\phi_1}\,d\phi_1,$$

$$\frac{d\phi}{\sqrt{1-k^2\sin^2\phi}} = \frac{2\,d\phi_1}{\sqrt{1+k^2+2k\cos 2\phi_1}} = \frac{2\,d\phi_1}{\sqrt{1+k^2+2k-4k\sin^2\phi_1}}$$

$$= \frac{2}{1+k}\,\frac{d\phi_1}{\sqrt{1-\dfrac{4k}{(1+k)^2}\sin^2\phi_1}}.$$

$$\phi_1 = 0 \quad \text{when} \quad \phi = 0,$$

hence
$$\int_0^{\phi}\frac{d\phi}{\sqrt{1-k^2\sin^2\phi}} = \frac{2}{1+k}\int_0^{\phi_1}\frac{d\phi_1}{\sqrt{1-\dfrac{4k}{(1+k)^2}\sin^2\phi_1}},$$

and
$$F(k,\phi) = \frac{2}{1+k}F(k_1,\phi_1),$$

where
$$k_1 = \frac{2\sqrt{k}}{1+k},$$

and
$$\sin(2\phi_1-\phi) = k\sin\phi. \qquad [4]$$

k_1 is less than 1 and greater than k; for $\frac{2\sqrt{k}}{1+k}<1$ reduces to $0<(1-\sqrt{k})^2$, which is obviously true, and $\frac{2\sqrt{k}}{1+k}>k$ reduces to $4>k(1+k)^2$, which is true, since k is less than 1. If ϕ is not greater than $\frac{\pi}{2}$, and the smallest value of ϕ_1 consistent with the relation $\sin(2\phi_1-\phi)=k\sin\phi$ is taken, $0<\phi_1<\phi$. Hence (4) is a reduction formula by which we can raise the modulus and lower the amplitude of our given function.

By applying the formula (4) n times, we get

$$F(k,\phi)=\frac{2}{1+k}\cdot\frac{2}{1+k_1}\cdot\frac{2}{1+k_2}\cdots\frac{2}{1+k_{n-1}}F(k_n,\phi_n);$$

or, since $\qquad \dfrac{2}{1+k}=\dfrac{k_1}{\sqrt{k}}$, etc.,

$$F(k,\phi)=k_n\sqrt{\frac{k_1 . k_2 . k_3 \ldots k_{n-1}}{k}}\,F(k_n,\phi_n),$$

where

$$k_p=\frac{2\sqrt{k_{p-1}}}{1+k_{p-1}},\text{ and }\sin(2\phi_p-\phi_{p-1})=k_{p-1}\sin\phi_{p-1}. \qquad [5]$$

If we suppose n in (5) to be indefinitely increased, we shall have $\underset{n=\infty}{\text{limit}}[k_n]=1$; for if we form the series

$$(1-k)+(1-k_1)+(1-k_2)+\cdots+(1-k_p)+\cdots,$$

we shall have

$$\frac{1-k_{p+1}}{1-k_p}=\frac{1-\dfrac{2\sqrt{k_p}}{1+k_p}}{1-k_p}=\frac{(1-\sqrt{k_p})^2}{1-k_p{}^2}=\frac{1-\sqrt{k_p}}{1+\sqrt{k_p}}\,\frac{1}{1+k_p},$$

which is always less than unity; hence the terms in the series must decrease indefinitely as p is increased and $\underset{n=\infty}{\text{limit}}[1-k_n]=0$. Since, as we have seen above, ϕ_n continually diminishes as n increases, but does not reach the value zero, it must have some limiting value Φ. Hence

$$\underset{n=\infty}{\text{limit}}[F(k_n,\phi_n)]=F(1,\Phi)=\int_0^\Phi\frac{d\phi}{\sqrt{1-\sin^2\phi}}$$

$$=\int_0^\Phi\sec\phi\,d\phi=\log\tan\left[\frac{\pi}{4}+\frac{\Phi}{2}\right];$$

and

$$F(k,\phi)=\log\tan\left[\frac{\pi}{4}+\frac{\Phi}{2}\right]\sqrt{\frac{k_1 . k_2 . k_3\ldots}{k}}. \qquad [6]$$

Formulas [5] and [6] lend themselves very readily to numerical computation.

181. Formula [4], Art. 180, may be used to decrease the modulus and increase the amplitude of a given Elliptic Integral. Interchanging the subscripts, and using (3) Art. 180 instead of (2) Art. 180, we have

$$\left.\begin{aligned} F(k, \phi) &= \frac{1+k_1}{2} F(k_1, \phi_1), \\ \text{where} \qquad k_1 &= \frac{1-\sqrt{1-k^2}}{1+\sqrt{1-k^2}}, \\ \text{and} \qquad \tan(\phi_1 - \phi) &= \sqrt{1-k^2}\tan\phi. \end{aligned}\right\} \qquad [1]$$

By repeated application of [1] we get

$$\left.\begin{aligned} F(k, \phi) &= (1+k_1)(1+k_2)\dots(1+k_n)\frac{F(k_n, \phi_n)}{2^n}, \\ \text{where} \qquad k_p &= \frac{1-\sqrt{1-k^2_{p-1}}}{1+\sqrt{1-k^2_{p-1}}}, \\ \text{and} \qquad \tan(\phi_p - \phi_{p-1}) &= \sqrt{1-k^2_{p-1}}\tan\phi_{p-1}. \end{aligned}\right\} \qquad [2]$$

It is easily shown, as in Art. 180, that $\lim\limits_{n=\infty} [k_n] = 0$, and consequently that $\lim\limits_{n=\infty} F(k_n, \phi_n) = \int_0^{\Phi} d\phi = \Phi$, where Φ is the limiting value approached by ϕ as n is increased.

If $\phi = \frac{\pi}{2}$, we get from [2], $\phi_1 = \pi$, $\phi_2 = 2\pi$, $\dots \phi_n = 2^{n-1}\pi$;

hence $$K = F\left(k, \frac{\pi}{2}\right) = \frac{\pi}{2}(1+k_1)(1+k_2)(1+k_3)\cdots. \qquad [3]$$

Formulas [2] and [3], like formulas [5] and [6] of Art. 180, lend themselves readily to computation.

With a large modulus, it is generally best to use [5] and [6] of Art. 180; with a small modulus, [2] or [3] of the present article will generally work more rapidly.

We give in the next article the whole work of computing the Elliptic Integral $F\left(\frac{\sqrt{2}}{2}, \frac{\pi}{4}\right)$ by each of the two methods, and of computing $K\left(\frac{\sqrt{2}}{2}\right) = F\left(\frac{\sqrt{2}}{2}, \frac{\pi}{2}\right)$ by the second method, employing five-place logarithms.

182. $F\left(\frac{\sqrt{2}}{2}, \frac{\pi}{4}\right)$. METHOD OF ART. 180.

$$k = 0.70712 \qquad \log k = 9.84949$$
$$1 + k = 1.70712 \qquad \log(1+k) = 0.23226$$

$$\begin{aligned} \log\sqrt{k} &= 9.92474 \\ \log 2 &= 0.30103 \\ \text{colog}\,(1+k) &= 9.76774 \\ \hline \log k_1 &= 9.99351 \end{aligned}$$

$$k_1 = 0.98518 \qquad \log k_1 = 9.99351$$
$$1 + k_1 = 1.98518 \qquad \log(1+k_1) = 0.29780$$

$$\begin{aligned} \log\sqrt{k_1} &= 9.99676 \\ \log 2 &= 0.30103 \\ \text{colog}\,(1+k_1) &= 9.70220 \\ \hline \log k_2 &= 9.99999 \end{aligned}$$

$$\underline{k_2 = 1}$$

$$\begin{aligned} \log k &= 9.84949 \\ \log\sin\frac{\pi}{4} &= 9.84949 \\ \hline \log\sin(2\phi_1 - \phi) &= 9.69898 \end{aligned}$$

$$\begin{aligned} 2\phi_1 - \phi &= 30^\circ\ 0'\ 3'' \\ 2\phi_1 &= 75^\circ\ 0'\ 3'' \\ \phi_1 &= 37^\circ\ 30'\ 2'' \end{aligned}$$

$$\begin{aligned} \log k_1 &= 9.99351 \\ \log\sin\phi_1 &= 9.78445 \\ \hline \log\sin(2\phi_2 - \phi_1) &= 9.77796 \end{aligned}$$

$$2\phi_2 - \phi_1 = 36^\circ\ 51'\ \ 3''$$
$$2\phi_2 = 74^\circ\ 21'\ \ 5''$$
$$\Phi = \phi_2 = 37^\circ\ 10'\ 32''$$
$$\tfrac{1}{2}\Phi + \frac{\pi}{4} = 63^\circ\ 35'\ 16''$$
$$\log\tan\left(\frac{\pi}{4} + \tfrac{1}{2}\Phi\right) = 0.30393$$

$$\log\sqrt{k_1} = 9.99676$$
$$\operatorname{colog}\sqrt{k} = 0.07526$$
$$\log\log\tan\left(\frac{\pi}{4} + \tfrac{1}{2}\Phi\right) = 9.48277$$
$$\operatorname{colog}\mu = 0.36222$$
$$\log F\left(\frac{\sqrt{2}}{2}, \frac{\pi}{4}\right) = 9.91701$$

$$F\left(\frac{\sqrt{2}}{2}, \frac{\pi}{4}\right) = 0.82605$$

$\mu = 0.43429$ is the modulus of the common system of logarithms.

$F\left(\frac{\sqrt{2}}{2}, \frac{\pi}{4}\right)$. METHOD OF ART. 181.

$$\sqrt{1-k^2} = k' = 0.70712$$

$1 - k' = 0.29288$	$\log(1-k') = 9.46669$
$1 + k' = 1.70712$	$\operatorname{colog}(1+k') = 9.76774$
$k_1 = 0.17157$	$\log k_1 = 9.23443$
$1 - k_1 = 0.82843$	$\log(1-k_1) = 9.91826$
$1 + k_1 = 1.17157$	$\log(1+k_1) = 0.06878$
	$\log k_1'^2 = 9.98704$
$k_1' = 0.98520$	$\log k_1' = 9.99352$
$1 - k_1' = 0.01480$	$\log(1-k_1') = 8.17026$
$1 + k_1' = 1.98520$	$\operatorname{colog}(1+k_1') = 9.70220$
$k_2 = 0.00746$	$\log k_2 = 7.87246$

$$1 - k_2 = 0.99254 \qquad \log(1 - k_2) = 9.99675$$
$$1 + k_2 = 1.00746 \qquad \log(1 + k_2) = 0.00323$$
$$\log k_2'^2 = 9.99998$$

$$k_2' = 1 \qquad \log k_2' = 9.99999$$
$$k_3 = 0$$

$$\log k' = 9.84949$$
$$\log \tan \phi = 0.00000$$

$$\log \tan(\phi_1 - \phi) = 9.84949$$

$$\phi_1 - \phi = 35^\circ\ 15'\ 53''$$
$$\phi_1 = 80^\circ\ 15'\ 53''$$

$$\log k_1' = 9.99352$$
$$\log \tan \phi_1 = 0.76557$$

$$\log \tan(\phi_2 - \phi_1) = 0.75909$$

$$\phi_2 - \phi_1 = 80^\circ\ 7'\ 17''$$
$$\phi_2 = 160^\circ\ 23'\ 10''$$

$$\tan(\phi_3 - \phi_2) = \tan \phi_2$$

$$\Phi = \phi_3 = 2\phi_2 = 320^\circ\ 46'\ 20''$$
$$\frac{1}{2^3}\Phi = 40^\circ\ 5'\ 48''$$
$$= 144348''$$
$$\pi = 648000''$$

$$\log\left(\frac{1}{2^3}\Phi\right)'' = 5.15942$$
$$\operatorname{colog} \pi'' = 4.18842$$
$$\log \pi = 0.49715$$

$$\log\left(\frac{1}{2^3}\Phi\right) = 9.84499$$

$$\log(1+k_1) = 0.06878$$
$$\log(1+k_2) = 0.00323$$
$$\log\left(\frac{1}{2^3}\Phi\right) = 9.84499$$
$$\log F\left(\frac{\sqrt{2}}{2}, \frac{\pi}{4}\right) = 9.91700$$
$$F\left(\frac{\sqrt{2}}{2}, \frac{\pi}{4}\right) = 0.82605$$

For $F\left(\frac{\sqrt{2}}{2}, \frac{\pi}{2}\right)$ we have by (3), Art. 181,

$$\log(1+k_1) = 0.06878$$
$$\log(1+k_2) = 0.00323$$
$$\log \pi = 0.49715$$
$$\text{colog}\, 2 = 9.69897$$
$$\log F\left(\frac{\sqrt{2}}{2}, \frac{\pi}{2}\right) = 0.26813$$
$$F\left(\frac{\sqrt{2}}{2}, \frac{\pi}{2}\right) = 1.8541$$

183. Landen's Transformation can also be applied to the computation of Elliptic Integrals of the second class, but the task is a more difficult one; we shall, however, give a brief sketch of the method; and in so doing we shall apply it to a more general form

$$G(k, \phi) = \int_0^\phi \frac{\alpha + \beta \sin^2\phi}{\sqrt{1 - k^2 \sin^2\phi}}\, d\phi, \qquad [1]$$

of which $E(k, \phi)$ is a special case.

From Art. 180 we have

$$\sqrt{1 - k^2 \sin^2\phi} = \frac{1 + k\cos 2\phi_1}{\sqrt{1 + k^2 + 2k\cos 2\phi_1}},$$

$$\cos\phi = \frac{k + \cos 2\phi_1}{\sqrt{1 + k^2 + 2k\cos 2\phi_1}},$$

$$d\phi = \frac{2(1 + k\cos 2\phi_1)}{1 + k^2 + 2k\cos 2\phi_1}\, d\phi_1.$$

Hence $\sqrt{1-k^2\sin^2\phi}+k\cos\phi=\sqrt{1+k^2+2k\cos 2\phi_1}.$

$$G(k,\phi)=\int_0^{\phi}\frac{\alpha+\beta\sin^2\phi}{\sqrt{1-k^2\sin^2\phi}}\,d\phi$$

$$=\int_0^{\phi}\left[\frac{\alpha}{\sqrt{1-k^2\sin^2\phi}}+\frac{\beta}{k^2}\cdot\frac{1-(1-k^2\sin^2\phi)}{\sqrt{1-k^2\sin^2\phi}}\right]d\phi$$

$$=\int_0^{\phi}\left[\frac{\alpha+\frac{\beta}{k^2}}{\sqrt{1-k^2\sin^2\phi}}-\frac{\beta}{k^2}\sqrt{1-k^2\sin^2\phi}\right]d\phi,$$

and

$$G(k,\phi)-\frac{\beta}{k}\sin\phi$$

$$=\int_0^{\phi}\left[\frac{\alpha+\frac{\beta}{k^2}}{\sqrt{1-k^2\sin^2\phi}}-\frac{\beta}{k^2}(\sqrt{1-k^2\sin^2\phi}+k\cos\phi)\right]d\phi.$$

Substituting ϕ_1 for ϕ, this becomes

$$G(k,\phi)-\frac{\beta}{k}\sin\phi=2\int_0^{\phi_1}\frac{\alpha-\frac{\beta}{k}\cos 2\phi_1}{\sqrt{1+k^2+2k\cos 2\phi_1}}\,d\phi_1$$

$$=\frac{2}{1+k}\int_0^{\phi_1}\frac{\alpha-\frac{\beta}{k}+\frac{2\beta}{k}\sin^2\phi_1}{\sqrt{1-\frac{4k}{(1+k)^2}\sin^2\phi_1}}\,d\phi_1.$$

Hence $$G(k,\phi)=\frac{\beta}{k}\sin\phi+\frac{2}{1+k}G_1(k_1,\phi_1), \qquad [2]$$

where

$$k_1=\frac{2\sqrt{k}}{1+k},\quad \sin(2\phi_1-\phi)=k\sin\phi,\quad \alpha_1=\alpha-\frac{\beta}{k},\quad \beta_1=\frac{2\beta}{k}. \qquad [3]$$

Formulas [2] and [3] enable us to make our given function depend upon one of the same form, but having a greater modulus and a less amplitude. A repeated use of [2], together with the reductions employed in Art. 180, gives us

$$G(k, \phi) = \frac{\beta}{k}\sin\phi + \frac{\beta_1}{\sqrt{k}}\sin\phi_1 + \sqrt{\frac{k_1}{k}}\cdot\beta_2\sin\phi_2$$

$$+\sqrt{\frac{k_1k_2}{k}}\cdot\beta_3\sin\phi_3 + \cdots + \sqrt{\frac{k_1k_2\cdots k_{n-2}}{k}}\beta_{n-1}\sin\phi_{n-1}$$

$$+k_n\sqrt{\frac{k_1k_2\cdots k_{n-1}}{k}}\cdot G_n(k_n, \phi_n), \qquad [4]$$

where $$\beta_p = \frac{2^p\beta}{kk_1k_2\cdots k_{p-1}}$$

and $$\alpha_p = \alpha - \frac{\beta}{k}\left(1 + \frac{2}{k_1} + \frac{2^2}{k_1k_2} + \cdots + \frac{2^{p-1}}{k_1k_2k_3\cdots k_{p-1}}\right).$$

Just as in Art. 180 k_n rapidly approaches 1 as n is increased; the limiting value of $G_n(k_n, \phi_n)$ is then

$$\text{limit } G_n(k_n, \phi_n) = \int_0^{\phi_n}\frac{\alpha_n + \beta_n\sin^2\phi}{\cos\phi}d\phi$$

$$= (\alpha_n + \beta_n)\log\tan\left(\frac{\pi}{4} + \frac{\phi_n}{2}\right) - \beta_n\sin\phi_n. \qquad [6]$$

By Art. 180, [5] and [6],

$$\text{limit } k_n\sqrt{\frac{k_1k_2\cdots k_{n-1}}{k}}\log\tan\left[\frac{\pi}{4} + \frac{\phi_n}{2}\right] = F(k, \phi).$$

[4] can thus be written

$$G(k, \phi) = F(k, \phi)\left[\alpha - \frac{\beta}{k}\left(1 + \frac{2}{k_1} + \frac{2^2}{k_1k_2} + \cdots\right.\right.$$

$$\left.\left. + \frac{2^{n-1}}{k_1k_2\ldots k_{n-1}} - \frac{2^n}{k_1k_2k_3\ldots k_{n-1}}\right)\right]$$

$$+\frac{\beta}{k}\left[\sin\phi + \frac{2}{\sqrt{k}}\sin\phi_1 + \frac{2^2}{\sqrt{kk_1}}\sin\phi_2 + \frac{2^3}{\sqrt{kk_1k_2}}\sin\phi_3 + \cdots\right.$$

$$\left. + \frac{2^{n-1}}{\sqrt{kk_1\ldots k_{n-2}}}\sin\phi_{n-1} - \frac{2^n}{\sqrt{kk_1\ldots k_{n-1}}}\sin\phi_n\right]. \qquad [7]$$

If $\alpha = 1$, and $\beta = -k^2$, [7] reduces to

$$E(k, \phi) = F(k, \phi)\left[1 + k\left(1 + \frac{2}{k_1} + \frac{2^2}{k_1 k_2} + \cdots + \frac{2^{n-1}}{k_1 k_2 \dots k_{n-1}} - \frac{2^n}{k_1 k_2 \dots k_{n-1}}\right)\right]$$
$$- k\left[\sin\phi + \frac{2}{\sqrt{k}}\sin\phi_1 + \frac{2^2}{\sqrt{k k_1}}\sin\phi_2 + \cdots + \frac{2^{n-1}}{\sqrt{k k_1 \dots k_{n-2}}}\sin\phi_{n-1} - \frac{2^n}{\sqrt{k k_1 \dots k_{n-1}}}\sin\phi_n\right], \qquad [8]$$

where $k_p = \dfrac{2\sqrt{k_{p-1}}}{1 + k_{p-1}}$, and $\sin(2\phi_p - \phi_{p-1}) = k_{p-1}\sin\phi_{p-1}$. [9]

By Formulas [8] and [9] an Elliptic Integral of the Second Class may be computed without difficulty.

184. Formula [2], Art. 183, may be used to decrease the modulus and increase the amplitude of an Elliptic Integral. Interchanging the subscripts, we have

$$G(k, \phi) = \frac{1 + k_1}{2}\left[G_1(k_1, \phi_1) - \frac{\beta_1}{k_1}\sin\phi_1\right];$$

or, since $\dfrac{\beta_1}{k_1} = \dfrac{\beta}{2}$, (Art. 183 [3]),

$$G(k, \phi) = \frac{1 + k_1}{2}\left[G_1(k_1, \phi_1) - \frac{\beta}{2}\sin\phi_1\right], \qquad [1]$$

where

$$k_1 = \frac{1 - \sqrt{1 - k^2}}{1 + \sqrt{1 - k^2}}, \ \tan(\phi_1 - \phi) = \sqrt{1 - k^2}\tan\phi, \ \alpha_1 = \alpha + \frac{\beta}{2}, \ \beta_1 = \frac{k_1\beta}{2}. \qquad [2]$$

A repeated use of [1] gives

$$G(k, \phi)=\frac{1+k_1}{2}\cdot\frac{1+k_2}{2}\cdot\frac{1+k_3}{2}\cdots\frac{1+k_n}{2}G_n(k_n, \phi_n)$$

$$-\tfrac{1}{2}\left[\frac{1+k_1}{2}\beta\sin\phi_1+\frac{1+k_1}{2}\cdot\frac{1+k_2}{2}\beta_1\sin\phi_2+\cdots\right.$$

$$\left.+\frac{1+k_1}{2}\cdot\frac{1+k_2}{2}\cdots\frac{1+k_n}{2}\beta_{n-1}\sin\phi_n\right], \qquad [3]$$

where $$\beta_p=\beta\frac{k_1k_2k_3\ldots k_p}{2^p},$$

and $$\alpha_p=\alpha+\tfrac{1}{2}\beta\left(1+\frac{k_1}{2}+\frac{k_1k_2}{2^2}+\frac{k_1k_2k_3}{2^3}+\cdots+\frac{k_1k_2\ldots k_{p-1}}{2^{p-1}}\right).$$

Just as in Art. 181 limit $k_n=0$, therefore limit $\beta_n=0$ and limit $G_n(k_n, \phi_n)=\int_0^{\phi_n}\alpha_n\,d\phi=\alpha_n\phi_n$.

By Art. 181, [2], $\frac{1+k_1}{2}\cdot\frac{1+k_2}{2}\cdots\frac{1+k_n}{2}\phi_n=F(k, \phi)$.

By Art. 180, [5], $\frac{1+k_1}{2}=\frac{\sqrt{k_1}}{k}$, $\frac{1+k_2}{2}=\frac{\sqrt{k_2}}{k_1}$, etc.

Hence [3] becomes

$$G(k, \phi)=F(k, \phi)\left[\alpha+\tfrac{1}{2}\beta\left(1+\frac{k_1}{2}+\frac{k_1k_2}{2^2}+\frac{k_1k_2k_3}{2^3}+\cdots\right)\right]$$

$$-\frac{\beta}{k}\left[\frac{\sqrt{k_1}}{2}\sin\phi_1+\frac{\sqrt{k_1k_2}}{2^2}\sin\phi_2+\frac{\sqrt{k_1k_2k_3}}{2^3}\sin\phi_3+\cdots\right]. \qquad [4]$$

If $\alpha=1$ and $\beta=-k^2$, [4] reduces to

$$E(k,\phi)=F(k,\phi)\left[1-\frac{k^2}{2}\left(1+\frac{k_1}{2}+\frac{k_1k_2}{2^2}+\frac{k_1k_2k_3}{2^3}+\cdots\right)\right]$$

$$+k\left[\frac{\sqrt{k_1}}{2}\sin\phi_1+\frac{\sqrt{k_1k_2}}{2^2}\sin\phi_2+\frac{\sqrt{k_1k_2k_3}}{2^3}\sin\phi_3+\cdots\right], \qquad [5]$$

where

$$k_p=\frac{1-\sqrt{1-k_{p-1}^2}}{1+\sqrt{1-k_{p-1}^2}}, \text{ and } \tan(\phi_p-\phi_{p-1})=\sqrt{1-k_{p-1}^2}\cdot\tan\phi_{p-1}. \qquad [6]$$

We have seen in Art. 181 that if $\phi=\frac{\pi}{2}$, $\phi_p=2^{p-1}\pi$.

Therefore, for a complete Elliptic Integral of the second class we have

$$E\left(k,\frac{\pi}{2}\right)=F\left(k,\frac{\pi}{2}\right)\left[1-\frac{k^2}{2}\left(1+\frac{k_1}{2}+\frac{k_1k_2}{2^2}+\frac{k_1k_2k_3}{2^3}+\cdots\right)\right]. \qquad [7]$$

Formulas [5] and [7] are admirably adapted to computation.

We give in the next article the work of computing $E\left(\frac{\sqrt{2}}{2},\frac{\pi}{4}\right)$ by each of the methods just given, and of computing $E\left(\frac{\sqrt{2}}{2},\frac{\pi}{2}\right)$ by the second method; using, as far as possible, the values already employed or obtained in Art. 182.

185. $E\left(\frac{\sqrt{2}}{2},\frac{\pi}{4}\right)$. METHOD OF ART. 183.

Here, as we have seen in Art. 182, if we carry the work only to five decimal places, $k_2=1$, and our working formula will be

$$E(k,\phi)=F(k,\phi)\left[1+k\left(1-\frac{2}{k_1}\right)\right]$$

$$-k\left[\sin\phi+\frac{2}{\sqrt{k}}\sin\phi_1-\frac{2^2}{\sqrt{kk_1}}\sin\phi_2\right].$$

$$\begin{aligned}\log 2 &= 0.30103\\ \log k &= 9.84949\\ \text{colog}\, k_1 &= \underline{0.00649}\\ &\ 0.15701\end{aligned} \qquad \frac{2k}{k_1} = 1.43553$$

$$1 + k = \underline{1.70712}$$

$$\log\left(1 + k - \frac{2k}{k_1}\right) = 9.43391 \qquad 1 + k - \frac{2k}{k_1} = 0.27159$$

$$\begin{aligned}\log F\left(\frac{\sqrt{2}}{2}, \frac{\pi}{4}\right) &= \underline{9.91701}\\ &\ 9.35092\end{aligned} \qquad F\left(\frac{\sqrt{2}}{2}, \frac{\pi}{4}\right)\left(1+k-\frac{2k}{k_1}\right) = 0.22435$$

$$\begin{aligned}\log k &= 9.84949\\ \log \sin\phi &= \underline{9.84949}\\ &\ 9.69898\end{aligned} \qquad k \sin\phi = 0.5$$

$$\begin{aligned}\log 2 &= 0.30103\\ \log\sqrt{k} &= 9.92474\\ \log \sin\phi_1 &= \underline{9.78445}\\ &\ 0.01022\end{aligned} \qquad \frac{2k}{\sqrt{k}} \sin\phi_1 = 1.0238$$

$$\begin{aligned}\log 2^2 &= 0.60206\\ \log\sqrt{k} &= 9.92474\\ \text{colog}\sqrt{k_1} &= 0.00324\\ \log \sin\phi_2 &= \underline{9.78122}\\ &\ 0.31126\end{aligned} \qquad \frac{2^2 k}{\sqrt{k k_1}} \sin\phi_2 = \underline{2.0477}$$

$$-k\left(\sin\phi + \frac{2}{\sqrt{k}}\sin\phi_1 - \frac{2^2}{\sqrt{k k_1}}\sin\phi_2\right) = 0.5239$$

$$F\left(\frac{\sqrt{2}}{2}, \frac{\pi}{4}\right)\left(1+k-\frac{2k}{k_1}\right) = \underline{0.22435}$$

$$E\left(\frac{\sqrt{2}}{2}, \frac{\pi}{4}\right) = \underline{\underline{0.74825}}$$

$E\left(\frac{\sqrt{2}}{2}, \frac{\pi}{4}\right)$. METHOD OF ART. 184.

$k_3 = 0.$ Therefore our formula is

$$E(k, \phi) = F(k, \phi)\left[1 - \frac{k^2}{2}\left(1 + \frac{k_1}{2} + \frac{k_1 k_2}{2^2}\right)\right]$$
$$+ k\left(\frac{\sqrt{k_1}}{2}\sin\phi_1 + \frac{\sqrt{k_1 k_2}}{2^2}\sin\phi_2\right).$$

$$\begin{aligned} \log k_1 &= 9.23443 \\ \log k_2 &= 7.87246 \\ \text{colog}\, 4 &= 9.39794 \\ \hline & 6.50483 \end{aligned} \qquad \frac{k_1 k_2}{2^2} = 0.00032$$

$$\frac{k_1}{2} = 0.08578$$

$$1 + \frac{k_1}{2} + \frac{k_1 k_2}{2^2} = 1.08610$$

$$\frac{k^2}{2}\left(1 + \frac{k_1}{2} + \frac{k_1 k_2}{2^2}\right) = 0.271525$$

$$\log 0.728475 = 9.862415 \qquad 1 - \frac{k^2}{2}\left(1 + \frac{k_1}{2} + \frac{k_1 k_2}{2^2}\right) = 0.728475$$

$$\begin{aligned} \log F\left(\frac{\sqrt{2}}{2}, \frac{\pi}{4}\right) &= 9.91700 \\ \hline & 9.779415 \end{aligned} \qquad F\left(\frac{\sqrt{2}}{2}, \frac{\pi}{4}\right)(0.728475) = 0.60178$$

$$\begin{aligned} \log k &= 9.84949 \\ \log \sqrt{k_1} &= 9.61722 \\ \text{colog}\, 2 &= 9.69897 \\ \log \sin\phi_1 &= 9.99370 \\ \hline & 9.15938 \end{aligned} \qquad \frac{k\sqrt{k_1}}{2}\sin\phi_1 = 0.14434$$

$$\log k = 9.84949$$
$$\log \sqrt{k_1} = 9.61722$$
$$\log \sqrt{k_2} = 8.93623$$
$$\operatorname{colog} 4 = 9.39794$$
$$\log \sin \phi_2 = 9.52592$$
$$7.32680 \qquad \frac{k\sqrt{k_1 k_2}}{2^2}\sin\phi_2 = 0.00212$$

$$k\left(\frac{\sqrt{k_1}}{2^2}\sin\phi_1 + \frac{\sqrt{k_1 k_2}}{2^2}\sin\phi_2\right) = 0.14646$$

$$F\left(\frac{\sqrt{2}}{2}, \frac{\pi}{4}\right)(0.728475) = 0.60178$$

$$E\left(\frac{\sqrt{2}}{2}, \frac{\pi}{4}\right) = 0.74824$$

$E\left(\frac{\sqrt{2}}{2}, \frac{\pi}{2}\right)$. METHOD OF ART. 184.

$$E\left(k, \frac{\pi}{2}\right) = F\left(k, \frac{\pi}{2}\right)\left[1 - \frac{k^2}{2}\left(1 + \frac{k_1}{2} + \frac{k_1 k_2}{2^2}\right)\right].$$

$$1 - \frac{k^2}{2}\left(1 + \frac{k_1}{2} + \frac{k_1 k_2}{2^2}\right) = 0.728475 \qquad \log 0.728475 = 9.862415$$

$$\log F\left(\frac{\sqrt{2}}{2}, \frac{\pi}{2}\right) = 0.26813$$

$$E\left(\frac{\sqrt{2}}{2}, \frac{\pi}{2}\right) = 1.3507 \qquad \log E\left(\frac{\sqrt{2}}{2}, \frac{\pi}{2}\right) = 0.13054$$

186. An Elliptic Integral of the first or second class, whose amplitude is greater than $\frac{\pi}{2}$, can be made to depend upon one whose amplitude is less than $\frac{\pi}{2}$, and upon the corresponding Complete Elliptic Integral.

We have

$$F(k,\pi)=\int_0^{\pi}\frac{d\phi}{\Delta\phi}=\int_0^{\frac{\pi}{2}}\frac{d\phi}{\Delta\phi}+\int_{\frac{\pi}{2}}^{\pi}\frac{d\phi}{\Delta\phi}=K+\int_{\frac{\pi}{2}}^{\pi}\frac{d\phi}{\Delta\phi},\text{ by [4], Art. 179.}$$

In $$\int_{\frac{\pi}{2}}^{\pi}\frac{d\phi}{\Delta\phi}\text{ let }\phi=\pi-\psi;$$

then $d\phi=-d\psi$ and $\Delta\phi=\sqrt{1-k^2\sin^2\phi}=\sqrt{1-k^2\sin^2\psi}=\Delta\psi$,

and we have $$\int_{\frac{\pi}{2}}^{\pi}\frac{d\phi}{\Delta\phi}=-\int_{\frac{\pi}{2}}^{0}\frac{d\psi}{\Delta\psi}=-\int_{\frac{\pi}{2}}^{0}\frac{d\phi}{\Delta\phi}=\int_0^{\frac{\pi}{2}}\frac{d\phi}{\Delta\phi}=K.$$

Hence $$F(k,\pi)=\int_0^{\pi}\frac{d\phi}{\Delta\phi}=2K. \qquad [1]$$

$$F(k,n\pi+\rho)=\int_0^{n\pi+\rho}\frac{d\phi}{\Delta\phi}$$

$$=\int_0^{\pi}\frac{d\phi}{\Delta\phi}+\int_{\pi}^{2\pi}\frac{d\phi}{\Delta\phi}+\int_{2\pi}^{3\pi}\frac{d\phi}{\Delta\phi}+\cdots+\int_{p\pi}^{(p+1)\pi}\frac{d\phi}{\Delta\phi}+\cdots+\int_{n\pi}^{n\pi+\rho}\frac{d\phi}{\Delta\phi}.$$

In $\int_{p\pi}^{(p+1)\pi}\frac{d\phi}{\Delta\phi}$ let $\phi=p\pi+\psi$; then $d\phi=d\psi$, and $\Delta\phi=\Delta\psi$,

and we have $$\int_{p\pi}^{(p+1)\pi}\frac{d\phi}{\Delta\phi}=\int_0^{\pi}\frac{d\psi}{\Delta\psi}=\int_0^{\pi}\frac{d\phi}{\Delta\phi}=2K.$$

The substitution of ψ for $\phi - n\pi$ in $\int_{n\pi}^{n\pi+\rho}\frac{d\phi}{\Delta\phi}$ gives us

$$\int_{n\pi}^{n\pi+\rho}\frac{d\phi}{\Delta\phi}=\int_0^{\rho}\frac{d\psi}{\Delta\psi}=\int_0^{\rho}\frac{d\phi}{\Delta\phi}=F(k,\rho).$$

Therefore $$F(k, n\pi+\rho)=2nK+F(k,\rho). \qquad [2]$$

In like manner it can be proved that

$$F(k, n\pi-\rho)=2nK-F(k,\rho), \qquad [3]$$

$$E(k, n\pi+\rho)=2nE+E(k,\rho), \qquad [4]$$

$$E(k, n\pi-\rho)=2nE-E(k,\rho), \qquad [5]$$

where $E=E\left(k,\frac{\pi}{2}\right)$ is the complete Elliptic Integral of the second class.

A table giving the values of the Elliptic Integrals of the first and second classes for values of the amplitude between 0 and $\frac{\pi}{2}$ is, then, a complete table.

Such a table, carried out to ten decimal places, is given by Legendre in his "Traité des Fonctions Elliptiques." We give in the next article a small three-place table.

It must be noted that the first column gives $F(0,\phi)$ and $E(0,\phi)$, that is, $\int_0^{\phi}d\phi=\phi$; and that the last column gives $F(1,\phi)$ and $E(1,\phi)$, that is, $\log\tan\left(\frac{\pi}{4}+\frac{\phi}{2}\right)$ and $\sin\phi$.

The complete Elliptic Integrals,

$$K=F\left(k,\frac{\pi}{2}\right) \text{ and } E=E\left(k,\frac{\pi}{2}\right),$$

are given in the last line of each table.

$F(k, \phi)$.

ϕ	$k=0$ sin 0°	$k=0.1$ sin 6°	$k=0.2$ sin 12°	$k=0.3$ sin 18°	$k=0.4$ sin 24°	$k=0.5$ sin 30°	$k=0.6$ sin 37°	$k=0.7$ sin 45°	$k=0.8$ sin 53°	$k=0.9$ sin 64°	$k=1.$ sin 90°
0°	0.000	0.000	0.000	0.000	0.000	0.000	0.000	0.000	0.000	0.000	0.000
5	0.087	0.087	0.087	0.087	0.087	0.087	0.087	0.087	0.087	0.087	0.087
10	0.175	0.175	0.175	0.175	0.175	0.175	0.175	0.175	0.175	0.175	0.175
15°	0.262	0.262	0.262	0.262	0.262	0.263	0.263	0.263	0.264	0.264	0.265
20	0.349	0.349	0.349	0.350	0.350	0.351	0.352	0.353	0.354	0.355	0.356
25	0.436	0.436	0.437	0.438	0.439	0.440	0.441	0.443	0.445	0.448	0.451
30°	0.524	0.524	0.525	0.526	0.527	0.529	0.532	0.536	0.539	0.544	0.549
35	0.611	0.611	0.612	0.614	0.617	0.620	0.624	0.630	0.636	0.644	0.653
40	0.698	0.699	0.700	0.703	0.707	0.712	0.718	0.727	0.736	0.748	0.763
45°	0.785	0.786	0.789	0.792	0.798	0.804	0.814	0.826	0.839	0.858	0.881
50	0.873	0.874	0.877	0.882	0.889	0.898	0.911	0.928	0.947	0.974	1.011
55	0.960	0.961	0.965	0.972	0.981	0.993	1.010	1.034	1.060	1.099	1.154
60°	1.047	1.049	1.054	1.062	1.074	1.090	1.112	1.142	1.178	1.233	1.317
65	1.134	1.137	1.143	1.153	1.168	1.187	1.215	1.254	1.302	1.377	1.506
70	1.222	1.224	1.232	1.244	1.262	1.285	1.320	1.370	1.431	1.534	1.735
75°	1.309	1.312	1.321	1.336	1.357	1.385	1.426	1.488	1.566	1.703	2.028
80	1.396	1.400	1.410	1.427	1.452	1.485	1.534	1.608	1.705	1.885	2.436
85	1.484	1.487	1.499	1.519	1.547	1.585	1.643	1.731	1.848	2.077	3.131
K. 90°	1.571	1.575	1.588	1.610	1.643	1.686	1.752	1.854	1.993	2.275	∞

$E(k, \phi)$.

ϕ	$k=0$ sin 0°	$k=0.1$ sin 6°	$k=0.2$ sin 12°	$k=0.3$ sin 18°	$k=0.4$ sin 24°	$k=0.5$ sin 30°	$k=0.6$ sin 37°	$k=0.7$ sin 45°	$k=0.8$ sin 53°	$k=0.9$ sin 64°	$k=1.$ sin 90°
0°	0.000	0.000	0.000	0.000	0.000	0.000	0.000	0.000	0.000	0.000	0.000
5	0.087	0.087	0.087	0.087	0.087	0.087	0.087	0.087	0.087	0.087	0.087
10	0.175	0.175	0.174	0.174	0.174	0.174	0.174	0.174	0.174	0.174	0.174
15°	0.262	0.262	0.262	0.262	0.261	0.261	0.261	0.260	0.260	0.259	0.259
20	0.349	0.349	0.349	0.348	0.348	0.347	0.347	0.346	0.345	0.343	0.342
25	0.436	0.436	0.436	0.435	0.434	0.433	0.431	0.430	0.428	0.425	0.423
30°	0.524	0.523	0.523	0.521	0.520	0.518	0.515	0.512	0.509	0.505	0.500
35	0.611	0.610	0.609	0.607	0.605	0.602	0.598	0.593	0.588	0.581	0.574
40	0.698	0.698	0.695	0.693	0.690	0.685	0.679	0.672	0.664	0.654	0.643
45°	0.785	0.785	0.782	0.779	0.773	0.767	0.759	0.748	0.737	0.723	0.707
50	0.873	0.872	0.869	0.864	0.857	0.848	0.837	0.823	0.808	0.789	0.766
55	0.960	0.959	0.955	0.948	0.939	0.928	0.914	0.895	0.875	0.850	0.819
60°	1.047	1.046	1.041	1.032	1.021	1.008	0.989	0.965	0.940	0.907	0.866
65	1.134	1.132	1.126	1.116	1.103	1.086	1.063	1.033	1.001	0.960	0.906
70	1.222	1.219	1.212	1.200	1.184	1.163	1.135	1.099	1.060	1.008	0.940
75°	1.309	1.306	1.297	1.283	1.264	1.240	1.207	1.163	1.117	1.053	0.966
80	1.396	1.393	1.383	1.367	1.344	1.316	1.277	1.227	1.172	1.095	0.985
85	1.484	1.480	1.468	1.450	1.424	1.392	1.347	1.289	1.225	1.135	0.996
E. 90°	1.571	1.566	1.554	1.533	1.504	1.467	1.417	1.351	1.278	1.173	1.000

Addition Formulas.

188. The Elliptic Integrals, $F(k, x)$ and $E(k, x)$, may be regarded as new functions of x, defined by the aid of definite integrals; namely,

$$F(k, x) = \int_0^x \frac{dx}{\sqrt{(1-x^2)(1-k^2x^2)}},$$

$$E(k, x) = \int_0^x \sqrt{\frac{1-k^2x^2}{1-x^2}} \cdot dx;$$

see Art. 178, [1] and [2].

We have seen how we may compute their values to any required degree of approximation when k and x are given. It remains to study their properties.

We are familiar with other and much simpler functions which may be defined as definite integrals, and whose most important properties can be deduced from these definitions.

For example, we may define $\log x$ as $\int_1^x \frac{dx}{x}$, $\sin^{-1}x$ as $\int_0^x \frac{dx}{\sqrt{1-x^2}}$, $\tan^{-1}x$ as $\int_0^x \frac{dx}{1+x^2}$, and the theory of these functions may be based upon these definitions. For instance, the fundamental property of the logarithm is expressed by what is called the *addition formula*,

$$\log x + \log y = \log(xy),$$

and the whole theory of logarithms may be based on this property; and there are addition formulas for the other functions defined above; namely,

$$\sin^{-1}x + \sin^{-1}y = \sin^{-1}(x\sqrt{1-y^2} + y\sqrt{1-x^2}),$$

$$\tan^{-1}x + \tan^{-1}y = \tan^{-1}\left(\frac{x+y}{1-xy}\right).$$

These three important formulas are usually obtained by more or less elaborate methods involving the theory of the functions which are the inverse or anti-functions of the $\log x$, the $\sin^{-1}x$, and the $\tan^{-1}x$, that is, of e^x, $\sin x$, and $\tan x$; but they may be obtained without difficulty from the definitions of $\log x$, $\sin^{-1}x$, and $\tan^{-1}x$, as definite integrals.

Take first $$\log x = \int_1^x \frac{dx}{x}.$$

Let us determine y in terms of x, so that

$$\log x + \log y = \log c, \tag{1}$$

where c is a given constant.

Since $$\log y = \int_1^y \frac{dy}{y},$$

if we differentiate (1), we have

$$\frac{dx}{x} + \frac{dy}{y} = 0,$$

or $$ydx + xdy = 0. \tag{2}$$

Integrate (2), and we get

$$\int ydx + \int xdy = C. \tag{3}$$

Simplify the first member of (3) by *integration by parts;*

$$xy - \int xdy + xy - \int ydx = C,$$

or $$2xy - \int (xdy + ydx) = C.$$

Reducing by the aid of (2), $\quad 2xy = C,$

or $$xy = C_1, \tag{4}$$

where C_1 is an undetermined constant. To determine C_1, let $x=1$ in (4), and we have $y=C_1$ when $x=1$; let $x=1$ in (1), then $\log x=\int_1^1\frac{dx}{x}=0$, $\log y=\log c$, and $y=c$, when $x=1$. Therefore $C_1=c$, and $xy=c$. Consequently $y=\frac{c}{x}$ is the required value of y, and we have (1)

$$\log x+\log\frac{c}{x}=\log c.$$

We can express this relation more neatly by replacing c by its value xy, and thus we reach our required addition formula

$$\log x+\log y=\log(xy). \qquad [5]$$

189. The addition formula for the $\sin^{-1}$ can be deduced in exactly the same way. We wish to determine y so that

$$\sin^{-1}x+\sin^{-1}y=\sin^{-1}c. \qquad (1)$$

We have $\sin^{-1}x=\int_0^x\frac{dx}{\sqrt{1-x^2}}$, $\sin^{-1}y=\int_0^y\frac{dy}{\sqrt{1-y^2}}$.

Differentiate (1).

$$\frac{dx}{\sqrt{1-x^2}}+\frac{dy}{\sqrt{1-y^2}}=0, \qquad (2)$$

or

$$\sqrt{1-y^2}\cdot dx+\sqrt{1-x^2}\cdot dy=0,$$

$$\int\sqrt{1-y^2}\cdot dx+\int\sqrt{1-x^2}\cdot dy=C.$$

Integrate by parts, and

$$x\sqrt{1-y^2}+y\sqrt{1-x^2}+\int xy\left(\frac{dx}{\sqrt{1-x^2}}+\frac{dy}{\sqrt{1-y^2}}\right)=C;$$

or, reducing by (2),

$$x\sqrt{1-y^2}+y\sqrt{1-x^2}=C. \qquad (3)$$

To determine C, we have from (3) $y = C$ when $x = 0$, and from (1) $y = c$ when $x = 0$, since $\sin^{-1}x = \int_0^0 \frac{dx}{\sqrt{1-x^2}} = 0$, when $x = 0$. Hence $C = c$, and $x\sqrt{1-y^2} + y\sqrt{1-x^2} = c$, and, finally,

$$\sin^{-1}x + \sin^{-1}y = \sin^{-1}\left(x\sqrt{1-y^2} + y\sqrt{1-x^2}\right). \qquad [4]$$

To get an addition formula for the $\tan^{-1}$, a slight device is required, that of dividing the differential equation corresponding to (2) by $1 - x^2y^2$.

As before, let

$$\tan^{-1}x + \tan^{-1}y = \tan^{-1}c, \qquad (5)$$

where $$\tan^{-1}x = \int_0^x \frac{dx}{1+x^2},$$

and $$\tan^{-1}y = \int_0^y \frac{dy}{1+y^2}.$$

$$\frac{dx}{1+x^2} + \frac{dy}{1+y^2} = 0,$$

or $$(1+y^2)\,dx + (1+x^2)\,dy = 0. \qquad (6)$$

Divide by $1 - x^2y^2$ and integrate.

$$\int \frac{1+y^2}{1-x^2y^2} \cdot dx + \int \frac{1+x^2}{1-x^2y^2} \cdot dy = C.$$

Integrate by parts. We have

$$d \cdot \frac{1+y^2}{1-x^2y^2} = \frac{2y}{(1-x^2y^2)^2}\left[(1+x^2)\,dy + xy\,(1+y^2)\,dx\right],$$

$$d \cdot \frac{1+x^2}{1-x^2y^2} = \frac{2x}{(1-x^2y^2)^2}\left[(1+y^2)\,dx + xy\,(1+x^2)\,dy\right],$$

and

$$x \cdot \frac{1+y^2}{1-x^2y^2} + y \cdot \frac{1+x^2}{1-x^2y^2} = \frac{x+y}{1-xy}.$$

Hence

$$\frac{x+y}{1-xy}-\int\frac{2xy(1+xy)}{(1-x^2y^2)^2}[(1+y^2)dx+(1+x^2)\,dy]=C.$$

Therefore, by (6), $$\frac{x+y}{1-xy}=C. \qquad (7)$$

To determine C, we have from (7) $y=C$ when $x=0$, and from (5) $y=c$ when $x=0$, since $\tan^{-1}x=\int_0^0\frac{dx}{1-x^2}=0$ when $x=0$.

Hence $C=c$, and $\frac{x+y}{1-xy}=c$,

and, finally, $$\tan^{-1}x+\tan^{-1}y=\tan^{-1}\left(\frac{x+y}{1-xy}\right). \qquad [8]$$

190. To get an addition formula for $F(k, x)$, as before

let $$F(k, x)+F(k, y)=F(k, c), \qquad (1)$$

where $$F(k, x)=\int_0^x\frac{dx}{\sqrt{(1-x^2)(1-k^2x^2)}},$$

and $$F(k, y)=\int_0^y\frac{dy}{\sqrt{(1-y^2)(1-k^2y^2)}}.$$

$$\frac{dx}{\sqrt{(1-x^2)(1-k^2x^2)}}+\frac{dy}{\sqrt{(1-y^2)(1-k^2y^2)}}=0, \qquad (2)$$

or

$$\sqrt{(1-y^2)(1-k^2y^2)}\cdot dx+\sqrt{(1-x^2)(1-k^2x^2)}\cdot dy=0. \qquad (3)$$

Divide by $1-k^2x^2y^2$ and integrate.

$$\int\frac{\sqrt{(1-y^2)(1-k^2y^2)}}{1-k^2x^2y^2}\cdot dx+\int\frac{\sqrt{(1-x^2)(1-k^2x^2)}}{1-k^2x^2y^2}\cdot dy=C.$$

Integrate by parts. We have

$$d\cdot\frac{\sqrt{(1-y^2)(1-k^2y^2)}}{1-k^2x^2y^2}=\frac{y}{(1-k^2x^2y^2)^2}\{[2k^2(x^2+y^2)$$

$$-(1+k^2)(1+k^2x^2y^2)]\frac{dy}{\sqrt{(1-y^2)(1-k^2y^2)}}$$

$$+2k^2xy\sqrt{(1-y^2)(1-k^2y^2)}\cdot dx\};$$

$$d\cdot\frac{\sqrt{(1-x^2)(1-k^2x^2)}}{1-k^2x^2y^2}=\frac{x}{(1-k^2x^2y^2)^2}\{[2k^2(x^2+y^2)$$

$$-(1+k^2)(1+k^2x^2y^2)]\frac{dx}{\sqrt{(1-x^2)(1-k^2x^2)}}$$

$$+2k^2xy\sqrt{(1-x^2)(1-k^2x^2)}\cdot dy\}.$$

Hence

$$\frac{x\sqrt{(1-y^2)(1-k^2y^2)}+y\sqrt{(1-x^2)(1-k^2x^2)}}{1-k^2x^2y^2}$$

$$-\int\frac{xy}{(1-k^2x^2y^2)^2}\{[2k^2(x^2+y^2)-(1+k^2)(1+k^2x^2y^2)]$$

$$\left[\frac{dx}{\sqrt{(1-x^2)(1-k^2x^2)}}+\frac{dy}{\sqrt{(1-y^2)(1-k^2y^2)}}\right]$$

$$+2k^2xy[\sqrt{(1-y^2)(1-k^2y^2)}\cdot dx$$

$$+\sqrt{(1-x^2)(1-k^2x^2)}\cdot dy]\}=C.$$

Reducing, by the aid of (2) and (3), we have

$$\frac{x\sqrt{(1-y^2)(1-k^2y^2)}+y\sqrt{(1-x^2)(1-k^2x^2)}}{1-k^2x^2y^2}=C. \qquad (4)$$

To determine C, from (4) $y = C$ when $x = 0$, and from (1) $y = c$ when $x = 0$. Therefore $C = c$, and we get

$$F(k, x) + F(k, y)$$
$$= F\left(k, \frac{x\sqrt{(1-y^2)(1-k^2y^2)} + y\sqrt{(1-x^2)(1-k^2x^2)}}{1-k^2x^2y^2}\right), \qquad [5]$$

our required addition formula.

An addition formula for $E(k, x)$ can be obtained in very much the same way, but the work is rather complicated, and it is better to use a method which will be explained later.

THE ELLIPTIC FUNCTIONS.

191. We have just seen that there is an analogy between the Elliptic Integral $F(k, x)$, and the familiar functions $\log x$, $\sin^{-1} x$, and $\tan^{-1} x$; and we know that the theory of these functions is ultimately connected with that of their inverse functions, $\log^{-1} u$ or e^u, $\sin u$, and $\tan u$; and, indeed, that the latter are so much simpler than the former that it is customary to regard them as the direct functions, and the logarithm, the anti-sine, and the anti-tangent as the inverse functions.

For example: the first three addition formulas just obtained are much simpler when we express them in terms of the direct functions, and they become

$$\log^{-1}(u+v) = \log^{-1}u \cdot \log^{-1}v,$$

or
$$e^{(u+v)} = e^u \cdot e^v, \qquad [1]$$

$$\sin(u+v) = \sin u\sqrt{1-\sin^2 v} + \sin v\sqrt{1-\sin^2 u};$$

or
$$\sin(u+v) = \sin u\cos v + \cos u \sin v, \qquad [2]$$

$$\tan(u+v) = \frac{\tan u + \tan v}{1 - \tan u \cdot \tan v}; \qquad [3]$$

and in this form they seem to better deserve the name of addition formulas.

In the same way the addition formula for $F(k, x)$ can be more simply written in terms of the function which we might naturally represent by $F^{-1}u$ (mod. k); and, as we might expect, this function has many interesting and important properties which well deserve investigation.

Since in most of the work which follows we shall generally employ the same modulus throughout, we shall not take the trouble to write it except in the few cases where its omission might give rise to confusion, and then we shall put (mod. k) after the function, as above with $F^{-1}u$ (mod. k).

192. In Arts. 178 and 179 we have adopted two forms of notation for an Elliptic Integral of the first class, $F(k, x)$ and $F(k, \phi)$;

$$F(k, x) = \int_0^x \frac{dx}{\sqrt{(1-x^2)(1-k^2x^2)}},$$

$$F(k, \phi) = \int_0^\phi \frac{d\phi}{\sqrt{1-k^2\sin^2\phi}} = \int_0^\phi \frac{d\phi}{\Delta\phi},$$

where $$x = \sin\phi,\ \sqrt{1-x^2} = \cos\phi,$$

and $$\sqrt{1-k^2x^2} = \sqrt{1-k^2\sin^2\phi} = \Delta\phi.$$

If we let $$u = F(k, x) = F(k, \phi),$$

we have in Art. 179 called ϕ the amplitude of u, and $\sin\phi$, $\cos\phi$, and $\Delta\phi$ may be called the sine, the cosine, and the delta of the amplitude of u; and ϕ, $\sin\phi$, $\cos\phi$, and $\Delta\phi$ may be written $\text{am}\,u$, $\text{sinam}\,u$, $\text{cosam}\,u$, and $\Delta\,\text{am}\,u$, or, more briefly, $\text{am}\,u$, $\text{sn}\,u$, $\text{cn}\,u$, and $\text{dn}\,u$; and may be read amplitude u, sine amplitude u, cosine amplitude u, and delta amplitude u. Formulating, we have

$$\left.\begin{aligned} u &= F(k,x) = F(k, \phi), \\ \phi &= \text{am}\,u, \\ x &= \sin\phi = \text{sn}\,u, \\ \sqrt{1-x^2} &= \cos\phi = \text{cn}\,u, \\ \sqrt{1-k^2x^2} &= \Delta\phi = \text{dn}\,u, \end{aligned}\right\}. \qquad [1]$$

$\operatorname{sn} u$, $\operatorname{cn} u$, $\operatorname{dn} u$, are trigonometric functions of ϕ, the amplitude of u, but they may be regarded as new and somewhat complicated functions of u itself, and from this point of view they are called *Elliptic Functions* of u.

$\operatorname{am} u$ also is sometimes called an Elliptic Function; and there are various allied functions that are sometimes included under the general title of Elliptic Functions. We shall, however, restrict the name to $\operatorname{sn} u$, $\operatorname{cn} u$, and $\operatorname{dn} u$. They have an analogy with trigonometric functions, and have a theory which closely resembles that of trigonometric functions, and which we shall proceed to develop. It must, however, be kept in mind that the independent variable u is not an angle, as in the case of the trigonometric functions.

Of course, with our notation, $u = F(k, x) = \operatorname{sn}^{-1} x \pmod k$, or $u = F(k, \phi) = \operatorname{am}^{-1} \phi \pmod k$.

The fundamental formulas connecting the Elliptic Functions of a single quantity follow immediately from the definitions [1], and are

$$\operatorname{sn}^2 u + \operatorname{cn}^2 u = 1, \qquad [2]$$

$$\operatorname{dn}^2 u + k^2 \operatorname{sn}^2 u = 1, \qquad [3]$$

$$\frac{d \operatorname{am} u}{du} = \operatorname{dn} u, \qquad [4]$$

$$\frac{d \operatorname{sn} u}{du} = \operatorname{cn} u \,.\, \operatorname{dn} u, \qquad [5]$$

$$\frac{d \operatorname{cn} u}{du} = - \operatorname{sn} u \,.\, \operatorname{dn} u, \qquad [6]$$

$$\frac{d \operatorname{dn} u}{du} = - k^2 \operatorname{sn} u \,.\, \operatorname{cn} u, \qquad [7]$$

The only one of this set which needs any explanation is [4].

We have $$u = \int_0^\phi \frac{d\phi}{\Delta\phi};$$

hence
$$du = \frac{d\phi}{\Delta\phi},\ du = \frac{d\,\text{am}\,u}{\text{dn}\,u};$$

and, finally,
$$\frac{d\,\text{am}\,u}{du} = \text{dn}\,u.$$

Since
$$\int_0^{-\phi}\frac{d\phi}{\Delta\phi} = \int_0^{\phi}\frac{d(-\phi)}{\Delta(-\phi)} = -\int_0^{\phi}\frac{d\phi}{\Delta\phi},$$

we see that
$$\left.\begin{aligned} \text{am}(-u) &= -\text{am}\,u,\\ \text{sn}(-u) &= -\text{sn}\,u,\\ \text{cn}(-u) &= \text{cn}\,u,\\ \text{dn}(-u) &= \text{dn}\,u, \end{aligned}\right\}. \qquad [8]$$

That is, $\text{sn}\,u$ is an odd function of u, and $\text{cn}\,u$ and $\text{dn}\,u$ are even functions of u.

Since
$$\int_0^0\frac{d\phi}{\Delta\phi} = 0,$$

we have
$$\left.\begin{aligned} \text{am}(0) &= 0,\\ \text{sn}(0) &= 0,\\ \text{cn}(0) &= 1,\\ \text{dn}(0) &= 1, \end{aligned}\right\}. \qquad [9]$$

193. Our addition formula for the sine amplitude flows immediately from [5], Art. 190. Let $u = F(k, x)$ and $v = F(k, y)$, and take the sine amplitude of each member of [5], Art. 190; we get

$$\text{sn}(u+v) = \frac{\text{sn}\,u\,.\,\text{cn}\,v\,.\,\text{dn}\,v + \text{cn}\,u\,.\,\text{sn}\,v\,.\,\text{dn}\,u}{1 - k^2\,.\,\text{sn}^2u\,.\,\text{sn}^2v}.$$

If now we replace v by $-v$, and simplify by [8], Art. 192, we have

$$\text{sn}(u-v) = \frac{\text{sn}\,u\,.\,\text{cn}\,v\,.\,\text{dn}\,v - \text{cn}\,u\,.\,\text{sn}\,v\,.\,\text{dn}\,u}{1 - k^2\,.\,\text{sn}^2u\,.\,\text{sn}^2v},$$

and the two formulas can be combined if we use the sign $\pm$;

$$\operatorname{sn}(u \pm v) = \frac{\operatorname{sn} u \,.\, \operatorname{cn} v \,.\, \operatorname{dn} v \pm \operatorname{cn} u \,.\, \operatorname{sn} v \,.\, \operatorname{dn} u}{1 - k^2 \,.\, \operatorname{sn}^2 u \,.\, \operatorname{sn}^2 v}. \qquad [1]$$

From [1], with the aid of [2] and [3], Art. 192, we can get, after a rather elaborate reduction, the addition formulas for cn and dn.

$$\operatorname{cn}(u \pm v) = \frac{\operatorname{cn} u \,.\, \operatorname{cn} v \mp \operatorname{sn} u \,.\, \operatorname{sn} v \,.\, \operatorname{dn} u \,.\, \operatorname{dn} v}{1 - k^2 \,.\, \operatorname{sn}^2 u \,.\, \operatorname{sn}^2 v}. \qquad [2]$$

$$\operatorname{dn}(u \pm v) = \frac{\operatorname{dn} u \,.\, \operatorname{dn} v \mp k^2 \,.\, \operatorname{sn} u \,.\, \operatorname{sn} v \,.\, \operatorname{cn} u \,.\, \operatorname{cn} v}{1 - k^2 \,.\, \operatorname{sn}^2 u \,.\, \operatorname{sn}^2 v}. \qquad [3]$$

From [1], [2], and [3] a large number of formulas can be readily obtained. We give only those for sn; there are similar ones for cn and dn.

$$\operatorname{sn}(u+v) + \operatorname{sn}(u-v) = \frac{2 \operatorname{sn} u \,.\, \operatorname{cn} v \,.\, \operatorname{dn} v}{1 - k^2 \,.\, \operatorname{sn}^2 u \,.\, \operatorname{sn}^2 v}. \qquad [4]$$

$$\operatorname{sn}(u+v) - \operatorname{sn}(u-v) = \frac{2 \operatorname{cn} u \,.\, \operatorname{sn} v \,.\, \operatorname{dn} u}{1 - k^2 \,.\, \operatorname{sn}^2 u \,.\, \operatorname{sn}^2 v}. \qquad [5]$$

$$\operatorname{sn}(u+v) \,.\, \operatorname{sn}(u-v) = \frac{\operatorname{sn}^2 u - \operatorname{sn}^2 v}{1 - k^2 \cdot \operatorname{sn}^2 u \cdot \operatorname{sn}^2 v}. \qquad [6]$$

$$1 + \operatorname{sn}(u+v) \,.\, \operatorname{sn}(u-v) = \frac{\operatorname{cn}^2 v + \operatorname{sn}^2 u \,.\, \operatorname{dn}^2 v}{1 - k^2 \,.\, \operatorname{sn}^2 u \,.\, \operatorname{sn}^2 v}. \qquad [7]$$

$$1 + k^2 \operatorname{sn}(u+v) \,.\, \operatorname{sn}(u-v) = \frac{\operatorname{dn}^2 v + k^2 \,.\, \operatorname{sn}^2 u \,.\, \operatorname{cn}^2 v}{1 - k^2 \,.\, \operatorname{sn}^2 u \,.\, \operatorname{sn}^2 v}. \qquad [8]$$

$$[1 + \operatorname{sn}(u+v)][1 + \operatorname{sn}(u-v)] = \frac{(\operatorname{cn} v + \operatorname{sn} u \,.\, \operatorname{dn} v)^2}{1 - k^2 \,.\, \operatorname{sn}^2 u \,.\, \operatorname{sn}^2 v}. \qquad [9]$$

From [2] and [3] comes the useful formula

$$\operatorname{cn}(u+v) = \operatorname{cn} u \,.\, \operatorname{cn} v - \operatorname{sn} u \,.\, \operatorname{sn} v \,.\, \operatorname{dn}(u+v). \qquad [10]$$

194. If in formulas [1], [2], and [3] of Art. 193 we let $v = u$, we get the following formulas for $\operatorname{sn} 2u$, $\operatorname{cn} 2u$, and $\operatorname{dn} 2u$:

$$\operatorname{sn} 2u = \frac{2 \operatorname{sn} u \,.\, \operatorname{cn} u \,.\, \operatorname{dn} u}{1 - k^2 \operatorname{sn}^4 u}, \qquad [1]$$

$$\operatorname{cn} 2u = \frac{\operatorname{cn}^2 u - \operatorname{sn}^2 u \,.\, \operatorname{dn}^2 u}{1 - k^2 \operatorname{sn}^4 u} = \frac{1 - 2 \operatorname{sn}^2 u + k^2 \operatorname{sn}^4 u}{1 - k^2 \operatorname{sn}^4 u}, \qquad [2]$$

$$\operatorname{dn} 2u = \frac{\operatorname{dn}^2 u - k^2 \,.\, \operatorname{sn}^2 u \,.\, \operatorname{cn}^2 u}{1 - k^2 \operatorname{sn}^4 u} = \frac{1 - 2k^2 \operatorname{sn}^2 u + k^2 \operatorname{sn}^4 u}{1 - k^2 \operatorname{sn}^4 u}. \qquad [3]$$

From these come readily

$$1 - \operatorname{cn} 2u = \frac{2 \operatorname{sn}^2 u \,.\, \operatorname{dn}^2 u}{1 - k^2 \operatorname{sn}^4 u}, \qquad [4]$$

$$1 + \operatorname{cn} 2u = \frac{2 \operatorname{cn}^2 u}{1 - k^2 \operatorname{sn}^4 u}, \qquad [5]$$

$$1 - \operatorname{dn} 2u = \frac{2k^2 \operatorname{sn}^2 u \,.\, \operatorname{cn}^2 u}{1 - k^2 \operatorname{sn}^4 u}, \qquad [6]$$

$$1 + \operatorname{dn} 2u = \frac{2 \operatorname{dn}^2 u}{1 - k^2 \operatorname{sn}^4 u}. \qquad [7]$$

195. Replacing u by $\frac{u}{2}$, and dividing [4] by [7] and [6] by [5], Art. 194, we have

$$\operatorname{sn}^2 \frac{u}{2} = \frac{1 - \operatorname{cn} u}{1 + \operatorname{dn} u} = \frac{1 - \operatorname{dn} u}{k^2(1 + \operatorname{cn} u)}, \qquad [1]$$

$$\operatorname{cn}^2 \frac{u}{2} = \frac{\operatorname{dn} u + \operatorname{cn} u}{1 + \operatorname{dn} u} = \frac{-k'^2 + k^2 \operatorname{cn} u + \operatorname{dn} u}{k^2(1 + \operatorname{cn} u)}, \qquad [2]$$

$$\operatorname{dn}^2 \frac{u}{2} = \frac{k'^2 + \operatorname{dn} u + k^2 \operatorname{cn} u}{1 + \operatorname{dn} u} = \frac{(\operatorname{cn} u + \operatorname{dn} u)}{(1 + \operatorname{cn} u)}, \qquad [3]$$

where $k'^2 = 1 - k^2$, and is the square of the complementary modulus.

From [1], [2], and [3], we can get without difficulty the set

$$\text{sn}^2\frac{u}{2} = \frac{\text{dn}\,u - \text{cn}\,u}{k'^2 + \text{dn}\,u - k^2\,\text{cn}\,u}, \qquad [4]$$

$$\text{cn}^2\frac{u}{2} = \frac{k'^2(1 + \text{cn}\,u)}{k'^2 + \text{dn}\,u - k^2\,\text{cn}\,u}, \qquad [5]$$

$$\text{dn}^2\frac{u}{2} = \frac{k'^2(1 + \text{dn}\,u)}{k'^2 + \text{dn}\,u - k^2\,\text{cn}\,u}. \qquad [6]$$

Numerous additional formulas can be obtained by the exercise of a little ingenuity, but we have given the most useful and important ones, and they form a set as complete as the usual collections of trigonometric formulas.

Periodicity of the Elliptic Functions.

196. We have seen (Art. 186, [2]) that

$$F(k, n\pi + \rho) = 2nK + F(k, \rho), \qquad [1]$$

where K is the complete Elliptic Integral of the first class.

Let $u = F(k, \rho)$, and take the amplitude of each member of [1]; we get

$$\text{am}\,(u + 2nK) = n\pi + \text{am}\,u\,; \qquad [2]$$

or, replacing n by $2n$,

$$\text{am}\,(u + 4nK) = 2n\pi + \text{am}\,u\,; \qquad [3]$$

whence

$$\left.\begin{aligned} \text{sn}\,(u + 4nK) &= \text{sn}\,u, \\ \text{cn}\,(u + 4nK) &= \text{cn}\,u, \\ \text{dn}\,(u + 4nK) &= \text{dn}\,u, \end{aligned}\right\}; \qquad [4]$$

and $\text{sn}\,u$, $\text{cn}\,u$, $\text{dn}\,u$ are periodic functions, and have the real period $4K$. $\text{dn}\,u$ actually has the smaller period $2K$, as may be seen by taking the delta of both members of [2].

Since the amplitude of K is $\frac{\pi}{2}$, we have

$$\left.\begin{aligned}\operatorname{sn} K &= 1,\\ \operatorname{cn} K &= 0,\\ \operatorname{dn} K &= k',\end{aligned}\right\}, \qquad [5]$$

and our addition formulas [1], [2], [3], Art. 193, give us readily

$$\left.\begin{aligned}\operatorname{sn}(u+K) &= \frac{\operatorname{cn} u}{\operatorname{dn} u},\\ \operatorname{cn}(u+K) &= -\frac{k' \operatorname{sn} u}{\operatorname{dn} u},\\ \operatorname{dn}(u+K) &= \frac{k'}{\operatorname{dn} u},\end{aligned}\right\}, \qquad [6]$$

$$\left.\begin{aligned}\operatorname{sn}(u+2K) &= -\operatorname{sn} u,\\ \operatorname{cn}(u+2K) &= -\operatorname{cn} u,\\ \operatorname{dn}(u+2K) &= \operatorname{dn} u,\end{aligned}\right\}, \qquad [7]$$

$$\left.\begin{aligned}\operatorname{sn}(u+3K) &= -\frac{\operatorname{cn} u}{\operatorname{dn} u},\\ \operatorname{cn}(u+3K) &= \frac{k' \operatorname{sn} u}{\operatorname{dn} u},\\ \operatorname{dn}(u+3K) &= \frac{k'}{\operatorname{dn} u},\end{aligned}\right\}, \qquad [8]$$

$$\left.\begin{aligned}\operatorname{sn}(u+4K) &= \operatorname{sn} u,\\ \operatorname{cn}(u+4K) &= \operatorname{cn} u,\\ \operatorname{dn}(u+4K) &= \operatorname{dn} u,\end{aligned}\right\}, \qquad [9]$$

a confirmation of [4].

197. It is easy to get formulas for the sn, cn, and dn of an imaginary variable, $u\sqrt{-1}$, by the aid of a transformation due to Jacobi.

Let
$$v = F(k, \phi) = \int_0^\phi \frac{d\phi}{\Delta\phi}, \tag{1}$$

so that $\phi = \operatorname{am} v$, $\sin\phi = \operatorname{sn} v$, and $\cos\phi = \operatorname{cn} v$. In (1), replace ϕ by ψ, ϕ and ψ being connected by the relation

$$\sin\phi = \sqrt{-1} \,.\, \tan\psi, \tag{2}$$

whence
$$\cos\phi = \sec\psi, \tag{3}$$

$$\Delta\phi = \sqrt{1 - k^2\sin^2\phi} = \sqrt{1 + k^2\tan^2\psi}, \tag{4}$$

and
$$d\phi = \sqrt{-1} \,.\, \sec\psi \,.\, d\psi.$$

Since ψ and ϕ equal zero together,

$$v = \int_0^\phi \frac{d\phi}{\Delta\phi} = \sqrt{-1}\int_0^\psi \frac{\sec\psi\, d\psi}{\sqrt{1 + k^2\tan^2\psi}}$$

$$= \sqrt{-1}\int_0^\psi \frac{d\psi}{\sqrt{1 - k'^2\sin^2\psi}} = \sqrt{-1} \,.\, F(k', \psi).$$

If now we let $u = F(k', \psi)$,

we have
$$v = u\sqrt{-1}. \tag{5}$$

Hence, since $\psi = \operatorname{am} u \,(\operatorname{mod} k')$, we have from (2), (3), and (4),

$$\operatorname{sn} v \,(\operatorname{mod} k) = \sqrt{-1}\,\frac{\operatorname{sn} u \,(\operatorname{mod} k')}{\operatorname{cn} u \,(\operatorname{mod} k')},$$

$$\operatorname{cn} v \,(\operatorname{mod} k) = \frac{1}{\operatorname{cn} u \,(\operatorname{mod} k')},$$

$$\operatorname{dn} v \,(\operatorname{mod} k) = \frac{\operatorname{dn} u \,(\operatorname{mod} k')}{\operatorname{cn} u \,(\operatorname{mod} k')};$$

or, as $v = u\sqrt{-1}$,

$$\left.\begin{aligned} \operatorname{sn} u\sqrt{-1}\,(\operatorname{mod} k) &= \sqrt{-1}\,\frac{\operatorname{sn} u\,(\operatorname{mod} k')}{\operatorname{cn} u\,(\operatorname{mod} k')},\\ \operatorname{cn} u\sqrt{-1}\,(\operatorname{mod} k) &= \frac{1}{\operatorname{cn} u\,(\operatorname{mod} k')},\\ \operatorname{dn} u\sqrt{-1}\,(\operatorname{mod} k) &= \frac{\operatorname{dn} u\,(\operatorname{mod} k')}{\operatorname{cn} u\,(\operatorname{mod} k')},\end{aligned}\right\}. \qquad [6]$$

It is interesting to note that if u is replaced in (6) by $u\sqrt{-1}$, the formulas reduce to

$$\begin{aligned} \operatorname{sn}(-u) &= -\operatorname{sn} u,\\ \operatorname{cn}(-u) &= \operatorname{cn} u,\\ \operatorname{dn}(-u) &= \operatorname{dn} u,\end{aligned}$$

and are still true. Consequently, in (6), u may be either a real or a pure imaginary.

Let $$\int_0^{\frac{\pi}{2}} \frac{d\psi}{\Delta\psi\,(\operatorname{mod} k')} = \int_0^{\frac{\pi}{2}} \frac{d\psi}{\sqrt{1-k'^2\sin^2\psi}} = K'.$$

Then, by Art. 196, $4K'$ is a period for $\operatorname{sn} u\,(\operatorname{mod} k')$, $\operatorname{cn} u\,(\operatorname{mod} k')$, and $\operatorname{dn} u\,(\operatorname{mod} k')$.

Hence

$$\begin{aligned} \operatorname{sn}(u\sqrt{-1} + 4nK'\sqrt{-1}) &= \operatorname{sn} u\sqrt{-1},\\ \operatorname{cn}(u\sqrt{-1} + 4nK'\sqrt{-1}) &= \operatorname{cn} u\sqrt{-1},\\ \operatorname{dn}(u\sqrt{-1} + 4nK'\sqrt{-1}) &= \operatorname{dn} u\sqrt{-1};\end{aligned}$$

or, replacing $u\sqrt{-1}$ by v,

$$\left.\begin{aligned} \operatorname{sn}(v + 4nK'\sqrt{-1}) &= \operatorname{sn} v,\\ \operatorname{cn}(v + 4nK'\sqrt{-1}) &= \operatorname{cn} v,\\ \operatorname{dn}(v + 4nK'\sqrt{-1}) &= \operatorname{dn} v,\end{aligned}\right\}, \qquad [7]$$

and $4K'\sqrt{-1}$ is a period for sn, cn, and dn.

We see, then, that our Elliptic Functions, like Trigonometric Functions, have a real period, and, like Exponential Functions, have a pure imaginary period. They are, then, what may be called

Doubly Periodic Functions, and they are often studied from the point of view of their double periodicity.

Like Trigonometric Functions, the Elliptic Functions may be developed in series, and from these series their values may be computed, and tables resembling Trigonometric tables may be prepared.

A partial three-place table is here presented as a sample. It is complete for Elliptic Functions having the modulus $\frac{\sqrt{2}}{2}$; that is, 0.7.

MODULUS $\frac{\sqrt{2}}{2} = 0.7$.

u	sn u	cn u	dn u
0.00	0.000	1.000	1.000
0.05	0.051	0.999	0.999
0.15	0.150	0.989	0.994
0.25	0.247	0.969	0.985
0.35	0.340	0.940	0.971
0.45	0.429	0.903	0.953
0.55	0.512	0.859	0.932
0.65	0.589	0.808	0.909
0.75	0.659	0.752	0.885
0.85	0.722	0.692	0.860
0.95	0.778	0.628	0.835
1.05	0.827	0.562	0.811
1.15	0.869	0.494	0.789
1.25	0.906	0.424	0.768
1.35	0.935	0.353	0.750
1.45	0.959	0.284	0.735
1.55	0.977	0.213	0.723
1.65	0.990	0.143	0.714
1.75	0.997	0.072	0.709
K. 1.85	1.000	0.000	0.707

From this table, by the aid of formulas [4], [6], [7], and (8) of Art. 196, sn u, cn u, and dn u may be readily obtained for any value of u if the modulus is $\frac{\sqrt{2}}{2}$.

As a matter of fact no complete set of tables for the Elliptic Functions has been published, and their values are usually obtained indirectly from Legendre's Tables of Elliptic Integrals (*v.* Arts. 186, 187), unless especial accuracy is required, in which case they must be computed by methods which we have not space to give.

198. The Elliptic Integral of the second class $E(k, \phi)$ can be expressed in terms of Elliptic Functions, and for some purposes there is a decided advantage in the new form.

We have
$$E(k, \phi) = \int_0^\phi \Delta\phi \,.\, d\phi.$$

Let $u = F(k, \phi)$, then $\phi = \text{am}\, u$, and $E(k, \phi)$ may be written $E(k, \text{am}\, u)$, or, more simply, $E(\text{am}\, u)$, if the modulus can be omitted without danger of confusion.

Then
$$E(\text{am}\, u) = \int_0^{\text{am}\, u} \text{dn}\, u \,.\, d\,\text{am}\, u;$$
or, since by (4), Art. 192,
$$d\,\text{am}\, u = \text{dn}\, u \,.\, du,$$
$$E(\text{am}\, u) = \int_0^u \text{dn}^2 u \,.\, du. \qquad [1]$$

As an example of the usefulness of the form just given in [1], we will employ it in getting an *addition formula* for Elliptic Integrals of the second class.

$$\begin{aligned}
E(\text{am}\, u) + E(\text{am}\, v) &= \int_0^u \text{dn}^2 u \,.\, du + \int_0^v \text{dn}^2 v \,.\, dv \\
&= \int_0^u \text{dn}^2 z \,.\, dz + \int_0^v \text{dn}^2 z \,.\, dz \\
&= \int_0^{u+v} \text{dn}^2 z \,.\, dz + \int_0^v \text{dn}^2 z \,.\, dz - \int_u^{u+v} \text{dn}^2 z \,.\, dz \\
&= E[\text{am}\,(u+v)] + \int_0^v \text{dn}^2 z \,.\, dz - \int_u^{u+v} \text{dn}^2 z \,.\, dz.
\end{aligned}$$

Replacing z by $u+z$, and remembering that u and v are given constants,

$$\int_u^{u+v} \mathrm{dn}^2 z \, . \, dz = \int_0^v \mathrm{dn}^2 (u+z)\, dz,$$

and

$$E(\mathrm{am}\, u) + E(\mathrm{am}\, v) = E[\mathrm{am}(u+v)] - \int_0^v [\mathrm{dn}^2(u+z) - \mathrm{dn}^2 z]\, dz. \qquad (2)$$

$$\mathrm{dn}^2(u+z) - \mathrm{dn}^2 z = [\mathrm{dn}(u+z) + \mathrm{dn}\, z][\mathrm{dn}(u+z) - \mathrm{dn}\, z]. \quad (3)$$

We can obtain from [3], Art. 193, the following formulas analogous to [4] and [5], Art. 193,

$$\mathrm{dn}(u+v) + \mathrm{dn}(u-v) = \frac{2\,\mathrm{dn}\, u \, . \, \mathrm{dn}\, v}{1 - k^2 \mathrm{sn}^2 u \, . \, \mathrm{sn}^2 v}, \qquad (4)$$

$$\mathrm{dn}(u+v) - \mathrm{dn}(u-v) = -\frac{2k^2 \mathrm{sn}\, u \, . \, \mathrm{sn}\, v \, . \, \mathrm{cn}\, u \, . \, \mathrm{cn}\, v}{1 - k^2 \mathrm{sn}^2 u \, . \, \mathrm{sn}^2 v}. \qquad (5)$$

If in (4) and (5) we let $u+v = u+z$, and $u-v = z$, and substitute the results in (3), we get

$$\mathrm{dn}^2(u+z) - \mathrm{dn}^2 z = -\frac{4k^2 \mathrm{sn}\left(\frac{u}{2}+z\right) \mathrm{cn}\left(\frac{u}{2}+z\right) \mathrm{dn}\left(\frac{u}{2}+z\right) \mathrm{sn}\frac{u}{2}\, \mathrm{cn}\frac{u}{2}\, \mathrm{dn}\frac{u}{2}}{\left[1 - k^2 \mathrm{sn}^2\frac{u}{2}\, \mathrm{sn}^2\left(\frac{u}{2}+z\right)\right]^2};$$

and

$$\int [\mathrm{dn}^2(u+z) - \mathrm{dn}^2 z]\, dz$$

$$= -2\,\mathrm{sn}\frac{u}{2}\, \mathrm{cn}\frac{u}{2}\, \mathrm{dn}\frac{u}{2} \int \frac{2k^2 \mathrm{sn}\left(\frac{u}{2}+z\right) \mathrm{cn}\left(\frac{u}{2}+z\right) \mathrm{dn}\left(\frac{u}{2}+z\right) dz}{\left[1 - k^2 \mathrm{sn}^2\frac{u}{2}\, \mathrm{sn}^2\left(\frac{u}{2}+z\right)\right]^2}$$

$$= -\frac{2\,\mathrm{sn}\frac{u}{2}\, \mathrm{cn}\frac{u}{2}\, \mathrm{dn}\frac{u}{2}}{\mathrm{sn}^2\frac{u}{2}} \cdot \frac{1}{1 - k^2 \mathrm{sn}^2\frac{u}{2}\, \mathrm{sn}^2\left(\frac{u}{2}+z\right)}$$

since $-2k^2\,\mathrm{sn}^2\frac{u}{2}\,\mathrm{sn}\left(\frac{u}{2}+z\right)\mathrm{cn}\left(\frac{u}{2}+z\right)\mathrm{dn}\left(\frac{u}{2}+z\right)dz$ is the differential of $1-k^2\,\mathrm{sn}^2\frac{u}{2}\,\mathrm{sn}^2\left(\frac{u}{2}+z\right)$.

$$\int_0^v[\mathrm{dn}^2(u+z)-\mathrm{dn}^2 z]\,dz$$

$$=-\frac{2\,\mathrm{sn}\frac{u}{2}\,\mathrm{cn}\frac{u}{2}\,\mathrm{dn}\frac{u}{2}}{\mathrm{sn}^2\frac{u}{2}}\left[\frac{1}{1-k^2\,\mathrm{sn}^2\frac{u}{2}\,\mathrm{sn}^2\left(\frac{u}{2}+v\right)}-\frac{1}{1-k^2\,\mathrm{sn}^4\frac{u}{2}}\right]$$

$$=-\frac{k^2\,2\,\mathrm{sn}\frac{u}{2}\,\mathrm{cn}\frac{u}{2}\,\mathrm{dn}\frac{u}{2}}{1-k^2\,\mathrm{sn}^4\frac{u}{2}}\cdot\frac{\mathrm{sn}^2\left(\frac{u}{2}+v\right)-\mathrm{sn}^2\frac{u}{2}}{1-k^2\,\mathrm{sn}^2\frac{u}{2}\,\mathrm{sn}^2\left(\frac{u}{2}+v\right)}$$

$$=-k^2\,.\,\mathrm{sn}\,u\,.\,\mathrm{sn}\,v\,.\,\mathrm{sn}(u+v),$$

by (1), Art. 194, and [6], Art. 193.

Hence by (2),

$$E(\mathrm{am}\,u)+E(\mathrm{am}\,v)=E[\mathrm{am}(u+v)]+k^2\,\mathrm{sn}\,u\,.\,\mathrm{sn}\,v\,.\,\mathrm{sn}(u+v), \qquad [6]$$

our required addition formula.

APPLICATIONS.

Rectification of the Lemniscate.

199. From the polar equation of the Lemniscate, $r^2=a^2\cos 2\theta$, referred to its centre as origin and its axis as axis, we get as the length of the arc, measured from the vertex to any point, P, whose coördinates are r and θ.

$$s=a\int_0^\theta\frac{d\theta}{\sqrt{\cos 2\theta}}=a\int_0^\theta\frac{d\theta}{\sqrt{1-2\sin^2\theta}}, \qquad [1]$$

and for the arc of the quadrant of the Lemniscate, that is, the arc from vertex to centre,

$$s_q = a\int_0^{\frac{\pi}{4}} \frac{d\theta}{\sqrt{1-2\sin^2\theta}}. \qquad [2]$$

These differ from Elliptic Integrals of the first class only in that the coefficient of $\sin^2\theta$ is greater than unity, and they may be reduced to the standard form by a simple device.

Introduce in [1] ϕ in place of θ, ϕ and θ being connected by the relation $\sin^2\phi = 2\sin^2\theta$.

Then we have $$\sqrt{1-2\sin^2\theta} = \cos\phi,$$

and $$d\theta = \frac{\sqrt{2}}{2}\frac{\cos\phi\, d\phi}{\sqrt{1-\frac{1}{2}\sin^2\phi}}.$$

Hence $$s = \frac{a\sqrt{2}}{2}\int_0^{\phi} \frac{d\phi}{\sqrt{1-\frac{1}{2}\sin^2\phi}} = \frac{a\sqrt{2}}{2}F\left(\frac{\sqrt{2}}{2}, \phi\right), \qquad [3]$$

and $$s_q = \frac{a\sqrt{2}}{2}\int_0^{\frac{\pi}{2}} \frac{d\phi}{\sqrt{1-\frac{1}{2}\sin^2\phi}} = \frac{a\sqrt{2}}{2}F\left(\frac{\sqrt{2}}{2}, \frac{\pi}{2}\right). \qquad [4]$$

The auxiliary angle ϕ is very easily constructed when the point P of the Lemniscate is given. We have $r = a\sqrt{\cos 2\theta}$, and we have seen that $\sqrt{\cos 2\theta} = \cos\phi$; hence $r = a\cos\phi$. If, then, on a as a diameter we describe a semi-circumference, and with the centre O of the Lemniscate as a centre, and with a radius equal to r, we describe an arc, and join with O the point Q where this arc intersects the semi-circumference, the angle made by OQ with a is equal to ϕ. For $OQ = a\cos AOQ$ and $OP = a\sqrt{\cos 2\theta}$.

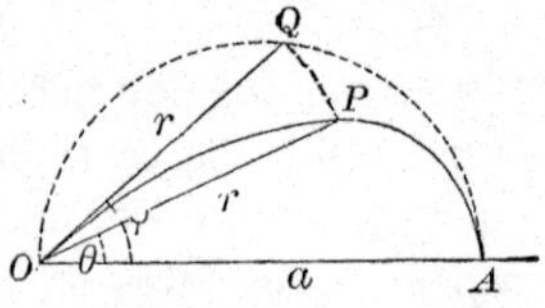

EXAMPLES.

(1) Find the numerical value of $\int_0^{\frac{\pi}{6}} \frac{d\phi}{\sqrt{1-4\sin^2\phi}}$.

Ans. 0.843.

(2) Reduce $\int_0^{\phi} \frac{d\phi}{\sqrt{1-n\sin^2\phi}}$ to an Elliptic Integral of the first class, when $n > 1$.

$$\textit{Ans.}\ \frac{1}{\sqrt{n}}\int_0^{\psi} \frac{d\psi}{\sqrt{1-\frac{1}{n}\sin^2\psi}} \text{ where } \sin^2\psi = n\sin^2\phi.$$

(3) The half-axis of a Lemniscate is 2. What is the length of the arc of a quadrant? of the arc from the vertex to the point whose polar angle is 30°? *Ans.* 2.622 ; 1.168.

In the inverse problem of cutting off an arc of given length the Elliptic Functions are of service. As an interesting example, let us find the point which bisects the quadrantal arc of the Lemniscate.

Here
$$s = \frac{a\sqrt{2}}{2}\,\tfrac{1}{2}\,F\left(\frac{\sqrt{2}}{2}, \frac{\pi}{2}\right),$$

and we wish to find ϕ and then θ.

Let
$$u = F\left(\frac{\sqrt{2}}{2}, \frac{\pi}{2}\right); \quad \text{we need am}\,\frac{u}{2}.$$

$$\text{am}\,u = \frac{\pi}{2},\ \text{sn}\,u = 1,\ \text{cn}\,u = 0, \text{ and dn}\,u = \frac{\sqrt{2}}{2}.$$

By [1] and [2], Art. 195,

$$\text{sn}^2\frac{u}{2} = \frac{1-\text{cn}\,u}{1+\text{dn}\,u}, \quad \text{cn}^2\frac{u}{2} = \frac{\text{dn}\,u + \text{cn}\,u}{1+\text{dn}\,u}.$$

Therefore,

$$\frac{\text{sn}^2\frac{u}{2}}{\text{cn}^2\frac{u}{2}} = \text{tn}^2\frac{u}{2} = \frac{1-\text{cn}\,u}{\text{dn}\,u+\text{cn}\,u} = \frac{1}{\frac{\sqrt{2}}{2}} = \sqrt{2}.$$

If, then, the required amplitude is ϕ,

$$\tan^2\phi = \sqrt{2},$$

and

$$\tan\phi = \sqrt[4]{2}.$$

Since $\sin^2\phi = 2\sin^2\theta$, we can compute θ without difficulty, and so get our required point. If, however, a construction will suffice, a very simple one gives the point.

.Erect at A a perpendicular whose length is a mean proportional between a and $a\sqrt{2}$. The angle subtended at O by this perpendicular is ϕ, and the corresponding point, P, is found by the method described on page 255.

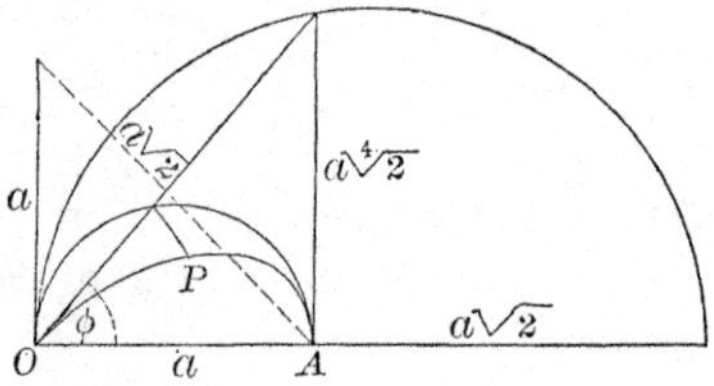

Rectification of the Ellipse.

200. We have seen in Art. 177 that the length of an arc of an Ellipse measured from the end of the minor axis is

$$s = \int_0^x \sqrt{\frac{a^2 - e^2x^2}{a^2 - x^2}} \cdot dx. \qquad [1]$$

If we let $x = a\sin\phi$, [1] becomes

$$s = a\int_0^\phi \sqrt{1 - e^2\sin^2\phi}\,.\,d\phi = aE(e, \phi), \qquad [2]$$

e, the modulus of the Elliptic Integral, being the eccentricity of the Ellipse. If $x = a$, $\phi = \frac{\pi}{2}$, and the length of the Elliptic quadrant is

$$s_q = a\int_0^{\frac{\pi}{2}} \sqrt{1 - e^2\sin^2\phi}\,.\,d\phi = aE\left(e, \frac{\pi}{2}\right). \qquad [3]$$

The length of an arc of the Elliptic quadrant, not measured from the extremity of the minor axis, can of course be expressed as the difference between two Elliptic Integrals of the second class.

The amplitude ϕ, corresponding to a given point P, of the Ellipse, is easily constructed as follows: On the major axis as diameter describe a circumference; extend the ordinate of P until it meets the circumference, and join the point of intersection with the centre of the ellipse. The angle the joining line makes with the minor axis is seen to be the required amplitude ϕ. If ϕ is given, P may be found by reversing the order of the steps of the construction.

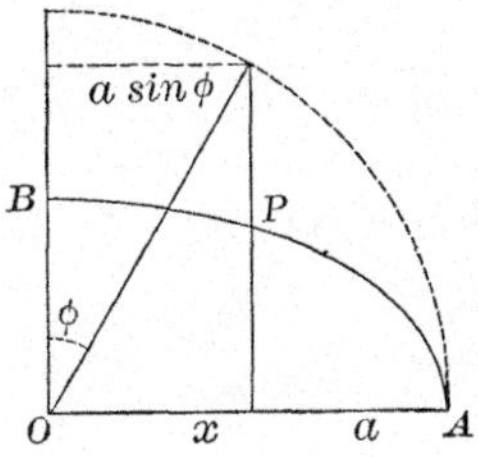

EXAMPLES.

The equation of an ellipse is $\frac{x^2}{16}+\frac{y^2}{8}=1$, required the length of the quadrantal arc; of the arc whose extremities have the abscissas 2 and $2\sqrt{2}$. *Ans.* 5.4; 0.944.

(2) Find the abscissa of the end of the unit arc measured from the extremity of the minor axis in the ellipse $\frac{x^2}{16}+\frac{y^2}{8}=1$; of the point which bisects the arc of the quadrant.

Ans. 0.996; 2.57.

201. By the aid of the addition formula

$$E(\operatorname{am} u)+E(\operatorname{am} v)=E\left[\operatorname{am}(u+v)\right]+k^2 \operatorname{sn} u \operatorname{sn} v \operatorname{sn}(u+v)$$

([6], Art. 198)

it is always possible to find an arc of an ellipse differing from the sum of two given arcs by an expression which is algebraic in terms of the abscissas of the extremities of the three arcs. This will be clearer if we modify slightly the form of our addition formula.

Let $\phi = \text{am}\, u, \quad \psi = \text{am}\, v, \quad \text{and } \sigma = \text{am}(u + v).$

Then the formula given above becomes

$$E(k, \phi) + E(k, \psi) = E(k, \sigma) + k^2 \sin\phi \sin\psi \sin\sigma, \qquad [1]$$

where ϕ, ψ, and σ are three angles connected by the relation

$$\cos\sigma = \cos\phi\cos\psi - \sin\phi\sin\psi\,\Delta\sigma, \qquad [2]$$

by [10], Art. 193.

If we multiply [1] by a and take k equal to e, we get

$$aE(e, \phi) + aE(e, \psi) = aE(e, \sigma) + \frac{e^2}{a^2} x_1 \cdot x_2 \cdot x_3,$$

if x_1, x_2, and x_3 are the abscissas of the points whose amplitudes are ϕ, ψ, and σ.

The most interesting case is when $\sigma = \frac{\pi}{2}$, in which case $aE(e, \sigma)$ is the arc of a quadrant. [2] then reduces to

$$0 = \cos\phi\cos\psi - \sin\phi\sin\psi\sqrt{1 - e^2},$$

or

$$\frac{b}{a}\sin\phi\sin\psi = \cos\phi\cos\psi,$$

or

$$\tan\phi\tan\psi = \frac{a}{b}, \qquad [3]$$

and we get from [1]

$$aE(e, \phi) - \left[aE\left(e, \frac{\pi}{2}\right) - aE(e, \psi)\right] = ae^2 \sin\phi\sin\psi. \qquad [4]$$

If, then, any point, P, is given, [3] will enable us to get the amplitude of a second point, Q, and thus to find Q, Q and P being so related that the arc BP, minus the arc AQ, shall be equal to a quantity which is proportional to the product of the abscissas of P and Q.

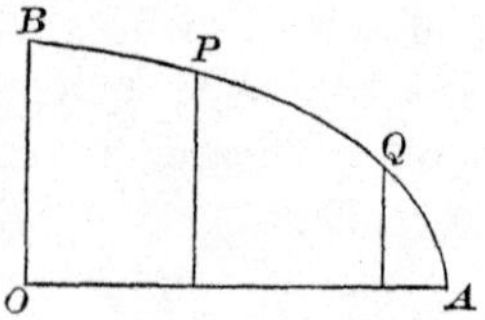

For the special case where ϕ and ψ are equal we have from [3],

$$\tan\phi = \sqrt{\frac{a}{b}},$$

and from [4],

$$BP - AP = ae^2 \sin^2\phi = \frac{a^2e^2}{a+b} = a - b.$$

This point, which divides the quadrant into two arcs whose difference is equal to the difference between the semi-axes, has a number of curious properties, and is known as *Fagnani's point.*

EXAMPLES.

(1) Show that the distance of the normal at Fagnani's point, from the centre of the ellipse, is equal to $a - b$.

(2) Show that the angle between the normals at P and Q in the figure is equal to $\psi - \phi$; that the normals are equidistant from O; that this distance is $BP - AQ$.

Rectification of the Hyperbola.

202. If the arc of the Hyperbola is measured from the vertex to any given point, P, whose coördinates are x and y, its length is easily found to be

$$s = \int_0^y \left(\frac{1 + \frac{a^2 + b^2}{b^4} y^2}{1 + \frac{y^2}{b^2}} \right)^{\frac{1}{2}} dy, \qquad [1]$$

or

$$s = \int_0^y \left(\frac{1 + \frac{a^2e^2}{b^4} y^2}{1 + \frac{y^2}{b^2}} \right)^{\frac{1}{2}} dy, \qquad [2]$$

if e is the eccentricity of the Hyperbola. Let

$$\frac{ae}{b^2} y = \tan\phi,$$

and [2] becomes

$$s = \frac{b^2}{ae}\int_0^\phi \frac{\sec^2\phi\, d\phi}{\sqrt{1 - \frac{1}{e^2}\sin^2\phi}};$$

hence
$$s = \frac{b^2}{ae}\int_0^\phi \frac{\sec^2\phi\, d\phi}{\sqrt{1 - k^2\sin^2\phi}} = \frac{b^2}{ae}\int_0^\phi \frac{\sec^2\phi\, d\phi}{\Delta\phi} \qquad [3]$$

if
$$k = \frac{1}{e}.$$

Now
$$\frac{1}{\Delta\phi} = \frac{1-k^2}{1-k^2}\cdot\frac{1}{\Delta\phi} = \frac{1}{1-k^2}\cdot\frac{1 - k^2\sin^2\phi - k^2\cos^2\phi}{\Delta\phi}$$

$$= \frac{1}{1-k^2}\left[\Delta\phi - \frac{k^2\cos^2\phi}{\Delta\phi}\right];$$

and
$$s = \frac{b^2}{ae}\cdot\frac{1}{1-k^2}\left[\int_0^\phi \sec^2\phi\,\Delta\phi\, d\phi - k^2\int_0^\phi\frac{d\phi}{\Delta\phi}\right]$$

$$= \frac{b^2}{ae}\cdot\frac{1}{1-k^2}\left[\int_0^\phi \sec^2\phi\,\Delta\phi\, d\phi - k^2 F(k, \phi)\right].$$

If we integrate by parts,

$$\int_0^\phi \sec^2\phi\,\Delta\phi\, d\phi = \tan\phi\,\Delta\phi + k^2\int_0^\phi\frac{\sin^2\phi}{\Delta\phi}d\phi;$$

but
$$k^2\frac{\sin^2\phi}{\Delta\phi} = \frac{1}{\Delta\phi} - \Delta\phi,$$

and
$$k^2\int_0^\phi\frac{\sin^2\phi}{\Delta\phi}\cdot d\phi = F(k, \phi) - E(k, \phi).$$

Hence

$$s = \frac{b^2}{ae}F(k, \phi) - \frac{b^2}{ae(1-k^2)}\left[E(k, \phi) - \tan\phi\,\Delta\phi\right].$$

But
$$1 - k^2 = \frac{b^2}{a^2e^2},$$

therefore $$s = \frac{b^2}{ae} F(k, \phi) - aeE(k, \phi) + ae \tan \phi \, \Delta\phi,$$

or $$s = \frac{b^2}{ae} F\left(\frac{1}{e}, \phi\right) - aeE\left(\frac{1}{e}, \phi\right) + ae \tan \phi \, \Delta\phi. \qquad [4]$$

The angle ϕ corresponding to a given point P is easily con-

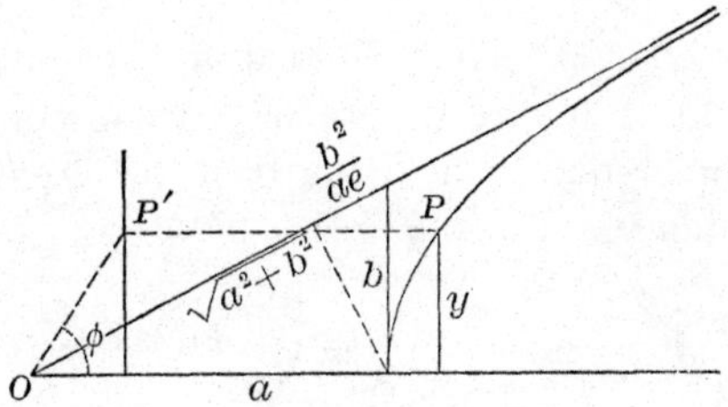

structed. We have only to erect a perpendicular to the transverse axis at a distance $\frac{b^2}{ae} = \frac{b^2}{\sqrt{a^2+b^2}}$ from the origin; that is, at a distance from the centre equal to the projection of b on the asymptote, and to join the projection of P on this line with the centre. The angle made by the joining line with the transverse axis is ϕ, for its tangent is clearly $\dfrac{y}{\dfrac{b^2}{\sqrt{a^2+b^2}}}$.

EXAMPLES.

(1) Find the length of the arc of the hyperbola $\frac{x^2}{16} - \frac{y^2}{9} = 1$, measured from the vertex to the point whose ordinate is 2.

Ans. 2.194.

(2) Show that $ae \tan \phi \, \Delta\phi$ is the distance from the centre to the normal at P.

(3) Show that the limiting value approached by the difference between the arc and the portion of the asymptote cut off by a perpendicular upon it from P, as P recedes indefinitely from

the origin, is $aeE\left(\frac{1}{e}, \frac{\pi}{2}\right) - \frac{b^2}{ae} F\left(\frac{1}{e}, \frac{\pi}{2}\right)$. This is generally referred to as the difference between the length of the infinite arc of the hyperbola and the length of the asymptote.

Show that in example (1) this difference is equal to 2.803.

The Pendulum.

203. We have seen in Art. 176 that if a pendulum starts from rest at a point of its arc whose distance above the lowest point is y_0, the time required in rising from the lowest point to a point whose distance above the lowest point is y, is

$$t = \sqrt{\frac{a}{g}} \int_0^\phi \frac{d\phi}{\sqrt{1 - k^2 \sin^2\phi}} = \sqrt{\frac{a}{g}} F(k, \phi), \qquad [1]$$

where $k = \sqrt{\frac{y_0}{2a}}$, and $\sin\phi = \sqrt{\frac{y}{y_0}}$.

In the figure let A be the lowest point of the arc, B the

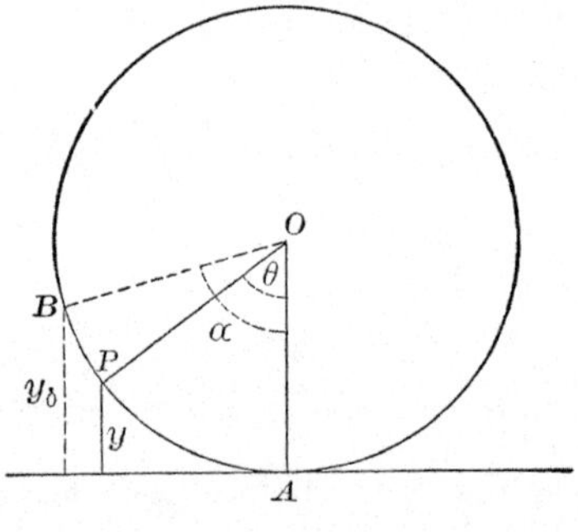

highest point reached by the pendulum, and P the point reached at the expiration of the time t. Call AOB α, and AOP θ.

Then $\frac{y_0}{a} = 1 - \cos\alpha$, and $\sqrt{\frac{y_0}{2a}} = \sqrt{\tfrac{1}{2}(1 - \cos\alpha)} = \sin\frac{\alpha}{2} = k$.

Consequently the modulus of the Elliptic Integral in [1] is

the sine of one-fourth the angle through which the pendulum swings.

$$\frac{y}{a} = 1 - \cos\theta,$$

and
$$\sqrt{\frac{y}{2a}} = \sqrt{\tfrac{1}{2}(1 - \cos\theta)} = \sin\frac{\theta}{2},$$

and
$$\sin\phi = \sqrt{\frac{y}{y_0}} = \frac{\sqrt{\frac{y}{2a}}}{\sqrt{\frac{y_0}{2a}}} = \frac{\sin\frac{\theta}{2}}{\sin\frac{\alpha}{2}},$$

and therefore the sine of the amplitude of the Elliptic Integral in [1] is easily computed when the angle through which the pendulum has risen is given. When $\theta = \alpha$, $\sin\phi = 1$, and $\phi = \frac{\pi}{2}$; so that the time of a half-oscillation is $\sqrt{\frac{a}{g}}F\left(\sin\frac{\alpha}{2}, \frac{\pi}{2}\right)$, a confirmation of [7], Art. 176. The construction indicated

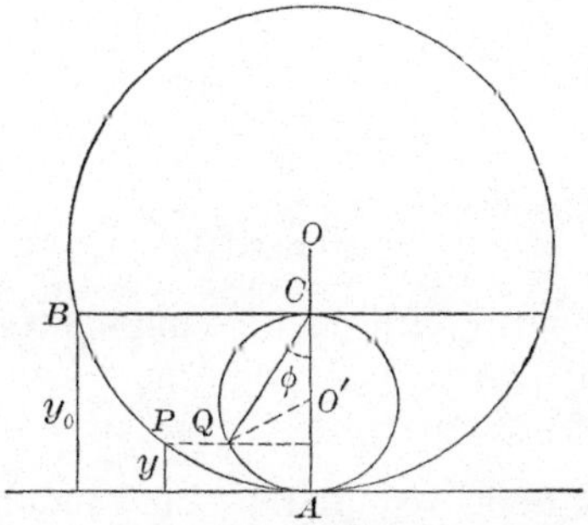

in the figure gives the angle ϕ, corresponding to any given arc AP. For

$$\frac{y}{\frac{1}{2}y_0} = 1 - \cos AO'Q,$$

and
$$\sqrt{\frac{y}{y_0}} = \sqrt{\tfrac{1}{2}(1 - \cos AO'Q)} = \sin\frac{AO'Q}{2} = \sin ACQ.$$

Therefore $ACQ = \phi$.

It is very easy to express the angle θ in terms of t.

We have $$t = \sqrt{\frac{a}{g}}\, F\left(\sin\frac{\alpha}{2}, \phi\right);$$

hence $$t\sqrt{\frac{g}{a}} = F\left(\sin\frac{\alpha}{2}, \phi\right),$$

$$\phi = \text{am}\left(t\sqrt{\frac{g}{a}}\right),$$

$$\sin\phi = \text{sn}\left(t\sqrt{\frac{g}{a}}\right),$$

and $$\sin\frac{\theta}{2} = \sin\frac{\alpha}{2}\,\text{sn}\left(t\sqrt{\frac{g}{a}}\right)\left(\text{mod}\sin\frac{\alpha}{2}\right);$$

then $$\cos\frac{\theta}{2} = \text{dn}\left(t\sqrt{\frac{g}{a}}\right)\left(\text{mod}\sin\frac{\alpha}{2}\right),$$

and $$\sin\theta = 2\sin\frac{\alpha}{2}\,\text{sn}\left(t\sqrt{\frac{g}{a}}\right)\text{dn}\left(t\sqrt{\frac{g}{a}}\right)\left(\text{mod}\sin\frac{\alpha}{2}\right).$$

EXAMPLES.

(1) A pendulum swings through an angle of 180°; required the time of oscillation.

Ans. $3.708\sqrt{\frac{a}{g}}$.

(2) Compare the times required by the pendulum in Ex. (1) to descend through the first 30°, the second 30°, and the third 30° of its arc respectively.

Ans. $1.028\sqrt{\frac{a}{g}}$; $0.446\sqrt{\frac{a}{g}}$; $0.380\sqrt{\frac{a}{g}}$.

(3) The time of vibration of a pendulum swinging in an arc of 72° is observed to be 2 seconds; how long does it take it to fall through an arc of 5°, beginning at a point 20° from the highest point of the arc of swing? *Ans.* 0.095 seconds.

(4) A pendulum for which $\sqrt{\frac{a}{g}}$ has been determined, and is equal to $\frac{1}{2}$, vibrates through an arc of 180°; through what arc does it rise in the first half-second after it has passed its lowest point? in the first $\frac{1}{8}$ of a second? *Ans.* 69°; 20° 6′.

(5) It has been shown in Art. 176 that if $y_0 > 2a$ the pendulum will make complete revolutions, and that the time required to pass from the lowest point to any point whose distance above the lowest point is y, is

$$t = a\sqrt{\frac{2}{gy_0}}\int_0^\phi \frac{d\phi}{\sqrt{1-k^2\sin^2\phi}} = a\sqrt{\frac{2}{gy_0}}F(k,\phi),$$

where $k = \sqrt{\frac{2a}{y_0}}$ and $\sin\phi = \sqrt{\frac{y}{2a}}$.

Show that in this case $\phi = \frac{\theta}{2}$, and that $\sin\frac{\theta}{2} = \text{sn}\left(\frac{t}{a}\sqrt{\frac{gy_0}{2}}\right)$.

NOTE.—In working with a pendulum it is often about as easy to compute $F(k, \phi)$ by developing by the binomial theorem and integrating two or three terms, as to use a table of Elliptic Integrals.

We have $$F(k,\phi) = \int_0^\phi \frac{d\phi}{\sqrt{1-k^2\sin^2\phi}},$$

$$(1-k^2\sin^2\phi)^{-\frac{1}{2}} = 1 + \tfrac{1}{2}k^2\sin^2\phi + \frac{1.3}{2.4}k^4\sin^4\phi + \cdots,$$

and $$F(k,\phi) = \int_0^\phi \frac{d\phi}{\sqrt{1-k^2\sin^2\phi}} = \phi + \frac{k^2}{4}(\phi - \sin\phi\cos\phi)$$

$$-\frac{3}{32}k^4\sin^3\phi\cos\phi + \frac{9}{64}k^4(\phi - \sin\phi\cos\phi)\cdots.$$

CHAPTER XVII.

INTRODUCTION TO THE THEORY OF FUNCTIONS.

204. A function having but a single value for any given value, real or imaginary, of the variable is called a *single-valued* function. Rational Algebraic Functions, Exponential Functions, the direct Trigonometric Functions, and the Elliptic Functions are single-valued.

A function which has in general two or more values for any given value of the variable is called a *multiple-valued* function. Irrational Algebraic Functions, Logarithmic Functions, the inverse or anti-Trigonometric Functions, and the Elliptic Integrals, are multiple-valued.

205. In Chapter II. we have explained the customary graphical method of representing an imaginary by the position of a point in a plane, the rectangular coördinates of the point being the real term and the real coefficient of the pure imaginary term of the imaginary in question.

In the ordinary treatment of the Theory of Functions this method of representation is of the greatest service, and enables us to bring the study of functions of imaginary variables within the province of Pure Geometry, and to give it great definiteness and precision.

For the sake of brevity we shall in future use the symbol i for $\sqrt{-1}$ and $\text{cis}\,\phi$ for $\cos\phi + \sqrt{-1}\sin\phi$, so that we shall write our typical imaginary as $x + yi$ or as $r\,\text{cis}\,\phi$, instead of using the longer forms $x + y\sqrt{-1}$, and $r(\cos\phi + \sqrt{-1}\sin\phi)$.

We shall also use the name *complex quantity* for an imaginary of the typical form when it is necessary to distinguish it from a pure imaginary.

206. A complex variable $z = x + yi$ is said to vary *continuously* when it varies in such a manner that the path traced by the point (x, y) representing it is a continuous line.

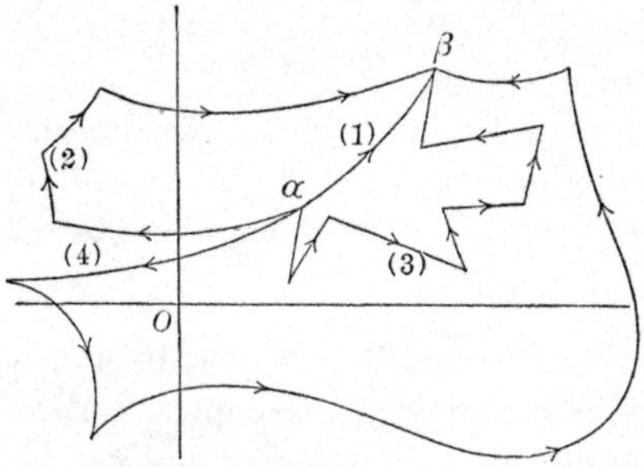

Thus if z changes from the value α to the value β, so that the point representing it traces any of the four lines in the figure, z varies continuously.

It will be seen that a variable can pass from the first to the second of two given values, real or imaginary, by any one of an infinite number of different paths without discontinuity if the variable in question is not restricted to real values; while a real variable can change continuously from one given value to another in but one way, since the point representing it is confined in its motion to the *axis of reals*.

207. A single-valued function w of a complex variable z is called a continuous function if the point representing it traces a continuous path whenever the point representing z traces a continuous path.

A multiple-valued function of z is continuous if each of the n points representing values corresponding to a value of z traces a continuous path whenever z traces a continuous path. These n paths are in general distinct, but two or more of them may intersect, a point of intersection corresponding to a value of z for which two or more of the n values of w, usually distinct, happen to coincide. Such a value of z is sometimes called a *critical value*, and the consideration of critical values plays an important part in the Theory of Functions.

In studying a multiple-valued function we may confine our attention to any one of its n values, and except for the possible presence of critical points this value may be treated just as we treat a single-valued function.

In representing graphically the changes produced in a function w by changing the variable z on which it depends, it is customary to avoid confusion by using separate sets of axes for w and z.

208. If we use the word *function* in its widest sense, $w = u + vi$ will be a function of a complex variable $z = x + yi$, if u and v are any given functions of x and y. For example,

$$xi,\ 6y,\ x^2 + y^2,\ x - yi,\ x^2 - y^2 + 2xyi,\ \frac{x - y + xi}{\sqrt{x^2 + y^2 + 4}},$$

may all be regarded as functions of z.

We have seen in Chapter II., Arts. 36–42, that with this definition of function the derivative with respect to z of a function of w is in general indeterminate; but that there are various functions of z, for instance, z^m, $\log z$, e^z, $\sin z$, where the derivative is not indeterminate. We are now ready to investigate more in detail the general question of the existence of a determinate derivative of a function of a complex variable.

Let $w = u + vi$ be a function of z; u and v, which are *real*, being functions of x and y.

Starting with the value $z_0 = x_0 + y_0 i$ of z and the corresponding value $w_0 = u_0 + v_0 i$ of w, let us change z by giving to x increment Δx without changing y.

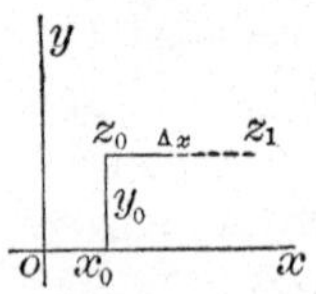

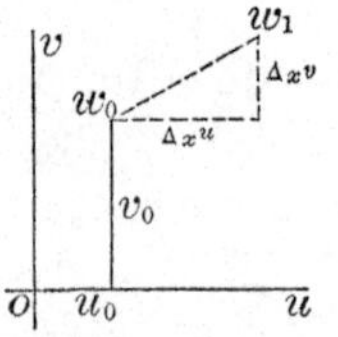

Let $\Delta_x u$ and $\Delta_x v$ be the corresponding increments of u and v; and z_1 and w_1 the new values of z and w.

We have $\qquad z_1 = z_0 + \Delta x, \quad w_1 = w_0 + \Delta_x u + i\Delta_x v.$

Then
$$\frac{w_1 - w_0}{z_1 - z_0} = \frac{\Delta_x u}{\Delta x} + \frac{i\Delta_x v}{\Delta x},$$

and the derivative of w with respect to z under the given circumstances is
$$\underset{\Delta z \doteq 0}{\text{limit}} \left[\frac{w_1 - w_0}{z_1 - z_0}\right] = D_x u + i D_x v. \qquad [1]$$

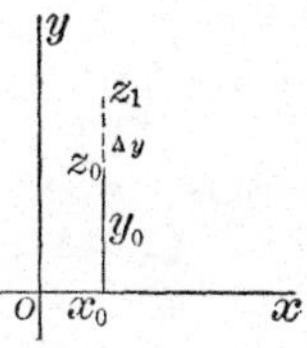

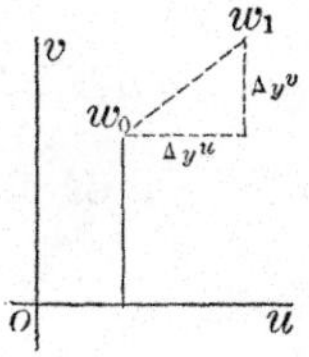

If, however, starting with the same value z_0 of z, we change z by giving y the increment Δy without changing x, we have

$$z_1 = x_0 + (y_0 + \Delta y)i = z_0 + i\Delta y,$$

$$w_1 = u_0 + \Delta_y u + (v_0 + \Delta_y v)i = w_0 + \Delta_y u + i\Delta_y v,$$

$$\frac{w_1 - w_0}{z_1 - z_0} = \frac{\Delta_y u}{i\Delta y} + \frac{i\Delta_y v}{i\Delta y},$$

and
$$\underset{\Delta z \doteq 0}{\text{limit}} \left[\frac{w_1 - w_0}{z_1 - z_0}\right] = D_y v - i D_y u, \qquad [2]$$

and this is the derivative of w with respect to z when we change y and do not change x.

Comparing [1] with [2], we see that if we start with a given value of z, and change z in the two different ways just considered, the limits of the ratios of the corresponding changes in w to the changes in z need not be the same. Indeed, the two values for $\frac{dw}{dz}$ given in [1] and [2] will not be the same unless $w = u + vi$ is such a function of $z = x + yi$ that

$$D_x u = D_y v \quad \text{and} \quad D_y u = - D_x v. \qquad [3]$$

We shall now show that if w is such a function of z that equations [3] are satisfied, $\lim_{\Delta z \doteq 0}\left[\frac{\Delta w}{\Delta z}\right]$ will be the same if we start with a given value z_0 of z, no mattter in what manner z may change; that is, no matter in what direction the point representing z may be supposed to move; or, in other words, no matter what may be the value of $\lim_{\Delta x \doteq 0}\left[\frac{\Delta y}{\Delta x}\right]$.

We have in general, since w is a function of the two variables x and y,

$$\Delta w = (D_x u + iD_x v)\,\Delta x + (D_y u + iD_y v)\,\Delta y + \epsilon,$$

where ϵ is an infinitesimal of higher order than Δx or Δy.

(I., Art. 198.)

$$\Delta z = \Delta x + i\,\Delta y.$$

Hence
$$\frac{\Delta w}{\Delta z} = \frac{D_x u\,.\,\Delta x + iD_y v\,.\,\Delta y + iD_x v\,.\,\Delta x + D_y u\,.\,\Delta y + \epsilon}{\Delta x + i\,\Delta y}$$

$$= \frac{D_x u + iD_y v\cdot\frac{\Delta y}{\Delta x} + iD_x v + D_y u\frac{\Delta y}{\Delta x} + \frac{\epsilon}{\Delta x}}{1 + i\frac{\Delta y}{\Delta x}},$$

and
$$\lim_{\Delta z \doteq 0}\left[\frac{\Delta w}{\Delta z}\right] = \frac{dw}{dz}$$

$$= \frac{D_x u + i D_y v\,.\lim_{\Delta x \doteq 0}\left[\frac{\Delta y}{\Delta x}\right] + i\left(D_x v - iD_y u\,.\lim_{\Delta x \doteq 0}\left[\frac{\Delta y}{\Delta x}\right]\right)}{1 + i\lim_{\Delta x \doteq 0}\left[\frac{\Delta y}{\Delta x}\right]}, \qquad [4]$$

a value involving $\lim_{\Delta x \doteq 0}\left[\frac{\Delta y}{\Delta x}\right]$, and therefore dependent upon the direction in which z is made to move.

If, however, [3] is satisfied, [4] reduces to

$$\frac{dw}{dz} = D_x u + iD_x v, \qquad [5]$$

and the derivative of w is independent of $\lim_{\Delta x \doteq 0}\left[\frac{\Delta y}{\Delta x}\right]$.

A function which satisfies equations [3], and which, therefore, has a derivative whose value depends only upon the value of the independent variable, and not upon the direction in which the point representing the variable is supposed to move, is called by some writers a *monogenic* function, by others a *function which has a derivative;* still others refuse to dignify with the name of function any other than *monogenic* functions.

209. Any function involving z as a whole, that is, any function which can be formed by performing an operation or a series of operations upon z as a whole, without introducing x and y except as they occur in z, is a monogenic function of z.

For if $$w = fz = f(x + yi),$$

where fz can be formed by operating upon z as a whole,

$$D_x w = f'z, \qquad \text{and} \qquad D_y w = if'z;$$

therefore $$iD_x w = D_y w, \qquad \text{or} \qquad iD_x(u + vi) = D_y(u + vi);$$

whence $$D_x u = D_y v, \qquad \text{and} \qquad D_y u = -D_x v;$$

and [3], Art. 208, is satisfied. Consequently w is monogenic. This accounts for the results of Arts. 38–42.

If w is a multiple-valued function of z, there may be several different values of $\frac{dw}{dz}$, corresponding to the same value of z; but if w is monogenic, each of these values depends only upon z, and not upon the way in which z is supposed to change.

In future, unless something is said to the contrary, we shall give the name function only to monogenic functions. Thus we shall not call such expressions as $x - yi$, or $x^2 + y^2 + 2xyi$, functions of z.

Conjugate Functions.

210. If u and v are functions of x and y, satisfying equations [3], Art. 208, it is easy to prove that

$$D_x^2 u + D_y^2 u = 0 \qquad \text{and} \qquad D_x^2 v + D_y^2 v = 0.$$

For since $D_x u = D_y v$ and $D_x v = - D_y u$,

we have $D_x^2 u = D_x D_y v$ and $D_y^2 u = - D_x D_y v$,

$D_x^2 v = - D_x D_y u$ and $D_y^2 v = D_x D_y u$;

u and v are then solutions of Laplace's equation,

$$D_x^2 V + D_y^2 V = 0. \qquad [1]$$

Any two functions ϕ and ψ of x and y, such that $\phi(x, y) + i\psi(x, y)$ is a monogenic function of $x + yi$, are called *conjugate* functions; and, by what has just been proved, each of a pair of conjugate functions is always a solution of Laplace's Equation [1].

Thus $x^2 - y^2$, $2xy$; $e^x \cos y$, $e^x \sin y$; $\frac{1}{2}\log(x^2 + y^2)$, $\tan^{-1}\frac{y}{x}$; are three pairs of conjugate functions, since $x^2 - y^2 + 2xyi = (x + yi)^2$, $e^x \cos y + ie^x \sin y = e^{x+yi}$, $\frac{1}{2}\log(x^2 + y^2) + i\tan^{-1}\frac{y}{x} = \log(x + yi)$, and consequently, by Art. 209, are all monogenic. Therefore each of the six functions at the beginning of this paragraph is a solution of Laplace's Equation [1].

It is clear that we can form pairs of conjugate functions at pleasure by merely forming functions of $x + yi$ and breaking them up into their real parts, and their pure imaginary parts; that is, throwing them into the typical form $u + vi$.

If each of a pair of conjugate functions, ϕ and ψ, is written equal to a constant, the equations thus formed will represent a pair of curves which intersect at right angles. For let (x, y) be a point of intersection of the curves $\phi = a$, $\psi = b$; the slopes of the two curves at (x, y) are respectively $-\frac{D_x\phi}{D_y\phi}$, $-\frac{D_x\psi}{D_y\psi}$ by I., Art. 202; and since $D_x\phi = D_y\psi$ and $D_x\psi = - D_y\phi$, the second slope is minus the reciprocal of the first, and the curves are perpendicular to each other at the point in question.

Thus $x^2 - y^2 = a$, $2xy = b$, cut each other orthogonally; as do also $\frac{1}{2}\log(x^2+y^2)=a$, $\tan^{-1}\frac{y}{x}=b$; or, what amounts to the same thing, $x^2+y^2=a_1$ $\frac{y}{x}=b_1$. It must be observed, however, that x^2+y^2 and $\frac{y}{x}$ are not conjugate functions, and that in general the converse of our proposition does not hold.

It may be easily proved that if ϕ and ψ are conjugate functions of x and y, and f and F are any second pair of conjugate functions of x and y, the new pair of functions formed by replacing x and y in ϕ and ψ by f and F respectively will be conjugate.

Thus $$(e^x\cos y)^2-(e^x\sin y)^2,\ 2e^x\cos y\,.\,e^x\sin y,$$

or, reducing, $$e^{2x}\cos 2y,\ e^{2x}\sin 2y,$$

are conjugate functions;

$$\tfrac{1}{2}\log\left[(x^2-y^2)^2+(2xy)^2\right],\ \tan^{-1}\left(\frac{2xy}{x^2-y^2}\right),$$

or, reducing, $$\log(x^2+y^2),\ \tan^{-1}\left(\frac{2xy}{x^2-y^2}\right),$$

are conjugate.

The properties of conjugate functions given in this article are of great importance in many branches of Mathematical Physics.

EXAMPLE.

Show that if x' and y' are conjugate functions of x and y, x and y are conjugate functions of x' and y'.

Preservation of Angles.

211. If w is a single-valued monogenic function of z, and the point representing z traces two arcs intersecting at a given angle, the corresponding arcs traced by the point representing w will in general intersect at the same angle.

For let z_0 be the point of intersection of the curves in the z plane, and w_0 the corresponding point in the w plane. Let z_1 be a point on the first curve, and z_2 a point on the second; and let

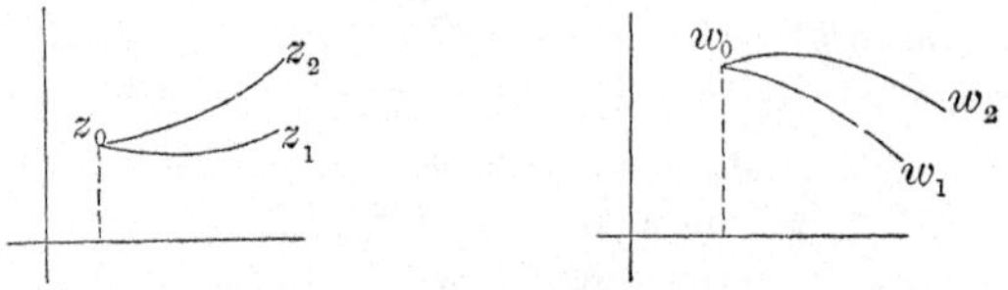

w_1 and w_2 be the corresponding points in the w figure.

Let r_1, r_2, s_1, and s_2 be the moduli of $z_1 - z_0$, $z_2 - z_0$, $w_1 - w_0$, and $w_2 - w_0$ respectively, ϕ_1, ϕ_2, ψ_1, and ψ_2 their arguments; then, since w is a monogenic function of z, we must have

$$\text{limit}\left[\frac{w_1 - w_0}{z_1 - z_0}\right] = \text{limit}\left[\frac{w_2 - w_0}{z_2 - z_0}\right],$$

or
$$\text{limit}\left[\frac{s_1 \text{ cis } \psi_1}{r_1 \text{ cis } \phi_1}\right] = \text{limit}\left[\frac{s_2 \text{ cis } \psi_2}{r_2 \text{ cis } \phi_2}\right];$$

whence, by Art. 23,

$$\text{limit}\left[\frac{s_1}{r_1} \text{ cis } (\psi_1 - \phi_1)\right] = \text{limit}\left[\frac{s_2}{r_2} \text{ cis } (\psi_2 - \phi_2)\right];$$

and since, when two imaginaries are equal, their moduli must be equal, and their arguments must be equal, unless the moduli are both zero or both infinite,

$$\text{limit } (\psi_2 - \psi_1) = \text{limit } (\phi_2 - \phi_1);$$

that is, the angle between the arcs in the w figure is equal to the angle between the corresponding arcs in the z figure; unless

$$\left[\frac{dw}{dz}\right]_{z=z_0} = 0, \quad \text{or} \quad \left[\frac{dw}{dz}\right]_{z=z_0} = \infty.$$

If w is a multiple-valued monogenic function of z, and if starting from any point z_0, the point which represents z traces out two curves intersecting at an angle α, each of the n points representing the corresponding values of w will trace out a pair of curves intersecting at the angle α; unless z_0 is a point at which $\frac{dw}{dz}$ is zero or infinite.

If, then, w is any monogenic function of z, and the point representing z is made to trace out any figure however complex, the point representing w will trace out a figure in which all the angles occurring in the z figure are preserved unchanged, except those having their vertices at points representing values of z which make $\frac{dw}{dz}$ zero or infinite.

This principle leads to the following working rule for transforming any given figure into another, in which the angles are preserved unchanged.

Substitute x' and y' for x and y in the equations of the curves which compose the given figure, x' and y' being any pair of conjugate functions (Art. 210) of x and y, and the new equations thus obtained will represent a set of curves forming a second figure in which all the angles of the given figure are preserved unchanged, except those having their vertices at points at which $D_x x'$ and $D_y y'$ are both zero, or at which one of them is infinite.

For example,
$$x - y = a, \tag{1}$$
$$x + y = b, \tag{2}$$
are a pair of perpendicular right lines. Replace x by $x^2 - y^2$ and y by $2xy$, and we get
$$x^2 - 2xy - y^2 = a, \tag{3}$$
$$x^2 + 2xy - y^2 = b, \tag{4}$$
a pair of hyperbolas that cut orthogonally.

212. If w is a *single-valued* continuous function of z, it is clear that if w_0 and w_1 are the values corresponding to z_0 and z_1,

and the point z moves from z_0 to z_1 by two different paths, the corresponding paths traced by w will begin at w_0 and end at w_1, and consequently that if z describes any closed contour, w also will describe a closed contour.

If w is a double-valued function of z, since to each value of z there will correspond two values of w, it is conceivable that if w_1 and w_1' are the values of w corresponding to z_1, and z moves from z_0 to z_1 by two different paths, w may in one case move from w_0 to w_1, and in the other case from w_0 to w_1'.

We can prove, however, that if the two paths traced by z do not enclose a *critical point* (Art. 207), and w is finite and continuous for the portion of the plane considered, this will not take place, and that the two paths starting from w_0 will terminate at the same point w_1.

For as z traces the first path, each of the two points representing the two values of w will trace a path, one starting at w_0, and the other at w_0', and unless the z path passes through a critical point, the two w paths will not intersect, but will be entirely separate and distinct, and will lead, one from w_0 to w_1, the other from w_0' to w_1'.

If, now, the z path be gradually swung into a second position without changing its beginning or its end, since w is a continuous function, the two w paths will be gradually swung into new positions; but, provided that the z path in its changing does not at any time pass through a critical point, the two w paths will at no time intersect, and consequently it will be impossible for the w points to pass over from one path to the other, and therefore the point which starts at w_0 must always come out at w_1, and not at w_1'.

It follows readily from this reasoning that if z describes a closed contour not embracing a critical point, each of the w points will describe a closed contour, and these contours will not intersect.

Of course, the proof given above holds for any multiple-valued function.

In any portion of the plane, then, not containing critical

points the separate values of a multiple-valued function may be separately considered, and may be regarded and treated as single-valued functions.

213. That in the case of a double-valued function two paths in the z plane, including between them a critical point, but having the same beginning and the same end, may lead to different values of the function, is easily shown by an example.

Let $w=\sqrt{z}$, and let z, starting with the value 1, move to the value -1 by the semi-circular path in the figure. That one of

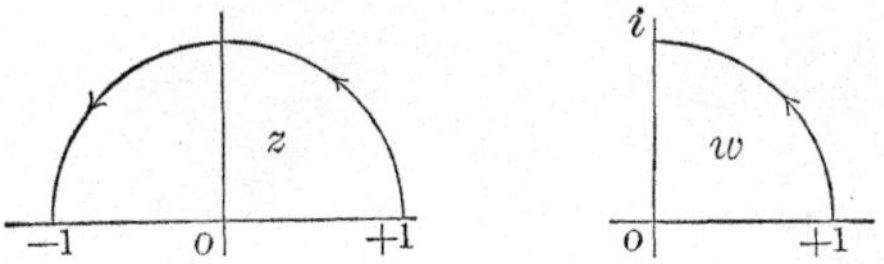

FIG. 1.

the corresponding values of w which starts with $+1$ will describe the quadrant shown in the figure, and will reach the point $1 . \text{cis}\frac{\pi}{2}$, or i. If, however, z moves from $+1$ to -1 by

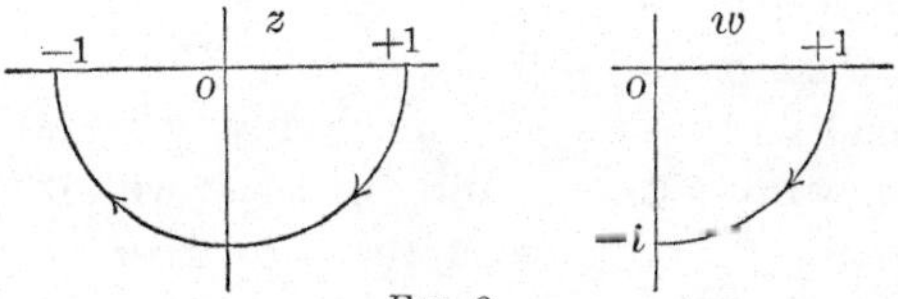

FIG. 2.

the semi-circular path in the second figure, the value of w which starts with $+1$ will describe the quadrant shown in the second figure, and will reach the value $1 . \text{cis}\left(-\frac{\pi}{2}\right)$, or $-i$. These two paths described by z, then, although beginning at the same point $+1$ and ending at the same point -1, cause that value of the function which begins with $+1$ to reach two different values; and the two paths in question embrace the point $z=0$, which is clearly a point at which the two values of w, ordinarily different, coincide; that is, a critical point.

It is easily seen that if z, starting with the value $+1$, describes a complete circumference about the origin, the value of w which starts from the point $+1$ will not describe a closed contour, but will move through a semi-circumference and end with the point $1 . \operatorname{cis} \pi$ or -1. Now, by Art. 212 any path

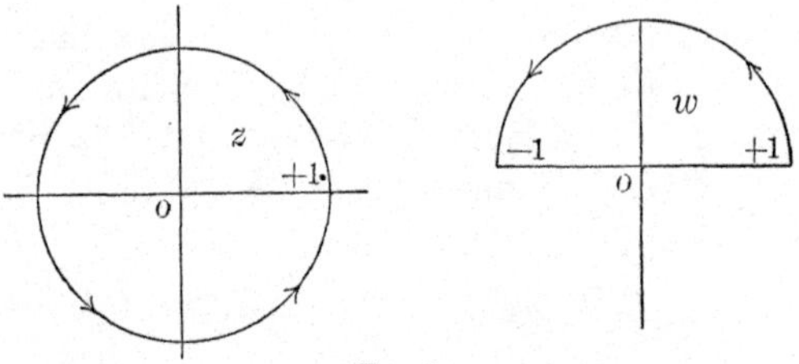

FIG. 3.

described by z beginning with $+1$ and ending with -1 and passing above the origin, since it can be deformed into the semi-circumference of Fig. 1 without passing through a critical point, will cause the value of w beginning with $+1$ to end with $+i$; and any path described by z beginning with $+1$ and ending with -1 and passing below the origin, since it can be deformed into the semi-circumference of Fig. 2 without passing through a critical point, will cause the value of w beginning with $+1$ to end with $-i$. Therefore any two paths described by z beginning with $+1$ and ending with -1 will, if they include the critical point $z=0$ between them, lead to different values of w, provided that the same value of w is taken at the start.

214. If w is a double-valued function of z, and z describes a closed contour about a single critical point, this contour may be deformed into a circle about the critical point, and a line leading from the starting point to the circumference of the circle, without affecting the final value of w (Art. 212). Thus, in the figure, the two paths $ABCDA$, $AB'C'D'B'A$ lead from the same initial to the same final value of w; and this is true no matter how small the radius of the circle $B'C'D'$.

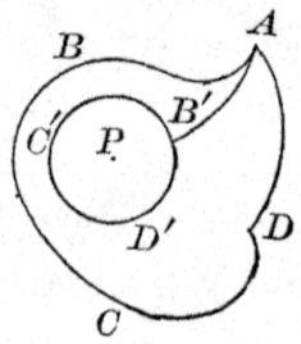

Let z_0 be the critical point, and let w_0 be the corresponding point in the w figure. As z moves from z_1 towards z_0, the points

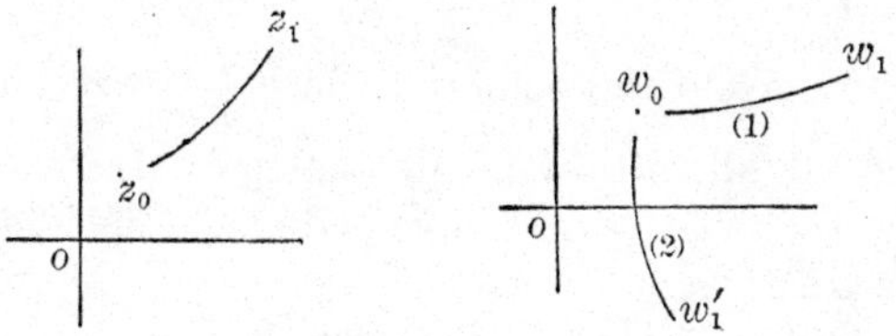

representing the corresponding values of w will start at w_1 and w_1' and move towards w_0, tracing distinct paths.

If, now, z describes a circumference about z_0, and then returns along its original path to z_1, the first value of w will either make a complete revolution about w_0 and return along the branch (1) to its initial value w_1, or it will describe about

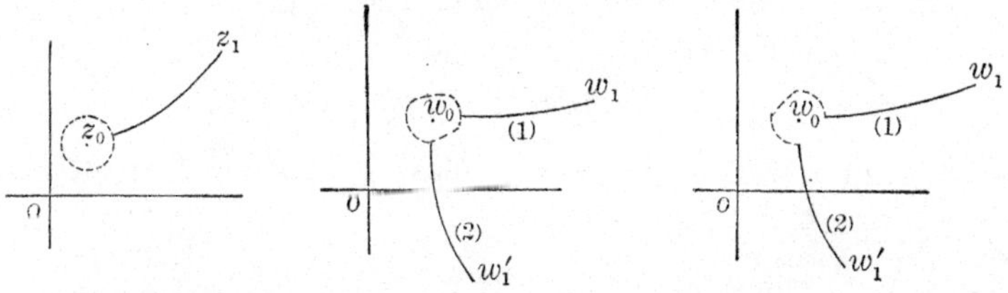

CASE I. CASE II.

w_0 a path ending with the branch (2) of the w curve, and move along that branch to the value w_1'.

In the first case, and in that case only, the value of w describes a closed contour when z describes a closed contour, and is practically a single-valued function.

If z_0 is a point at which $\frac{dw}{dz}$ is neither zero nor infinite (*v.* Art. 211), when z describes about z_0 a circle of infinitesimal radius, w will make about w_0 a complete revolution; for since if two radii are drawn from z_0, the curves corresponding to them will form at w_0 an angle equal to the angle between the radii, when a radius drawn to the moving point which is describing the circle about z_0 revolves through an angle of 360°, the corresponding line joining w_0 with the moving point representing w will revolve through 360°, and we shall have what we have called Case I.

If, then, we avoid the points at which $\frac{dw}{dz}$ is zero or infinite, we shall avoid all critical points that can vitiate the results obtained by treating our double-valued or multiple-valued functions as we treat single-valued functions.

A *critical point* of such a character that when z describes a closed contour about it the corresponding path traced by any one of the values of w is not closed, we shall call a *branch point.*

When a function is finite, continuous, and single-valued for all values of z lying in a given portion of the z plane, or when if multiple-valued it is finite and continuous, and has no *branch points* in the portion of the plane in question, it is said to be *holomorphic* in that portion of the plane.

Definite Integrals.

215. In the case of real variables, $\int_{z_0}^{Z} fz \,.\, dz$ was defined in Art. 80 in effect as follows:

$$\int_{z_0}^{Z} fz\,.\,dz = \underset{n=\infty}{\text{limit}} \left[fz_0(z_1 - z_0) + fz_1(z_2 - z_1) + fz_2(z_3 - z_2) + \cdots + fz_{n-1}(Z - z_{n-1}) \right], \qquad [1]$$

where $z_1, z_2, z_3, \ldots z_{n-1}$ are values of z dividing the interval between z_0 and Z into n parts, each of which is made to approach zero as its limit as n is indefinitely increased.

In other words, $\int_{z_0}^{Z}$ is the *line integral* of fz (Art. 163) taken along the straight line, joining z_0 and Z if z_0 and Z are represented as in the Calculus of Imaginaries.

It has been proved that if fz is finite and continuous between z_0 and Z, this integral depends merely upon the initial and final values of z, and is equal to $FZ - Fz_0$ where Fz is the indefinite integral $\int fz\,.\,dz$.

If z is a *complex* variable, and passes from z_0 to Z along any given path, we shall still define the definite integral $\int_{z_0}^{Z} fz\,.\,dz$ by [1] where now $z_1, z_2, z_3, \ldots z_{n-1}$ are points in the given path.

Two important results follow immediately from this definition:

1st. That
$$\int_{Z}^{z_0} fz\,.\,dz = -\int_{z_0}^{Z} fz\,.\,dz, \qquad [2]$$
if z traverses in each integral the same path connecting z_0 and Z.

2d. That the modulus of $\int_{z_0}^{Z} fz\,.\,dz$ is not greater than the line-integral of the modulus of fz taken along the given path joining z_0 and Z.

If we let
$$fz = w = u + vi,\ z = x + yi,\ u = \phi(x, y),\ \text{and}\ v = \psi(x, y),$$
then
$$\int_{z_0}^{Z} fz\,.\,dz = \int (u + vi)(dx + idy)$$
$$= \int \phi(x, y)\,dx + i\int \psi(x, y)\,dx - \int \psi(x, y)\,dy + i\int \phi(x, y)\,dy, \qquad [3]$$
each of the integrals in the last member being the line-integral of a real function of real variables, taken along the given path connecting z_0 and Z.

If the given path is changed, each of the integrals in the last member of [3] will in general change, and the value of $\int_{z_0}^{Z} fz\,.\,dz$ will change; and, since z may pass from z_0 to Z by an infinite number of different paths, we have no reason to expect that $\int_{z_0}^{Z} fz\,.\,dz$ will in general be determinate.

We shall, however, prove that in a large and important class of cases $\int_{z_0}^{Z} fz\,.\,dz$ is determinate, and depends for its value upon z_0 and Z, and not at all upon the nature of the path traversed by z in going from z_0 to Z.

216. If fz is *holomorphic* in a given portion of the plane,

$$\int_{z_0}^{z_0} fz\,.\,dz = 0 \qquad [1]$$

if z describes any closed contour lying wholly within that portion of the plane.

From [3], Art. 215, we have

$$\int_{z_0}^{z_0} fz\,.\,dz = \int w\,.\,dz = \int udx + i\int vdx - \int vdy + i\int udy, \qquad [2]$$

the integral in each case being the line-integral around the closed contour in question.

Since $w = fz$ is holomorphic, $u = \phi(x, y)$, and $v = \psi(x, y)$, and $D_x u$, $D_y u$, $D_x v$, and $D_y v$ are easily seen to be finite, continuous, and single-valued in the portion of the plane considered. Therefore, by Art. 170,

$$\int udx = \int\int D_y u dxdy; \qquad \int vdx = \int\int D_y v dxdy;$$

$$\int vdy = -\int\int D_x v dxdy; \quad \int udy = -\int\int D_x u dxdy;$$

the integral in the first member of each equation being taken around the contour, and that in the second member being a surface-integral taken over the surface bounded by the contour.

We have, then, from [2],

$$\int_{z_0}^{z_0} fz\,.\,dz = \int\int (D_y u + D_x v)dxdy + i\int\int(D_y v - D_x u)dxdy, \qquad [3]$$

but $D_x u = D_y v$, and $D_y u = - D_x v$ from [3], Art. 208. Therefore, [3] reduces to $\int_{z_0}^{z_0} fz\,.\,dz = 0$.

From this result we get easily the very important fact that if fz is holomorphic in a given portion of the plane, $\int_{z_0}^{Z} fz\,.\,dz$ will have the same value for all paths leading from z_0 to Z, provided they lie wholly in the given part of the plane. For let $z_0 aZ$ and $z_0 bZ$ be any two paths not intesecting between z_0 and Z. Then $z_0 aZbz_0$ is a closed contour, and

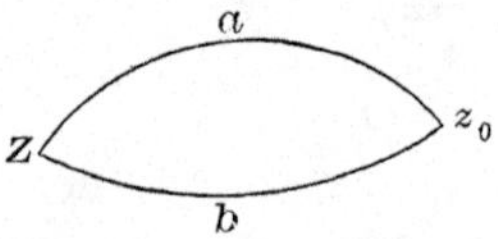

$$\int_{z_0}^{z_0} fz\,.\,dz \text{ (along } z_0aZbz_0)$$

$$= \int_{z_0}^{Z} fz\,.\,dz \text{ (along } z_0aZ) + \int_{Z}^{z_0} fz\,.\,dz \text{ (along } Zbz_0) = 0\,;$$

but $$\int_{Z}^{z_0} fz\,.\,dz \text{ (along } Zbz_0) = -\int_{z_0}^{Z} fz\,.\,dz \text{ (along } z_0bZ)$$

by Art. 215.

Therefore, $$\int_{z_0}^{Z} fz\,.\,dz \text{ (along } z_0aZ) = \int_{z_0}^{Z} fz\,.\,dz \text{ (along } z_0bZ).$$

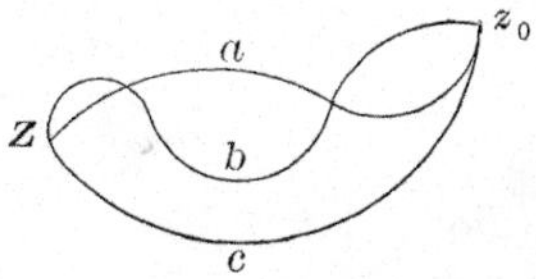

If the paths z_0aZ and z_0bZ intersect, a third path z_0cZ may be drawn not intersecting either of them, and by the proof just given

$$\int_{z_0}^{Z} fz\,.\,dz \text{ (along } z_0aZ) = \int_{z_0}^{Z} fz\,.\,dz \text{ (along } z_0cZ),$$

$$\int_{z_0}^{Z} fz\,.\,dz \text{ (along } z_0bZ) = \int_{z_0}^{Z} fz\,.\,dz \text{ (along } z_0cZ)\,;$$

therefore,

$$\int_{z_0}^{Z} fz\,.\,dz \text{ (along } z_0aZ) = \int_{z_0}^{Z} fz\,.\,dz \text{ (along } z_0bZ).$$

217. If fz, while in other respects holomorphic in a given portion of the plane, becomes infinite for a value T of z, then $\int fz\,.\,dz$ taken around a closed contour embracing T, while not zero, is, however, equal to the integral taken around any other closed path surrounding T.

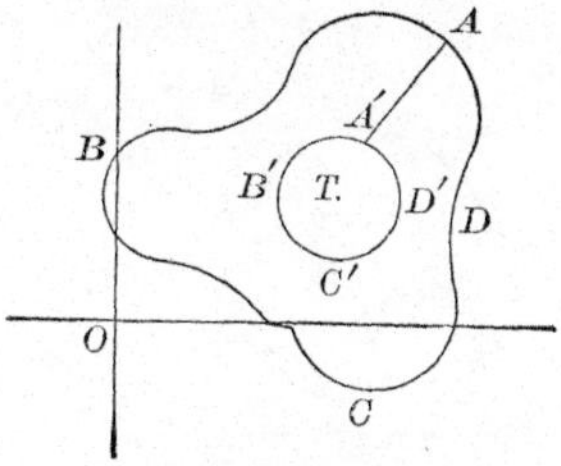

For let $ABCD$ be any closed contour about T. With T as a centre, and a radius ϵ, describe a circumference, taking ϵ so small that the circumference lies wholly within $ABCD$. Join the two contours by a line AA'. Then $ABCDAA'D'C'B'A'A$ is a closed path within which fz is holomorphic.

Therefore,

$$\int fz\,.\,dz \text{ (along } ABCDAA'D'C'B'A'A) = 0,$$

or $$\int fz\,.\,dz \text{ (along } ABCDA) \quad + \int fz\,.\,dz \text{ (along } AA')$$

$$+ \int fz\,.\,dz \text{ (along } A'D'C'B'A') + \int fz\,.\,dz \text{ (along } A'A) = 0\,;$$

but

$$\int fz\,.\,dz \text{ (along } AA') \qquad = - \int fz\,.\,dz \text{ along } (A'A),$$

and

$$\int fz\,.\,dz \text{ (along } A'D'C'B'A') = - \int fz\,.\,dz \text{ (along } A'B'C'D'A')$$

Hence

$$\int fz\,.\,dz \text{ (along } ABCDA) \quad = \quad \int fz\,.\,dz \text{ (along } A'B'C'D'A').$$

218. That the integral of a function of z around a closed contour embracing a point at which the function is infinite is not necessarily zero is easily shown by an example.

$fz = \dfrac{1}{z-t}$, t being a given constant, is single-valued, continuous, and finite throughout the whole of the plane except at the point t, at which $\dfrac{1}{z-t}$ becomes infinite, without, however, ceasing to be single-valued.

Let us take $\int \dfrac{dz}{z-t}$ around a circle whose centre is t, and whose radius is any arbitrarily chosen value ϵ. If z is on the circumference of this circle

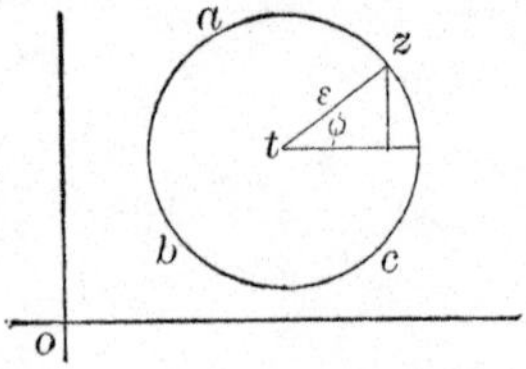

$$z - t = \epsilon\,(\cos\phi + i\sin\phi)$$

$$= \epsilon e^{\phi i} \qquad \text{by [5], Art. 31.}$$

$$z = t + \epsilon e^{\phi i}$$

and $$dz = i\epsilon e^{\phi i}\,d\phi.$$

Hence $$\int \frac{dz}{z-t} \text{ (around } abc) = \int_0^{2\pi} \frac{i\epsilon e^{\phi i}\,d\phi}{\epsilon e^{\phi i}} = 2\pi i.$$

From what has been proved in Art. 217, it follows that $\int \frac{dz}{z-t}$ around any closed contour embracing t must also be equal to $2\pi i$.

As another example let us consider $\int \frac{Fz}{z-t}dz$, when Fz is supposed to be holomorphic in the portion of the plane considered, and where the integral is to be taken around any closed contour embracing the point $z=t$.

$\frac{Fz}{z-t}$ is holomorphic except at the point $z=t$, where it becomes infinite. The required integral is, then, equal to the integral around a circumference described from the point t as a centre, with any given radius ϵ, that is, by the reasoning just used in the case of $\int \frac{dz}{z-t}$, to

$$\int_0^{2\pi} \frac{F(t+\epsilon e^{\phi i})\,i\epsilon e^{\phi i}\,d\phi}{\epsilon e^{\phi i}}, \quad \text{or} \quad i\int_0^{2\pi} F(t+\epsilon e^{\phi i})\,d\phi;$$

and in this expression ϵ may be taken at pleasure. If now ϵ is made infinitesimal $\epsilon e^{\phi i}$ is infinitesimal, and since Fz is continuous $F(t+\epsilon e^{\phi i})$ is equal to $Ft+\eta$ where η is some infinitesimal, and $F(t+\epsilon e^{\phi i})\,d\phi$ is equal to $Ft\,.\,d\phi+\eta\,.\,d\phi$.

Now, by I. Art. 161,

$$\int_0^{2\pi} (Ft\,.\,d\phi+\eta d\phi) = \int_0^{2\pi} Ft\,.\,d\phi.$$

Hence $$i\int_0^{2\pi} F(t+\epsilon e^{\phi i})\,d\phi = i\int_0^{2\pi} Ft\,.\,d\phi = 2\pi i Ft;$$

and we get the important result that $\int \frac{Fz}{z-t}dz$, taken around any contour including the point $z=t$, is equal to $2\pi i\,.\,Ft$.

From this we have $$Ft = \frac{1}{2\pi i}\int \frac{Fz}{z-t}dz;$$

and we see that *a holomorphic function is determined everywhere inside a closed contour if its value is given at every point of the contour.*

If in the formula $$Ft = \frac{1}{2\pi i}\int \frac{Fz}{z-t}\,dz \qquad [1]$$

we change t to $t+\Delta t$, we get

$$\Delta Ft = \frac{1}{2\pi i}\int Fz\,.\,dz\left(\frac{1}{z-t-\Delta t}-\frac{1}{z-t}\right) = \frac{1}{2\pi i}\int \frac{Fz\,.\,dz\,.\,\Delta t}{(z-t)(z-t-\Delta t)};$$

whence

$$\underset{\Delta t \doteq 0}{\text{limit}}\left[\frac{\Delta Ft}{\Delta t}\right] = \frac{1}{2\pi i}\int Fz\,.\,dz\,.\underset{\Delta t \doteq 0}{\text{limit}}\left[\frac{1}{(z-t)(z-t-\Delta t)}\right],$$

or $$F't = \frac{1}{2\pi i}\int \frac{Fz\,.\,dz}{(z-t)^2}; \qquad [2]$$

and in like manner we get

$$F''t = \frac{2\,!}{2\pi i}\int \frac{Fz\,.\,dz}{(z-t)^3}; \qquad [3]$$

and in general $$F^{(n)}t = \frac{n\,!}{2\pi i}\int \frac{Fz\,.\,dz}{(z-t)^{n+1}}, \qquad [4]$$

each of the integrals in these formulas being taken around a closed contour lying wholly in that portion of the plane in which Fz is holomorphic, and enclosing the point $z=t$.

219. *The integral of a holomorphic function along any given path is finite and determinate*, for, by [3], Art. 215, it is equal to the sum of four line integrals, each of which is finite and determinate (Art. 166).

If a series $w_0 + w_1 + w_2 + \cdots$, *where* w_0, w_1, $w_2 \cdots$ *are holomorphic functions of* z, *is convergent for all values of* z *in a certain portion of the plane, the integral of the series* along any given path lying in that portion of the plane *is the series formed of the integrals of the terms of the given series* along the path in question, *and the new series is convergent.*

For let $$S = w_0 + w_1 + w_2 + \cdots + w_n + w_{n+1} + \cdots$$
$$= w_0 + w_1 + w_2 + \cdots + w_n + R,$$
where $$R = w_{n+1} + w_{n+2} + \cdots,$$
and where by hypothesis $\underset{n=\infty}{\text{limit}} [R] = 0.$ (1)

$$\int Sdz = \int w_0\, dz + \int w_1\, dz + \cdots + \int w_n\, dz + \int R \,.\, dz$$

for any given value of n.

By the proposition at the beginning of this article, $\int Sdz$ along the given path is finite and determinate, as are also

$$\int w_0\, dz, \int w_1\, dz, \text{ etc.}$$

If, now, n is indefinitely increased,

$$\int Sdz = \int w_0\, dz + \int w_1\, dz + \int w_2\, dx + \cdots + \underset{n=\infty}{\text{limit}} \int Rdz.$$

The modulus of $\int Rdz$ is not greater than the line-integral along the given path of the modulus of R (v. Art. 215). As n increases each value of the modulus of R approaches zero as its limit (1); consequently each element of the cylindrical surface representing the line-integral of the modulus of R (v. Art. 166) approaches zero, and $\underset{n=\infty}{\text{limit}} \int Rdz = 0.$

Therefore, $$\int Sdz = \int w_0\, dz + \int w_1\, dz + \int w_2\, dz + \cdots; \qquad [2]$$

and, since the first member is finite and determinate, the second member is a *convergent* series.

Taylor's and Maclaurin's Theorems.

220. $$\frac{1-q^n}{1-q} = 1 + q + q^2 + q^3 + \cdots + q^{n-1}$$

identically, if n is a positive integer, even when q is imaginary.

If the modulus of q is less than 1,

$$\underset{n=\infty}{\text{limit}}\,[q^n]=0.$$

Hence $$1+q+q^2+q^3+\cdots=\underset{n=\infty}{\text{limit}}\left[\frac{1-q^n}{1-q}\right]=\frac{1}{1-q}, \quad [1]$$

even when q is imaginary, provided that the modulus of q is less than 1.

Suppose, now, that everywhere within and on a certain circumference described with the point $z=a$ as a centre Fz is holomorphic. Let $z=t$ be any point within this circumference, and $z=Z$ be a point on the circumference. Then the modulus of $Z-a$ is the distance from a to Z, and the modulus of $t-a$ is the distance from a to t;

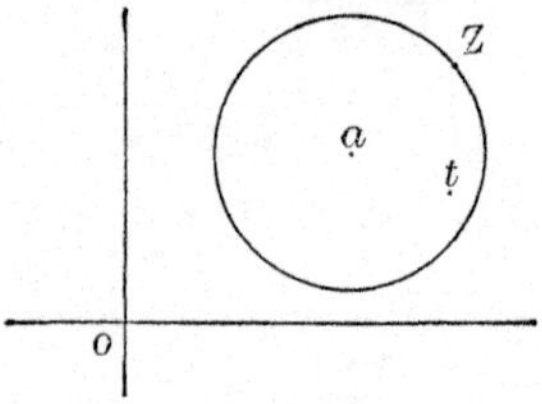

hence $\text{mod}\,(t-a)<\text{mod}\,(Z-a)$,

and $\text{mod}\left(\dfrac{t-a}{Z-a}\right)<1.$

$$\frac{1}{Z-t}=\frac{1}{Z-a-(t-a)}=\frac{1}{Z-a}\cdot\frac{1}{1-\dfrac{t-a}{Z-a}},$$

$$=\frac{1}{Z-a}\left[1+\frac{t-a}{Z-a}+\frac{(t-a)^2}{(Z-a)^2}+\frac{(t-a)^3}{(Z-a)^3}+\cdots\right],$$

by [1]

Hence

$$\frac{1}{Z-t}=\frac{1}{Z-a}+\frac{t-a}{(Z-a)^2}+\frac{(t-a)^2}{(Z-a)^3}+\frac{(t-a)^3}{(Z-a)^4}+\cdots, \quad [2]$$

and the second member of [2] is a convergent series.

Multiply [2] by $\dfrac{FZ}{2\pi i}$, and the series will still be convergent for each value of z which we have to consider; we get

$$\frac{1}{2\pi i}\,\frac{FZ}{Z-t}$$

$$=\frac{1}{2\pi i}\left[\frac{FZ}{Z-a}+(t-a)\,\frac{FZ}{(Z-a)^2}+(t-a)^2\frac{FZ}{(Z-a)^3}+\cdots\right]. \quad [3]$$

Integrate now both members of [3] around the circumference, and we have

$$\frac{1}{2\pi i}\int\frac{FZ}{Z-t}dZ=\frac{1}{2\pi i}\left[\int\frac{FZ}{Z-a}dZ+(t-a)\int\frac{FZ}{(Z-a)^2}dZ\right.$$

$$\left.+(t-a)^2\int\frac{FZ}{(Z-a)^3}dZ+\cdots\right]; \qquad [4]$$

and, since each of the functions to be integrated is holomorphic on the contour around which the integral is taken, and the second member of [3] is convergent, each integral will be finite and determinate, and the second member of [4] will be convergent.

Substituting in [4] the values obtained in Art. 218, [1], [2], [3], and [4], we have

$$Ft=Fa+(t-a)\,F'a+\frac{(t-a)^2}{2\,!}F''a+\frac{(t-a)^3}{3\,!}F'''a+\cdots$$

$$+\frac{(t-a)^n}{n\,!}F^{(n)}a+\cdots. \qquad [5]$$

If the point $z=a$ is at the origin, $a=0$ and [5] becomes

$$Ft=Fo+tF'o+\frac{t^2}{2\,!}F''o+\frac{t^3}{3\,!}F'''o+\cdots, \qquad [6]$$

which is Maclaurin's Theorem.

That [5] is merely a new form of Taylor's Theorem is easily seen if we let $t-a=h$, whence $t=a+h$, and [5] becomes

$$F(a+h)=Fa+h\,F'a+\frac{h^2}{2\,!}F''a+\frac{h^3}{3\,!}F'''a+\cdots. \qquad [7]$$

[6] can, of course, be written

$$Fz=Fo+zF'o+\frac{z^2}{2\,!}F''o+\frac{z^3}{3\,!}F'''o+\cdots, \qquad [8]$$

and [5] as

$$Fz=Fa+(z-a)\,F'a+\frac{(z-a)^2}{2\,!}F''a+\frac{(z-a)^3}{3\,!}F'''a+\cdots; \qquad [9]$$

and we get the very important result that *if a function of* z *is holomorphic within a circle whose centre is at the origin it may be developed by Maclaurin's Theorem, and the development will hold*, that is, *the series will be convergent, for all values of* z *lying within the circle.*

If a function of z is holomorphic within a circle described from $z=a$ as a centre it can be developed by Taylor's Theorem into a series arranged according to powers of $z-a$, and the development will hold for all values of z lying within the circle.

The question of the convergency of either Taylor's or Maclaurin's Series for the case when z lies on the circumference of the circle needs special investigation, and will not be considered here.

If the function which we wish to develop is single-valued, in drawing our circle of convergence we need avoid only those points at which the function becomes infinite; but if it is multiple-valued we must avoid also those at which its derivative is zero or infinite (v. Art. 214).

221. We are now able to investigate from a new point of view the question of the convergence of the series obtained by Taylor's and Maclaurin's Theorems in I. Chap. IX.

Let us begin with the *Binomial Theorem*,

$$(a)\quad (a+h)^n = a^n + na^{n-1}h + n\frac{(n-1)}{2\,!}a^{n-2}h^2 + \cdots, \qquad [1]$$

or, following the notation of [9], Art. 220,

$$z^n = a^n + na^{n-1}(z-a) + \frac{n(n-1)}{2\,!}a^{n-2}(z-a)^2 + \cdots. \qquad [2]$$

If n is a positive integer, z^n is holomorphic throughout the whole plane, and [2] holds for all values of z and a, and [1] for all values of a and h.

If n is a negative integer, z^n is single-valued, and it is finite and continuous except for $z=0$, where z^n becomes infinite. [2] is, then, convergent for all values of z lying within a circle described with a as a centre and passing through the origin;

that is, for all values of z, such that $mod\,(z-a) < mod\,a$; and consequently [1] holds if $mod\,h < mod\,a$.

If n is a fraction, z^n is multiple-valued, and our circle of convergence must avoid the points at which $\frac{dz^n}{dz}$ becomes zero or infinite; but as the origin is the only point of this character, the circle of convergence is the same as in the case last considered, and [1] holds for all cases where $mod\,h < mod\,a$.

When a and h are real our results agree with those obtained in I. Art. 131.

(b) $$e^z = e^{x+yi} = e^x(\cos y + i\sin y) \qquad \text{([4], Art. 31)}$$

is single-valued and continuous, and becomes infinite only when $x = \infty$. Maclaurin's development for e^z holds, then, for all finite values of z.

(c) $$\log z = \log(r\,\mathrm{cis}\,\phi) = \log r + \phi i \qquad \text{(Art. 33)}$$

is finite and continuous throughout the whole plane. It is, however, multiple-valued, but its derivative $\frac{1}{z}$ becomes infinite only when $z = 0$, and does not become zero for any finite value of z. $\log z$, then, can be developed into a convergent series, arranged according to powers of $z - a$, for all values of z within a circle having the centre a and passing through the origin; that is, for all cases where $mod\,(z-a) < mod\,a$.

If $z - a = h$, we get

$$\log(a+h) = \log a + \frac{h}{a} - \frac{h^2}{2a^2} + \frac{h^3}{3a^3} - \frac{h^4}{4a^4} + \cdots, \qquad [3]$$

[3] holding for all cases where $mod\,h < mod\,a$.

If $a = 1$ and $h = z$, we get

$$\log(1+z) = \frac{z}{1} - \frac{z^2}{2} + \frac{z^3}{3} - \frac{z^4}{4} + \cdots, \qquad [4]$$

which holds for all values of z where $mod\,z < 1$.

(d) $$\sin z = \sin(x+yi) = \sin x \cdot \frac{e^y + e^{-y}}{2} + i\cos x \cdot \frac{e^y - e^{-y}}{2},$$

and $\cos z = \cos(x + yi) = \cos x \cdot \frac{e^y + e^{-y}}{2} - i \sin x \cdot \frac{e^y - e^{-y}}{2}$,
(v. [3] and [4], Art. 35)

are single-valued, and are finite and continuous throughout the plane. Therefore, Maclaurin's developments for $\sin z$ and $\cos z$ hold for all values of z.

(*e*) $\tan z = \frac{\sin z}{\cos z}$, and $\sec z = \frac{1}{\cos z}$, are single-valued and continuous, and become infinite only when $\cos z = 0$; that is, when $z = \frac{\pi}{2}$. Therefore, Maclaurin's developments for $\tan z$ and $\sec z$ (I. Art. 138), hold for every value of z whose modulus is less than $\frac{\pi}{2}$.

(*f*) $\text{ctn}\, z = \frac{\cos z}{\sin z}$, and $\csc z = \frac{1}{\sin z}$ become infinite when $z = 0$, and cannot be developed by Maclaurin's Theorem.

(*g*) $\sin^{-1} z$ is finite and continuous throughout the plane; it is, however, multiple-valued, and its derivative $\frac{1}{\sqrt{1 - z^2}}$ becomes infinite when $z = 1$, and when $z = -1$. Therefore, the development for $\sin^{-1} z$ (I. Art. 135 [2]), holds for any value of z whose modulus is less than 1.

(*h*) $\tan^{-1} z$ is finite and continuous throughout the plane; it is multiple-valued, and its derivative $\frac{1}{1 + z^2}$ becomes infinite when $z = i$, and when $z = -i$. Therefore, the development for $\tan^{-1} z$ (I. Art. 135 [1]), holds if $mod\, z < mod\, i$; that is, if $mod\, z < 1$.

EXAMPLES.

(1) Show that the development of $\frac{z}{1 + z} + \log(1 + z)$, given in I. Art. 136, Ex. 1, holds if $\text{mod}\, z < 1$.

(2) Show that the development of $\log(1 + e^z)$, given in I. Art. 136, Ex. 2, holds if $\text{mod}\, z < \pi$.

(3) Obtain the following developments, and find for what real values of x they hold good:

(a) $\log(x+\sqrt{x^2+a^2}) = \log a + \frac{x}{a} - \frac{1}{2}\cdot\frac{x^3}{3a^3} + \frac{1.3}{2.4}\cdot\frac{x^5}{5a^5} - \cdots$

(b) $(e^x + e^{-x})^n = 2^n\left(1 + \frac{nx^2}{2!} + n(3n-2)\frac{x^4}{4!} + \cdots\right).$

(c) $e^x . \cos x = 1 + x - \frac{2x^3}{3!} - \frac{4x^4}{4!}\cdots$

(d) $e^{\cos x} = e\left(1 - \frac{x^2}{2!} + \frac{4x^4}{4!} - \frac{31x^6}{6!}\cdots\right).$

(e) $e^x \log(1+x) = x + \frac{x^2}{2!} + \frac{2x^3}{3!} + \frac{9x^5}{5!}\cdots$

(f) $\tan^4 x = x^4 + \frac{4x^6}{3} + \frac{6x^8}{5}\cdots$

(g) $(1+2x+3x^2)^{-\frac{1}{2}} = 1 - x + 2x^3 - \frac{7x^4}{4}\cdots$

(h) $e^{\tan^{-1}x} = 1 + x + \frac{x^2}{2} - \frac{x^3}{6}\cdots$

(i) $\log\left(\frac{1+x}{1-x}\right) = 2\left(x + \frac{x^3}{3} + \frac{x^5}{5}\cdots\right).$

(j) $\tan x = x + \frac{2}{3!}x^3 + \frac{2^4}{5!}x^5 + \frac{2^4.17}{7!}x^7 + \frac{2^8.31}{9!}x^9 + \frac{2^9.691}{11!}x^{11}\cdots$

(k) $x.\text{ctn}\, x = 1 - \frac{x^2}{3} - \frac{x^4}{45} - \frac{2x^6}{945} - \frac{x^8}{4725} - \frac{2x^{10}}{93555}\cdots$

(l) $\log\tan\left(\frac{\pi}{4}+x\right) = \log\tan\frac{\pi}{4} + 2x + \frac{2^3}{3!}x^3 + \frac{5.2^5}{5!}x^5 + \frac{61.2^7}{7!}x^7\cdots$

(m) $e^{\sin x} = 1 + \frac{x}{1!} + \frac{x^2}{2!} - \frac{3x^4}{4!} - \frac{8x^5}{5!} - \frac{3x^6}{6!} + \frac{56x^7}{7!} + \cdots$

(n) $e^{\tan x} = 1 + x + \frac{x^2}{2!} + \frac{3x^3}{3!} + \frac{9x^4}{4!} + \frac{37x^5}{5!} + \frac{177x^6}{6!}.$

(o) $(\text{ver}\sin^{-1}x)^2 = 2\left(x + \frac{x^2}{3.2} + \frac{1.2}{3.5}\,\frac{x^3}{3} + \frac{1.2.3}{3.5.7}\,\frac{x^4}{4} + \cdots\right).$

(p) $\frac{1}{2-2x+x^2} = \frac{2}{4} + \frac{2x}{4} + \frac{x^2}{4} - \frac{2x^4}{4^2} - \frac{2x^5}{4^2} - \frac{x^6}{4^2}\cdots$

(q) $\frac{1+x}{6-5x+x^2} = \frac{3}{2-x} - \frac{4}{3-x} = \left(\frac{3}{2} - \frac{4}{3}\right) + \left(\frac{3}{2^2} - \frac{4}{3^2}\right)x + \cdots$

Answers.

(a) $-a < x < a$;

(b) $-\infty < x < \infty$ if $n > 0$, $-\frac{\pi}{2} < x < \frac{\pi}{2}$ if $n < 0$;

(c) $-\infty < x < \infty$;

(d) $-\infty < x < \infty$;

(e) $-1 < x < 1$;

(f) $-\frac{\pi}{2} < x < \frac{\pi}{2}$;

(g) $-\frac{1}{3} < x < \frac{1}{3}$;

(h) $-1 < x < 1$;

(i) $-1 < x < 1$;

(j) $-\frac{\pi}{2} < x < \frac{\pi}{2}$;

(k) $-\pi < x < \pi$;

(l) $-\frac{\pi}{4} < x < \frac{\pi}{4}$;

(m) $-\infty < x < \infty$;

(n) $-\frac{\pi}{2} < x < \frac{\pi}{2}$;

(o) $-2 < x < 2$;

(p) $-\sqrt{2} < x < \sqrt{2}$;

(q) $-2 < x < 2$.

CHAPTER XVIII.

KEY TO THE SOLUTION OF DIFFERENTIAL EQUATIONS.

222. In this chapter an analytical key leads to a set of concise, practical rules, embodying most of the ordinary methods employed in solving differential equations; and the attempt has been made to render these rules so explicit that they may be understood and applied by any one who has mastered the Integral Calculus proper.

The key is based upon "Boole's Differential Equations" (London: Macmillan & Co.), to which or to "Forsyth's Differential Equations" (London: Macmillan & Co.), we refer the student who wishes to become familiar with the theoretical considerations upon which the working rules are based.

223. A *differential equation* is an expressed relation involving derivatives with or without the primitive variables from which they are derived.

For example:

$$(1+x)\,y+(1-y)\,x\frac{dy}{dx}=0, \tag{1}$$

$$x\frac{dy}{dx}-ay=x+1, \tag{2}$$

$$x\frac{dy}{dx}-y+x\sqrt{x^2-y^2}=0, \tag{3}$$

$$\left(\frac{dy}{dx}\right)^3=y^4\left(y+x\frac{dy}{dx}\right), \tag{4}$$

$$\frac{d^4y}{dx^4}-2\frac{d^3y}{dx^3}+2\frac{d^2y}{dx^2}-2\frac{dy}{dx}+y=1, \tag{5}$$

$$\sin^2 x \frac{d^2 y}{dx^2} + \sin x \cos x \frac{dy}{dx} - y = x - \sin x, \tag{6}$$

$$x(1-x)^2 \frac{d^2 y}{dx^2} - 2y = 0, \tag{7}$$

$$\frac{d^2 y}{dx^2} + \left(1 - \frac{2}{x^2}\right) y = 0, \tag{8}$$

$$D_x^2 z - a^2 D_y^2 z = 0, \tag{9}$$

are differential equations.

The *order* of a differential equation is the same as that of the derivative of highest order which appears in the equation.

Equations (1), (2), (3), and (4) are of the first order; (6), (7), (8), and (9) of the second order; and (5) of the fourth order.

The *degree* of a differential equation is the same as the power to which the derivative of highest order in the equation is raised, that derivative being supposed to enter into the equation in a rational form.

Equations (1), (2), (3), (5), (6), (7), (8), and (9) are all of the first degree; (4) is of the third degree.

A differential equation is *linear* when it would be of the first degree if the dependent variable and all its derivatives were regarded as unknown quantities.

Equations (2), (5), (6), (7), (8), and (9) are *linear*.

The equation not containing differentials or derivatives, and expressing the most general relation between the primitive variables consistent with the given differential equation, is called its *general solution* or *complete primitive*. A general solution will always contain arbitrary constants or arbitrary functions.

The differential equation is formed from the complete primitive by direct differentiation, or by differentiation and the subsequent elimination of constants or functions between the primitive and the derived equations.

If it has been formed by differentiation *only* without subsequent elimination or reduction, the differential equation is said to be *exact*.

A *singular solution* of a differential equation is a relation between the primitive variables which satisfies the differential equation by means of the values which it gives to the derivatives, but which cannot be obtained from the complete primitive by giving particular values to the arbitrary constants.

224. We shall illustrate the use of the key by solving equations (1), (2), (3), (4), (5), (6), (7), (8), and (9) of Art. 223 by its aid.

(1) $(1+x)y+(1-y)x\frac{dy}{dx}=0$, or $(1+x)ydx+(1-y)xdy=0$.

Beginning at the beginning of the key, we see that we have a single equation, and hence look under I., p. 310; it involves ordinary derivatives: we are then directed to II., p. 310; it contains two variables: we go to III., p. 310; it is of the first order, IV., p. 310, and of the first degree, V., p. 310.

It is reducible to the form

$$\frac{1+x}{x}dx+\frac{1-y}{y}dy=0,$$

which comes under $Xdx+Ydy=0$.

Hence we turn to (1), p. 314, and there find the specific directions for its solution. Integrating each term separately, we get

$$\log x+x+\log y-y=c, \quad \text{or} \quad \log(xy)+x-y=c,$$

the required primitive equation.

(2) $$x\frac{dy}{dx}-ay=x+1.$$

Beginning again at the beginning of the key, we are directed through I., II., III., IV., to V., p. 310. Looking under V., we see that it will come under either the third or the fourth head. Let us try the fourth; we are referred to (4), p. 314, for specific directions.

Obeying instructions, the work is as follows:

$$x\frac{dy}{dx}-ay=0,$$

$$xdy - aydx = 0,$$

$$\frac{dy}{y} - \frac{adx}{x} = 0,$$

$$\log y - a\log x = c,$$

$$\log\frac{y}{x^a} = c;$$

$$\frac{y}{x^a} = C,$$

$$y = Cx^a, \qquad (1)$$

$$\frac{dy}{dx} = aCx^{a-1} + x^a\frac{dC}{dx}.$$

Substitute in the given equation,

$$aCx^a + x^{a+1}\frac{dC}{dx} - aCx^a = x + 1,$$

$$x^{a+1}\frac{dC}{dx} - (x+1) = 0,$$

$$dC - \frac{x+1}{x^{a+1}}dx = 0,$$

$$C + \frac{1}{(a-1)x^{a-1}} + \frac{1}{ax^a} = C'.$$

Substitute this value for C in (1), and we get

$$y = C'x^a - \left(\frac{1}{a} + \frac{x}{a-1}\right),$$

the required primitive.

(3) $$x\frac{dy}{dx} - y + x\sqrt{x^2 - y^2} = 0.$$

Beginning at the beginning of the key, we are directed through I., II., III., IV., to V., page 310. Looking under V., we find that our equation does not come under any of the special forms there given. We are consequently driven to obtaining a solution in the form of a series, and for specific instructions we are referred to (13), page 316. Obeying these, our process is the following:

$$\frac{dy}{dx} = \frac{y}{x} - \sqrt{x^2 - y^2}, \qquad \frac{dy_0}{dx_0} = \frac{y_0}{x_0} - \sqrt{x_0^2 - y_0^2},$$

$$\frac{d^2 y}{dx^2} = -y - \frac{2}{x}\sqrt{x^2 - y^2}, \qquad \frac{d^2 y_0}{dx_0^2} = -y_0 - \frac{2}{x_0}\sqrt{x_0^2 - y_0^2},$$

$$\frac{d^3 y}{dx^3} = -\frac{3y}{x} + \sqrt{x^2 - y^2}, \qquad \frac{d^3 y_0}{dx_0^3} = -\frac{3y_0}{x_0} + \sqrt{x_0^2 - y_0^2},$$

$$\frac{d^4 y}{dx^4} = y + \frac{4}{x}\sqrt{x^2 - y^2}, \qquad \frac{d^4 y_0}{dx_0^4} = y_0 + \frac{4}{x_0}\sqrt{x_0^2 - y_0^2},$$

$$\frac{d^5 y}{dx^5} = \frac{5y}{x} - \sqrt{x^2 - y^2}, \qquad \frac{d^5 y_0}{dx_0^5} = \frac{5y_0}{x_0} - \sqrt{x_0^2 - y_0^2},$$

$$\frac{d^6 y}{dx^6} = -y - \frac{6}{x}\sqrt{x^2 - y^2}, \qquad \frac{d^6 y_0}{dx_0^6} = -y_0 - \frac{6}{x_0}\sqrt{x_0^2 - y_0^2},$$

$$\frac{d^7 y}{dx^7} = -\frac{7y}{x} + \sqrt{x^2 - y^2}, \qquad \frac{d^7 y_0}{dx_0^7} = -\frac{7y_0}{x_0} + \sqrt{x_0^2 - y_0^2};$$

and the general value of y is

$$y = y_0 + (x - x_0)\left(\frac{y_0}{x_0} - \sqrt{x_0^2 - y_0^2}\right) - \frac{(x-x_0)^2}{2\,!}\left(y_0 + \frac{2}{x_0}\sqrt{x_0^2 - y_0^2}\right)$$

$$- \frac{(x-x_0)^3}{3\,!}\left(\frac{3y_0}{x_0} - \sqrt{x_0^2 - y_0^2}\right)$$

$$+ \frac{(x-x_0)^4}{4\,!}\left(y_0 + \frac{4}{x_0}\sqrt{x_0^2 - y_0^2}\right) + \frac{(x-x_0)^5}{5\,!}\left(\frac{5y_0}{x_0} - \sqrt{x_0^2 - y_0^2}\right)$$

$$- \frac{(x-x_0)^6}{6\,!}\left(y_0 + \frac{6}{x_0}\sqrt{x_0^2 - y_0^2}\right)\cdots$$

This result can be very greatly simplified by breaking up the series; we have

$$y = y_0\left(1 - \frac{(x-x_0)^2}{2\,!} + \frac{(x-x_0)^4}{4\,!} - \frac{(x-x_0)^6}{6\,!}\cdots\right)$$

$$+ y_0\frac{(x-x_0)}{x_0}\left(1 - \frac{(x-x_0)^2}{2\,!} + \frac{(x-x_0)^4}{4\,!} - \frac{(x-x_0)^6}{6\,!}\cdots\right)$$

$$- \sqrt{x_0^2 - y_0^2}\left((x-x_0) - \frac{(x-x_0)^3}{3\,!} + \frac{(x-x_0)^5}{5\,!} - \frac{(x-x_0)^7}{7\,!}\cdots\right)$$

$$-\sqrt{x_0^2-y_0^2}\left(\frac{x-x_0}{x_0}\right)\left((x-x_0)-\frac{(x-x_0)^3}{3\,!}+\frac{(x-x_0)^5}{5\,!}\right.$$

$$\left.-\frac{(x-x_0)^7}{7\,!}\cdots\right),$$

or

$$y=x\left[\frac{y_0}{x_0}\cos(x-x_0)-\frac{\sqrt{x_0^2-y_0^2}}{x_0}\sin(x-x_0)\right]$$

$$=x\left[\frac{y_0}{x_0}\cos(x-x_0)-\sqrt{1-\frac{y_0^2}{x_0^2}}\cdot\sin(x-x_0)\right].$$

$\frac{y_0}{x_0}$ is entirely arbitrary; call it $\sin\alpha$, then

$$y=x[\sin\alpha\cos(x-x_0)-\cos\alpha\sin(x-x_0)]=x\sin[\alpha-(x-x_0)],$$

$y=x\sin(c-x)$, where c is any constant.

(4) $$\left(\frac{dy}{dx}\right)^3=y^4\left(y+x\frac{dy}{dx}\right).$$

Beginning at the beginning of the key, we are directed through I., II., III., IV., to VI., page 311. Looking under VI. we see that the equation is of the first degree in x; we are referred to (17), page 318, for our specific instructions.

Obeying these, we first replace $\frac{dy}{dx}$ by p; the equation becomes

$$p^3=y^4(y+xp).$$

Differentiate relatively to y, and we get

$$3p^2\frac{dp}{dy}=4y^3(y+xp)+2y^4+xy^4\frac{dp}{dy}.$$

Eliminate x,

$$3p^2\frac{dp}{dy}=4\frac{p^3}{y}+2y^4+\left(p^2-\frac{y^5}{p}\right)\frac{dp}{dy},$$

or $$\frac{2p^3+y^5}{p}\frac{dp}{dy}-2\frac{(2p^3+y^5)}{y}=0.$$

Striking out the factor $2p^3 + y^5$, we have

$$\frac{1}{p}\frac{dp}{dy} - \frac{2}{y} = 0,$$

a differential equation of the first order and degree in which the variables are separated, and which therefore can be solved by (1), page 314.

Its solution is $$\log p - \log y^2 = C,$$

or $$\frac{p}{y^2} = c.$$

Eliminating p between this and the given equation, and reducing, we have $cy(x - c^2) = 1$, as our required solution.

(5) $$\frac{d^4y}{dx^4} - 2\frac{d^3y}{dx^3} + 2\frac{d^2y}{dx^2} - 2\frac{dy}{dx} + y = 1. \qquad (1)$$

Beginning at the beginning of the key, we are directed through I., II., III., VII., to (22) (*a*), page 319, for our specific directions.

We see at once that $y = 1$ is a particular solution.

Obeying directions, we have now to solve

$$\frac{d^4y}{dx^4} - 2\frac{d^3y}{dx^3} + 2\frac{d^2y}{dx^2} - 2\frac{dy}{dx} + y = 0 \quad \text{by (21).} \qquad (2)$$

Let $y = e^{mx}$, and we have

$$m^4 - 2m^3 + 2m^2 - 2m + 1 = 0,$$

as our auxiliary algebraic equation in m. Its roots are

$$1,\ 1,\ \sqrt{-1},\ -\sqrt{-1}.$$

The solution of (2) is then

$$y = (A + Bx)e^x + C\cos x + D\sin x,$$

and of (1) is

$$y = (A + Bx)e^x + C\cos x + D\sin x + 1.$$

(6) $$\sin^2 x \frac{d^2y}{dx^2} + \sin x \cos x \frac{dy}{dx} - y = x - \sin x. \qquad (1)$$

Beginning at the beginning of the key, we are directed through I., II., III., VII., to (24), page 321, for our specific instructions.

Dividing through by $\sin^2 x$, the equation becomes

$$\frac{d^2y}{dx^2} + \operatorname{ctn} x \frac{dy}{dx} - \csc^2 x \cdot y = x \csc^2 x - \csc x. \qquad (2)$$

$y = \operatorname{ctn} x$ is found by inspection to be a solution of

$$\frac{d^2y}{dx^2} + \operatorname{ctn} x \frac{dy}{dx} - \csc^2 x \cdot y = 0;$$

(2) can then be solved by (24) (a).

Substitute $y = z \operatorname{ctn} x$ in (2), and it becomes

$$\operatorname{ctn} x \frac{d^2z}{dx^2} + (\operatorname{ctn}^2 x - 2\csc^2 x)\frac{dz}{dx} = x \csc^2 x - \csc x,$$

or $$\frac{d^2z}{dx^2} - (\tan x + \sec x \csc x)\frac{dz}{dx} = x \sec x \csc x - \sec x. \qquad (3)$$

Referring to (25), page 323, and obeying instructions, we let $z' = \frac{dz}{dx}$, and (3) becomes

$$\frac{dz'}{dx} - (\tan x + \sec x \csc x) z' = x \sec x \csc x - \sec x,$$

a linear differential equation of the first order in z'. whose solution by (4), page 314, is

$$z' = A \tan x \sec x - x \sec^2 x + \tan x \sec x \left(\log \tan \frac{x}{2} - \log \sin x\right);$$

but $z' = \frac{dz}{dx}$, whence integrating, we have

$$z = B + A \sec x - x \tan x - (1 + \sec x) \log (1 + \cos x),$$

and

$$y = A \csc x + B \operatorname{ctn} x - x - (\csc x + \operatorname{ctn} x) \log (1 + \cos x).$$

(7) $$x(1-x)^2\frac{d^2y}{dx^2}-2y=0.$$

Beginning at the beginning of the key, we are directed through I., II., III., VII., to (24), page 321, for our specific instructions.

Let us try the method of (24) (e), page 323.

Assume $y=\Sigma a_m x^m$, and substitute in the given equation; we have

$$\Sigma\left[m(m-1)a_m x^{m-1}-2m(m-1)a_m x^m\right.$$
$$\left.+m(m-1)a_m x^{m+1}-2a_m x^m\right]=0.$$

Writing the coefficient of x^m in this sum equal to zero, we have

$$m(m+1)a_{m+1}-2[m(m-1)+1]a_m+(m-1)(m-2)a_{m-1}=0,$$

and we wish to choose the simplest set of values that will satisfy this relation.

Substituting $m=0$, $m=-1$, $m=-2$, etc., in this relation, we find

$$a_{-1}=a_0,\ a_{-2}=a_{-1},\ a_{-3}=a_{-2},\ \cdots.$$

Hence if we take $a_0=0$, it follows that

$$a_{-1}=a_{-2}=a_{-3}\cdots=0,$$

and no negative powers of x will occur in our particular solution.

Substituting now $m=1$, $m=2$, $m=3$, etc., we have

$$a_1=a_2=a_3=a_4=\cdots.$$

Taking $a_1=1$, we get as our required particular solution of the given equation

$$y=x+x^2+x^3+x^4+\cdots.$$

This can be written in finite form, since we know that

$$1+x+x^2+x^3\cdots=\frac{1}{1-x}.$$

Hence $$y=\frac{x}{1-x}$$

is a particular solution.

Turning now to (24) (a), page 321, we find

$$y = \frac{cx}{1-x} + c'\left(1 + x + \frac{2x}{1-x}\log x\right).$$

(8) $$\frac{d^2y}{dx^2} + \left(1 - \frac{2}{x^2}\right)y = 0.$$

Beginning at the beginning of the key, we are directed through I., II., III., VII., to (24), page 321, for our specific instructions. Let us try again the method (24) (e), page 323.

Assume $y = \Sigma a_m x^m$, and substitute in the given equation,

$$\Sigma\left[m(m-1)a_m x^{m-2} + a_m x^m - 2a_m x^{m-2}\right] = 0.$$

The terms containing x^m are

$$(m+2)(m+1)a_{m+2}x^m + a_m x^m - 2a_{m+2}x^m;$$

writing the sum of the coefficients equal to zero, we have

$$m(m+3)a_{m+2} + a_m = 0. \tag{1}$$

Letting $m = 0$ and $m = -3$, we get $a_0 = 0$ and $a_{-3} = 0$; and all terms of y involving even negative powers of x disappear, as do all terms involving odd negative powers, except the -1st.

In general $$a_{m+2} = -\frac{a_m}{m(m+3)}. \tag{2}$$

From this we get

$$a_4 = -\frac{a_2}{2.5} = -\frac{1}{3!5}, \quad \text{if we take } a_2 = \tfrac{1}{3},$$

$$a_6 = \frac{a_2}{2.4.5.7} = \frac{1}{5!7},$$

$$a_8 = -\frac{a_2}{2.4.5.6.7.9} = -\frac{1}{7!9},$$

$$a_{10} = \frac{a_2}{2.4.5.6.7.8.9.11} = \frac{1}{9!11}.$$

Hence $$y = \frac{x^2}{3} - \frac{x^4}{3!5} + \frac{x^6}{5!7} - \frac{x^8}{7!9} + \frac{x^{10}}{9!11} \cdots$$

is a particular solution of the given equation. This can be thrown into finite form without much labor.

We have $$xy = \frac{x^3}{3} - \frac{x^5}{3!\,5} + \frac{x^7}{5!\,7} - \frac{x^9}{7!\,9} + \frac{x^{11}}{9!\,11}\cdots,$$

$$\frac{d(xy)}{dx} = x^2 - \frac{x^4}{3!} + \frac{x^6}{5!} - \frac{x^8}{7!} + \frac{x^{10}}{9!}\cdots,$$

$$= x\left(x - \frac{x^3}{3!} + \frac{x^5}{5!} - \frac{x^7}{7!} + \frac{x^9}{9!}\cdots\right),$$

$$= x \sin x;$$

whence $$xy = \sin x - x \cos x,$$

and $$y = \frac{1}{x}(\sin x - x \cos x).$$

By going back to (2), and using odd values of m, we get another solution of our given equation, namely,

$$y = \frac{1}{x} + \frac{x}{2} - \frac{x^3}{2!\,4} + \frac{x^5}{4!\,6} - \frac{x^7}{6!\,8},$$

which can be reduced to

$$y = \frac{1}{x}(\cos x + x \sin x).$$

Hence our complete solution is

$$y = \frac{1}{x}[A(\cos x + x \sin x) + B(\sin x - x \cos x)],$$

or $$y = A'\left[\frac{\cos(x-c)}{x} + \sin(x-c)\right],$$

if we let $\dfrac{B}{A} = \tan c$.

(9) $$D_x{}^2 z - a^2 D_y{}^2 z = 0.$$

Beginning at the beginning of the key, we are directed through I. and IX. to (45), p. 331, for our specific instructions.

Obeying these, our work is as follows:

$$dy^2 - a^2\,dx^2 = 0,$$

$$dy - adx = 0, \tag{1}$$

$$dy + adx = 0, \tag{2}$$

$$dpdy - a^2\,dqdx = 0. \tag{3}$$

Combining (1) and (3), we get

$$dpdy - adqdy = 0,$$

or

$$dp - adq = 0. \tag{4}$$

(1) gives $y - ax = \alpha.$

(4) gives $p - aq = \beta.$

(2) and (3) give us, in the same way,

$$y + ax = \alpha_1,$$

$$p + aq = \beta_1;$$

and our two first integrals are

$$p - aq = f_1(y - ax), \tag{5}$$

$$p + aq = f_2(y + ax), \tag{6}$$

f_1 and f_2 denoting arbitrary functions.

Determining p and q, from (5) and (6),

$$p = \tfrac{1}{2}[f_2(y + ax) + f_1(y - ax)],$$

$$q = \frac{1}{2a}[f_2(y + ax) - f_1(y - ax)];$$

$$dz = \tfrac{1}{2}[f_2(y+ax) + f_1(y-ax)]\,dx + \frac{1}{2a}[f_2(y+ax) - f_1(y-ax)]dy$$

$$= \frac{f_2(y + ax)(dy + adx) - f_1(y - ax)(dy - adx)}{2a}.$$

Hence, $z = F(y + ax) + F_1(y - ax),$

where F and F_1 denote arbitrary functions obtained by integrating f_1 and f_2, which are arbitrary.

225. When a differential equation does not come under any of the forms given in the key, a change of dependent or independent variable, or of both, will often reduce it to one of the standard forms. No general rule can be laid down for such a substitution. It will, however, often suffice to introduce a new letter for the sum, or the difference, or the product, or the quotient of the variables, or for a power of one or of both. Sometimes an ingenious trigonometric substitution is effective, or a change from rectangular to polar coördinates; that is, the introduction of $r\cos\phi$ for x and $r\sin\phi$ for y.

The following examples of such substitutions are instructive.

(A.) *Change of dependent variable.*

(1) $(x+y)^2\dfrac{dy}{dx}=a^2$, reduces to $\dfrac{z^2}{a^2+z^2}dz-dx=0$,

if we introduce $z=x+y$.

(2) $\dfrac{d\theta}{d\phi}=\sin(\phi-\theta)$, reduces to $\dfrac{d\omega}{1-\sin\omega}-d\phi=0$,

if $\omega=\phi-\theta$.

(3) $(x-y^2)dx+2xydy=0$, reduces to $(x-z)\,dx+xdz=0$,

if $z=y^2$.

(4) $x\dfrac{dy}{dx}-y+x\sqrt{x^2-y^2}=0$, reduces to $\dfrac{dz}{\sqrt{1-z^2}}+dx=0$,

if $z=\dfrac{y}{x}$.

(5) $\dfrac{d^2y}{dx^2}+\dfrac{2}{x}\dfrac{dy}{dx}-n^2y=0$, reduces to $\dfrac{d^2z}{dx^2}-n^2z=0$,

if $z=xy$.

(B.) *Change of independent variable.*

(1) $(1-x^2)^2\dfrac{d^2y}{dx^2}+y=0$, reduces to

$$\cos^2\theta\frac{d^2y}{d\theta^2}+\sin\theta\cos\theta\frac{dy}{d\theta}+y=0,\text{ if }x=\sin\theta.$$

(2) $\frac{d^2y}{dx^2} + \tan x \frac{dy}{dx} + \cos^2 x \cdot y = 0$, reduces to $\frac{d^2y}{dz^2} + y = 0$,

if $z = \sin x$.

(C.) *Change of both variables.*

(1) $\left(1 - \frac{dy^2}{dx^2}\right) xy = \frac{dy}{dx}(x^2 - y^2 - a^2)$, reduces to

$$v - z\frac{dv^2}{dz^2} - \frac{dv}{dz}(z - v - a^2) = 0, \text{ if } z = x^2 \text{ and } v = y^2.$$

(2) $(y - x)(1 + x^2)^{\frac{1}{2}} \frac{dy}{dx} = (1 + y^2)^{\frac{3}{2}}$, reduces to

$$\sin(\phi - \theta)\, d\phi = d\theta, \text{ if } x = \tan\theta \text{ and } y = \tan\phi.$$

(3) $\left(x\frac{dy}{dx} - y\right)^2 = a\left(1 + \frac{dy^2}{dx^2}\right)(x^2 + y^2)^{\frac{3}{2}}$, reduces to

$$\frac{dr}{\sqrt{r(1 - ar)}} - \frac{d\phi}{\sqrt{a}} = 0, \text{ if } x = r\cos\phi \text{ and } y = r\sin\phi.$$

KEY.

* Of course, X and Y may be constants.

		Page
Of the form $(ax + by + c)\,dx + (a'x + b'y + c')\,dy = 0$	(3)	314
Linear. General form, $\frac{dy}{dx} + X_1 y = X_2$, where X_1 and X_2 are functions of x alone* . . .	(4)	314
Of the form $\frac{dy}{dx} + X_1 y = X_2 y^n$, where X_1 and X_2 are functions of x alone*	(5)	315
$Mdx + Ndy$ an exact differential. Test, $D_y M = D_x N$	(6)	315
$Mx + Ny = 0$	(7)	315
$Mx - Ny = 0$	(8)	315
Of the form $F_1(xy)\,ydx + F_2(xy)\,xdy = 0$ . .	(9)	315
$\frac{D_y M - D_x N}{N}$, a function of x alone	(10)	315
$\frac{D_x N - D_y M}{M}$, a function of y alone	(11)	316
$\frac{D_y M - D_x N}{Ny - Mx}$, a function of (xy)	(12)	316
A solution in the form of a series can always be obtained	(13)	316

VI. Not of first degree.

Can be solved as an algebraic equation in p, where p stands for $\frac{dy}{dx}$	(14)	317
Involves only one of the variables and p, where p stands for $\frac{dy}{dx}$	(15)	317
Of the first degree in x and y; that is, of the form $xf_1p + yf_2p = f_3p$, where p stands for $\frac{dy}{dx}$	(16)	317
Of the first degree in x or y	(17)	318
Homogeneous relatively to x and y	(18)	318

* Of course, X_1 and X_2 may be constants.

* The first member is supposed to contain only those terms involving the dependent variable or its derivatives.

† See note, p. 310.

* See note, p. 310.

(1) Of or reducible to the form $Xdx + Ydy = 0$, where X is a function of x alone and Y is a function of y alone.

Integrate each term separately, and write the sum of their integrals equal to an arbitrary constant.

(2) M and N homogeneous functions of x and y of the same degree.

Introduce in place of y the new variable v defined by the equation $y = vx$, and the equation thus obtained can be solved by (1).

Or, multiply the equation through by $\frac{1}{Mx + Ny}$, and its first member will become an exact differential, and the solution may be obtained by (6).

(3) Of the form $(ax + by + c)\,dx + (a'x + b'y + c')\,dy = 0$.

If $ab' - a'b = 0$, the equation may be thrown into the form $(ax + by + c)\,dx + \frac{a'}{a}(ax + by + c)\,dy = 0$. If now $z = ax + by$ be introduced in place of either x or y, the resulting equation can be solved by (1).

If $ab' - a'b$ does not equal zero, the equation can be made homogeneous by assuming $x = x' - \alpha$, $y = y' - \beta$, and determining α and β so that the constant terms in the new values of M and N shall disappear, and it can then be solved by (2).

(4) Linear. General form $\frac{dy}{dx} + X_1 y = X_2$, where X_1 and X_2 are functions of x alone.

Solve on the supposition that $X_2 = 0$ by (1); and from this solution obtain a value for y, involving of course an arbitrary constant C. Substitute this value of y in the given equation, regarding C as a variable, and there will result a differential equation, involving C and x, whose solution by (1) will express C as a function of x. Substitute this value for C in the expression already obtained for y, and the result will be the required solution.

(5) Of the form $\frac{dy}{dx} + X_1 y = X_2 y^n$, where X_1 and X_2 are functions of x alone.

Divide through by y^n, and then introduce $z = y^{1-n}$ in place of y, and the equation will become linear and may be solved by (4).

(6) $Mdx + Ndy$ an exact differential. Test $D_y M = D_x N$.

Find $\int Mdx$, regarding y as constant, and add an arbitrary function of y. Determine this function of y by the fact that the differential of the result just mentioned, taken on the supposition that x is constant, must equal Ndy. Write equal to an arbitrary constant the $\int Mdx$ above mentioned plus the function of y just determined.

(7) $Mx + Ny = 0$.

Divide the first term of $Mdx + Ndy = 0$ by Mx, and the second by its equal $-Ny$, and integrate by (1).

(8) $Mx - Ny = 0$.

Divide the first term of $Mdx + Ndy = 0$ by Mx, and the second by its equal Ny, and integrate by (1).

(9) Of the form $f_1(xy)\,ydx + f_2(xy)\,xdy = 0$.

Multiply through by $\frac{1}{Mx - Ny}$, and the first member will become an exact differential. The solution may then be found by (6).

(10) $\frac{D_y M - D_x N}{N}$, a function of x alone.

Multiply the equation through by $e^{\int \frac{D_y M - D_x N}{N} \cdot dx}$, and the first member will become an exact differential. The solution may then be found by (6).

(11) $\dfrac{D_x N - D_y M}{M}$, a function of y alone.

Multiply the equation through by $e^{\int \frac{D_x N - D_y M}{M} \cdot dy}$, and the first member will become an exact differential. The solution may then be found by (6).

(12) $\dfrac{D_y M - D_x N}{Ny - Mx}$, a function of (xy).

Multiply the equation through by $e^{\int \frac{D_y M - D_x N}{Ny - Mx} \cdot dv}$ where $v = xy$, and the first member will become an exact differential. The solution may thus be found by (6).

(13) A solution of $Mdx + Ndy = 0$ in the form of a series can always be obtained.

Throw the given equation into the form $\dfrac{dy}{dx} = -\dfrac{M}{N}$, then differentiate, and in the result replace $\dfrac{dy}{dx}$ by $-\dfrac{M}{N}$, thus obtaining a value of $\dfrac{d^2y}{dx^2}$ in terms of x and y; by successive differentiations and substitutions get values of $\dfrac{d^3y}{dx^3}$, $\dfrac{d^4y}{dx^4}$, etc., in terms of x and y.

If y_0 is the value of y corresponding to any chosen value x_0 of x, y can now be developed by Taylor's Theorem.

We have $$y = fx = f(x_0 + x - x_0)$$

$$= fx_0 + (x - x_0)f'x_0 + \frac{(x - x_0)^2}{2!}f''x_0 + \frac{(x - x_0)^3}{3!}f'''x_0 + \cdots,$$

or

$$y = y_0 + (x - x_0)\frac{dy_0}{dx_0} + \frac{(x - x_0)^2}{2!}\frac{d^2y_0}{dx_0^{\,2}} + \frac{(x - x_0)^3}{3!}\frac{d^3y_0}{dx_0^{\,3}} + \cdots,$$

where $$\frac{dy_0}{dx_0}, \frac{d^2y_0}{dx_0^{\,2}}, \frac{d^3y_0}{dx_0^{\,3}}, \text{ etc.,}$$

are obtained by replacing x and y by x_0 and y_0 in the values of

$$\frac{dy}{dx}, \frac{d^2y}{dx^2}, \frac{d^3y}{dx^3}, \text{ etc.,}$$

described above.

In the general case y_0 is entirely arbitrary, and if the given equation is at all complicated, the solution is apt to be too complicated to be of much service. If, however, in a special problem the value of y corresponding to some value of x is given, and these values are taken as y_0 and x_0, the solution will generally be useful.

(14) Can be solved as an algebraic equation in p, where p stands for $\frac{dy}{dx}$.

Solve as an algebraic equation in p, and, after transposing all the terms to the first member, express the first member as the product of factors of the first order and degree. Write each of these factors separately equal to zero, and find its solution in the form $V - c = 0$ by (V.). Write the product of the first members of these solutions equal to zero, using the same arbitrary constant in each.

(15) Involves only one of the variables and p, where p stands for $\frac{dy}{dx}$.

By algebraic solution express the variable as an explicit function of p, and then differentiate through relatively to the other variable, regarding p as a new variable and remembering that $\frac{dx}{dy} = \frac{1}{p}$. There will result a differential equation of the first order and degree between the second variable and p which can be solved by (1). Eliminate p between this solution and the given equation, and the resulting equation will be the required solution.

(16) Of the form $xf_1p + yf_2p = f_3p$, where p stands for $\frac{dy}{dx}$.

Differentiate the equation relatively to one of the variables, regarding p as a new variable, and, with the aid of the given equation, eliminate the other original variable. There will result a linear differential equation of the first order between p and the remaining variable, which may be simplified by striking out any factor not containing $\frac{dp}{dx}$ or $\frac{dp}{dy}$, and can be solved by (4). Eliminate p between this solution and the given equation, and the result will be the required solution.

(17) Of the first degree in x or y.

The equation can sometimes be solved by the method of (16), differentiating relatively to the variable which does not enter to the first degree.

(18) Homogeneous relatively to x and y.

Let $y = vx$, and solve algebraically relatively to p or v, p standing for $\frac{dy}{dx}$. The result will be of the form $p = fv$, or $v = Fp$. If

$$p = fv,\quad \frac{dy}{dx} = fv,\quad \frac{d(vx)}{dx} = fv,\quad x\frac{dv}{dx} + v = fv,$$

an equation that can be solved by (1). If

$$v = Fp,\quad \frac{y}{x} = Fp,\quad y = xFp,$$

an equation that can be solved by (16).

(19) Of the form $F(\phi, \psi) = 0$, where ϕ and ψ are functions of x, y, and $\frac{dy}{dx}$, such that $\phi = a$ and $\psi = b$ will lead, on differentiation, to the same differential equations of the second order.

Eliminate $\frac{dy}{dx}$ between $\phi = a$ and $\psi = b$, where a and b are arbitrary constants subject to the relation that $F(a, b) = 0$, and the result will be the required solution.

(20) Singular solution will answer.

Let $\frac{dy}{dx} = p$, and express p as an explicit function of x and y. Take $\frac{dp}{dy}$, regarding x as constant, and see whether it can be made infinite by writing equal to zero any expression involving y. If so, and if the equation thus formed will satisfy the given differential equation, it is a singular solution.

Or take $\dfrac{d\left(\frac{1}{p}\right)}{dx}$, regarding y as constant, and see whether it can be made infinite by writing equal to zero any expression involving x. If so, and if the equation thus formed is consistent with the given equation, it is a singular solution.

(21) Linear, with constant coefficients. Second member zero.

Assume $y = e^{mx}$; m being constant, substitute in the given equation, and then divide through by e^{mx}. There will result an algebraic equation in m. Solve this equation, and the complete value of y will consist of a series of terms characterized as follows: For every distinct real value of m there will be a term Ce^{mx}; for each pair of imaginary values, $a + b\sqrt{-1}$, $a - b\sqrt{-1}$, a term $Ae^{ax}\cos bx + Be^{ax}\sin bx$; each of the coefficients A, B, and C being an arbitrary constant, if the root or pair of roots occurs but once; and an algebraic polynomial in x of the $(r-1)$st degree with arbitrary constant coefficients, if the root or pair of roots occurs r times.

(22) Linear, with constant coefficients. Second member not zero.

(a) If a particular solution of the given equation can be obtained by inspection, this value plus the value of y obtained by (21) on the hypothesis that the second member is zero, will be the complete value of the dependent variable.

(*b*) If the second member of the given equation can be got rid of by differentiation, or by differentiation and elimination between the given and the derived equations, solve the new differential equation thus obtained, by (21), and determine the superfluous arbitrary constants so that the given equation shall be satisfied.

In determining these superfluous constants, it will generally save labor to solve the original equation on the hypothesis that its second member is zero, and then to strike out from the preceding solution the terms which are duplicates of the ones in the second solution before proceeding to differentiate, as from the nature of the case they would drop out in the course of the work.

(*c*) If the given equation is of the second order, solve on the hypothesis that the second member is zero, by (21), obtain from this solution a simple particular solution by letting one of the arbitrary constants equal zero and the other equal unity, and let $y = v$ be this last solution; then substitute vz for y in the given equation; there will result a differential equation of the second order between x and z in which the dependent variable z will be wanting, and which can be completely solved by (25). Substitute the value of z thus obtained in $y = vz$ and there will result the required solution of the given equation.

(*d*) Solve, on the hypothesis that the second member is zero, and obtain the complete value of y by (21). Denoting the order of the given equation by n, form the $n-1$ successive derivatives $\frac{dy}{dx}, \frac{d^2y}{dx^2} \cdots \frac{d^{n-1}y}{dx^{n-1}}$. Then differentiate y and each of the values just obtained, regarding the arbitrary constants as new variables, and substitute the resulting values in the given equation; and by its aid, and that of the $n-1$ equations of condition formed by writing each of the derivatives of the second set, except the nth, equal to the derivative of the same order in the first set, determine the arbitrary coefficients and substitute their values in the original expression for y.

(23) Of the form

$$(a+bx)^n\frac{d^n y}{dx^n}+A\,(a+bx)^{n-1}\frac{d^{n-1}y}{dx^{n-1}}+\cdots+Ly=X,$$

where X is a function of x alone.

Assume $a+bx=e^t$, and change the independent variable in the given equation so as to introduce t in place of x. The solution can then be obtained by (22).

(24) Linear; of second order; coefficients not constants.

General form $\quad \dfrac{d^2 y}{dx^2}+P\dfrac{dy}{dx}+Qy=R.$

(a) If a particular solution $y=v$ of the equation

$$\frac{d^2 y}{dx^2}+P\frac{dy}{dx}+Qy=0$$

can be found by inspection or other means, substitute $y=vz$ in the given equation, which will then reduce to the form

$$v\frac{d^2 z}{dx^2}+\left(2\frac{dv}{dx}+Pv\right)\frac{dz}{dx}=R,$$

and can be solved by (25). Substitute the value of z thus found in $y=vz$, and the result will be the general solution of the given equation.

(b) The substitution of $y=vz$ in the given equation, where v is given by the auxiliary differential equation

$$2\frac{dv}{dx}+Pv=0,$$

and can be found by (1), and should be used in the simplest possible form, will lead to a differential equation in z of the form

$$\frac{d^2 z}{dx^2} + Iz = R,$$

which is often simpler than the original equation.

(*c*) The introduction of z in place of the independent variable x, z being a solution of the auxiliary differential equation

$$\frac{d^2 z}{dx^2} + P\frac{dz}{dx} = 0,$$

the simpler the better, will reduce the given equation to the form

$$\frac{d^2 y}{dz^2} + Iy = S,$$

which is often simpler than the original equation.

(*d*) If the first member of the given equation regarded as an operation performed on y can be resolved into the product of two operations, the equation can always be solved. The conditions of such a resolution are the following: let the given equation be

$$u\frac{d^2 y}{dx^2} + v\frac{dy}{dx} + wy = R,$$

where u, v, w, and R are functions of x; this can be resolved into

$$\left(p\frac{d}{dx} + q\right)\left(r\frac{d}{dx} + s\right)y = R,$$

where p, q, r, and s are functions of x, if

$$pr = u,\quad qr + p\left(\frac{dr}{dx} + s\right) = v,\ \text{and}\ qs + p\frac{ds}{dx} = w;$$

and the values of p, q, r, and s can usually be obtained by inspection. We have first to solve $p\frac{dz}{dx}+qz=R$ by (4), and then to solve $r\frac{dy}{dx}+sy=z$ by (4).

(e) A particular solution of the equation

$$\frac{d^2y}{dx^2}+P\frac{dy}{dx}+Qy=0,$$

can often be obtained by assuming that y is of the form $\Sigma a_m x^m$, m being an integer, substituting this value for y in the given equation, writing the sum of the coefficients of x^m equal to zero, since the equation must be identically true, and thus obtaining a relation between successive coefficients of the assumed series. The simplest set of values consistent with this relation should be substituted in the assumed value of y, which will then be a particular solution of the equation. If this solution can be expressed in finite form, the complete solution of the given equation can be obtained from it by the method described in (24) (a). If, however, two different particular solutions can be found by the method just described, each of them should be multiplied by an arbitrary constant, and the sum of these products will be the complete solution of the given equation.

(25) Either of the primitive variables wanting.

Assume z equal to the derivative of lowest order in the equation, and express the equation in terms of z and its derivatives with respect to the primitive variable actually present, and the order of the resulting equation will be lower than that of the given one.

(26) Of the form $\frac{d^ny}{dx^n}=X$. X being a function of x alone.

Solve by integrating n times successively with regard to x.

Or solve by (22).

(27) Of the form $\frac{d^2 y}{dx^2} = Y$. Y being a function of y alone.

Multiply by $2\frac{dy}{dx}$ and integrate relatively to x. There will result the equation $\left(\frac{dy}{dx}\right)^2 = 2\int Ydy + C$, whence $\frac{dy}{dx} = (2\int Ydy + C)^{\frac{1}{2}}$, an equation that may be solved by (1).

(28) Of the form $\frac{d^n y}{dx^n} = f\frac{d^{n-1} y}{dx^{n-1}}$.

Assume

$$\frac{d^{n-1}y}{dx^{n-1}} = z, \text{ then } \frac{dz}{dx} = fz \text{ or } dx = \frac{dz}{fz},\ x = \int\frac{dz}{fz} + C.$$

After effecting this integration, express z in terms of x and C. Then, since $z = \frac{d^{n-1}y}{dx^{n-1}}$, $\frac{d^{n-1}y}{dx^{n-1}} = F(x,\ C)$, an equation that may be treated by (26).

Or, since

$$\frac{d^{n-1}y}{dx^{n-1}} = z,\ \frac{d^{n-2}y}{dx^{n-2}} = \int zdx + c = \int\frac{zdz}{fz} + c, \text{ since } dx = \frac{dz}{fz}.$$

$$\frac{d^{n-3}y}{dx^{n-3}} = \int dx\left(\int\frac{zdz}{fz} + c\right) + c_1 = \int\frac{dz}{fz}\left(\int\frac{zdz}{fz} + c\right) + c_1, \cdots.$$

Continue this process until y is expressed in terms of z and $n-1$, arbitrary constants, and then eliminate z by the aid of the equation $x = \int\frac{dz}{fz} + C$.

(29) Of the form $\frac{d^n y}{dx^n} = f\frac{d^{n-2} y}{dx^{n-2}}$.

Let $\frac{d^{n-2}y}{dx^{n-2}} = z$, and the equation becomes $\frac{d^2 z}{dx^2} = fz$, and may be solved by (27).

(30) Homogeneous on the supposition that x and y are of the degree 1, $\frac{dy}{dx}$ of the degree 0, $\frac{d^2y}{dx^2}$ of the degree -1, ...

Assume $x = e^\theta$, $y = e^\theta z$, and by changing the variables introduce θ and z into the equation in the place of x and y. Divide through by e^θ and there will result an equation involving only z, $\frac{dz}{d\theta}$, $\frac{d^2z}{d\theta^2}$, ..., whose order may be depressed by (25).

(31) Homogeneous on the supposition that x is of the degree 1, y of the degree n, $\frac{dy}{dx}$ of the degree $n-1$, $\frac{d^2y}{dx^2}$ of the degree $n-2$, ...

Assume $x = e^\theta$, $y = e^{n\theta}z$, and by changing the variables introduce θ and z into the equation in the place of x and y. The resulting equation may be freed from θ by division and treated by (25).

(32) Homogeneous relatively to y, $\frac{dy}{dx}$, $\frac{d^2y}{dx^2}$, ...

Assume $y = e^z$, and substitute in the given equation. Divide through by e^z and treat by (25).

(33) Containing the first power only of the derivative of the highest order.

The equation may be *exact*.

Call its first member $\frac{dV}{dx}$. If n is the order of the equation, represent $\frac{d^{n-1}y}{dx^{n-1}}$ by p and $\frac{d^ny}{dx^n}$ by $\frac{dp}{dx}$. Multiply the term containing $\frac{dp}{dx}$ by dx and integrate it as if p were the only variable, calling the result U_1; then replacing p by $\frac{d^{n-1}y}{dx^{n-1}}$, find the complete derivative $\frac{dU_1}{dx}$, and form the expression $\frac{dV}{dx} - \frac{dU_1}{dx}$, representing it by $\frac{dV_1}{dx}$. If $\frac{dV_1}{dx}$ contains the first power only of the highest derivative of y, it may itself be an exact derivative, and is to be treated precisely as the first member of the given equation $\frac{dV}{dx}$ has been. Continue this process until a remainder $\frac{dV_{n-1}}{dx}$ of the first order occurs.

Write this equal to zero, and see if the equation thus formed is *exact*, see (6). If so, solve it by (6), throwing its solution into the form $V_{n-1} = C$. A complete first integral of the given equation will be $U_1 + U_2 + \cdots + V_{n-1} = C$. The occurrence at any step of the process of a remainder $\frac{dV_k}{dx}$, containing a higher power than the first of its highest derivative of y, or the failure of the resulting equation of the first order above described to be *exact*, shows that the first member of the given equation was not an exact derivative, and that this method will not apply.

(34) Of the form $\frac{d^2y}{dx^2} + X\frac{dy}{dx} + Y\left[\frac{dy}{dx}\right]^2 = 0$, where X is a function of x alone and Y a function of y alone. Multiply through by $\left[\frac{dy}{dx}\right]^{-1}$ and the equation will become exact, and may be solved by (33).

(35) Singular integral will answer.

Call $\frac{d^{n-1}y}{dx^{n-1}}$ p, and $\frac{d^ny}{dx^n}$ q, and find $\frac{dq}{dp}$, regarding p and q as the only variables, and see whether $\frac{dq}{dp}$ can be made infinite by writing equal to zero any factor containing p.

If so, eliminate q between this equation and the given equation, and if the result is a solution it will be a singular integral.

(36) General form, $Pdx + Qdy + Rdz = 0$.

If the equation can be reduced to the form $Xdx + Ydy + Zdz = 0$, where X is a function of x alone, Y a function of y alone, and Z a function of z alone, integrate each term separately, and write the sum of the integrals equal to an arbitrary constant.

If not, integrate the equation by (V.) on the supposition that one of the variables is constant and its differential zero, writing an arbitrary function of that variable in place of the arbitrary constant in the result. Transpose all the terms to the first member, and then take its complete differential, regarding all the original variables as variable, and write it equal to the first member of the given equation, and from this equation of condition determine the arbitrary function. Substitute for the arbitrary function in the first integral its value thus determined, and the result will be the solution required.

If the equation of condition contains any other variables than the one involved in the arbitrary function, they must be eliminated by the aid of the primitive equation already obtained; and if this elimination cannot be performed, the given equation is not derivable from a single primitive equation, but must have come from two simultaneous primitive equations.

In that case, assume any arbitrary equation $f(x,y,z) = 0$ as one primitive, differentiate it, and eliminate between it its derived equation and the given equation, one variable, and its differential. There will result a differential equation containing only two variables, which may be solved by (III.), and will lead to the second primitive of the given equation.

(37) General form, $Pdx_1 + Qdx_2 + Rdx_3 + \cdots\cdots = 0$.

If the equation can be reduced to the form $X_1dx_1 + X_2dx_2 + X_3dx_3 + \cdots\cdots = 0$, where X_1 is a function of x_1 alone, X_2 a function of x_2 alone, X_3 a function of x_3 alone, etc., integrate each term separately, and write the sum of their integrals equal to an arbitrary constant.

If not, integrate the equation by (V.), on the supposition that all the variables but two are constant and their differentials zero, writing an arbitrary function of these variables in place of the arbitrary constant in the result. Transpose all the terms to the first member, and then take its complete differential, regarding all of the original variables as variable, and write it equal to the first member of the given equation, and from this equation of condition determine the arbitrary function. Substitute for the arbitrary function in the first integral its value thus determined, and the result will be the solution required.

If the equation of condition cannot, even with the aid of the primitive equation first obtained, be thrown into a form where the complete differential of the arbitrary function is given equal to an exact differential, the function cannot be determined, and the given equation is not derivable from a single primitive equation.

(38) System of simultaneous equations of the first order.

If any of the equations of the set can be integrated separately by (II.) so as to lead to single primitives, the problem can be simplified; for by the aid of these primitives a number of variables equal to the number of solved equations can be eliminated from the remaining equations of the series, and there will be formed a simpler set of simultaneous equations whose primitives, together with the primitives already found, will form the primitive system of the given equations.

There must be n equations connecting $n+1$ variables, in order that the system may be determinate.

Let x, x_1, x_2, ..., x_n be the original variables. Choose any two, x and x_1, as the independent and the principal dependent variable, and by successive eliminations form the n equations $\frac{dx_1}{dx}=f_1(x,x_1,x_2,\ldots\ldots,x_n)$, $\frac{dx_2}{dx}=f_2(x,x_1,x_2,\ldots\ldots,x_n)$, ..., up to $\frac{dx_n}{dx}=f_n(x,x_1,x_2,\ldots\ldots,x_n)$. Differentiate the first of these with respect to x $n-1$ times, substituting for $\frac{dx_2}{dx}, \frac{dx_3}{dx}, \ldots\ldots, \frac{dx_n}{dx}$, after each step their values in terms of the original variables. There will result n equations, which will express each of the n successive derivatives $\frac{dx_1}{dx}, \frac{d^2x_1}{dx^2}, \frac{d^3x_1}{dx^3}, \ldots\ldots, \frac{d^nx_1}{dx^n}$, in terms of x, x_1, x_2, ..., x_n. Eliminate from these all the variables except x and x_1, obtaining a single equation of the nth order between x and x_1. Solve this by (VII.), and so get a value of x_1 in terms of x and n arbitrary constants. Find by differentiating this result values for $\frac{dx_1}{dx}, \frac{d^2x_1}{dx^2}, \ldots\ldots, \frac{d^{n-1}x_1}{dx^{n-1}}$, and write them equal to the ones already obtained for them in terms of the original variables. The $n-1$ equations thus formed, together with the equation expressing x_1 in terms of x and arbitrary constants, are the complete primitive system required.

(39) System of simultaneous equations not of the first order. Regard each derivative of each dependent variable, from the first to the next to the highest as a new variable, and the given equations, together with the equations defining these new variables, will form a system of simultaneous equations of the first order which may be solved by (38). Eliminate the new variables representing the various derivatives from the equations of the solution, and the equations obtained will be the complete primitive system required.

(40) All the partial derivatives taken with respect to one of the independent variables.

Integrate by (II.) as if that one were the only independent variable, replacing each arbitrary constant by an arbitrary function of the other independent variables.

(41) Of the first order and linear, containing three variables. General form, $PD_xz + QD_yz = R$.

Form the auxiliary system of ordinary differential equations $\frac{dx}{P} = \frac{dy}{Q} = \frac{dz}{R}$, and integrate by (38). Express their primitives in the form $u = a$, $v = b$, a and b being arbitrary constants; and $u = fv$, where f is an arbitrary function, will be the required solution.

(42) Of the first order and linear, containing more than three variables. General form, $P_1D_{x_1}z + P_2D_{x_2}z + \cdots\cdots = R$, where x_1, x_2, $\cdots\cdots$, x_n, are the independent and z the dependent variables.

Form the auxiliary system of ordinary differential equations $\frac{dx_1}{P_1} = \frac{dx_2}{P_2} \cdots\cdots = \frac{dx_n}{P_n} = \frac{dz}{R}$, and integrate them by (38). Express their primitives in the form $v_1 = a$, $v_2 = b$, $v_3 = c$, $\cdots\cdots$, and $v_1 = f(v_2, v_3, \cdots\cdots, v_n)$, where f is an arbitrary function, will be the required solution.

(43) Of the first order and not linear, containing three variables, $F(x,y,z,p,q) = 0$, where $p = D_xz$, $q = D_yz$.

Express q in terms of x, y, z and p from the given equation, and substitute its value thus obtained in the auxiliary system of ordinary differential equations $\frac{dx}{-D_pq} = dy$ $= \frac{dz}{q - pD_pq} = \frac{dp}{D_xq + pD_zq}$. Deduce by integration from these equations, by (36), a value of p involving an arbitrary constant, and substitute it with the corresponding value of q in the equation $dz = pdx + qdy$. Integrate this result by (36), if possible; and if a single primitive equation be obtained, it will be a complete primitive of the given equation.

A singular solution may be obtained by finding the partial derivatives $D_p z$ and $D_q z$ from the given equation, writing them separately equal to zero, and eliminating p and q between them and the given equation.

(44) Of the first order and not linear, containing more than three variables. $F(x_1, x_2, \cdots\cdots, x_n, z, p_1, p_2, \cdots\cdots, p_n) = 0$, where $p_1 = D_{x_1} z$, $p_2 = D_{x_2} z$, $\cdots\cdots$.

Form the linear partial differential equation $\Sigma_i[(D_{x_i}F + p_i D_z F) D_{p_i}\Phi - D_{p_i}F(D_{x_i}\Phi + p_i D_z \Phi)] = 0$, where Φ is an unknown function of $(x_1, \cdots\cdots, x_n, p_1, \cdots\cdots, p_n)$, and where Σ_i means the sum of all the terms of the given form that can be obtained by giving i successively the values 1, 2, 3, $\cdots\cdots$, n.

Form, by (42), its auxiliary system of ordinary differential equations, and from them get, by (38), $n-1$ integrals, $\Phi_1 = a_1$, $\Phi_2 = a_2$, $\cdots\cdots$, $\Phi_{n-1} = a_{n-1}$. By these equations and the given equation express $p_1, p_2, \cdots\cdots, p_n$ in terms of the original variables, and substitute their values in the equation $dz = p_1 dx_1 + p_2 dx_2 + \cdots\cdots + p_n dx_n$. Integrate this by (37), and the result will be the required complete primitive.

(45) Of the second order and containing the derivatives of the second order only in the first degree. General form, $RD_x^2 z + SD_x D_y z + TD_y^2 z = V$, where R, S, T, and V may be functions of x, y, z, $D_x z$, and $D_y z$.

Call $D_x z$ p and $D_y z$ q.

Form first the equation

$$Rdy^2 - Sdxdy + Tdx^2 = 0, \qquad [1]$$

and resolve it, supposing the first member not a complete square, into two equations of the form

$$dy - m_1 dx = 0, \qquad dy - m_2 dx = 0. \qquad [2]$$

From the first of these, and from the equation

$$Rdpdy + Tdqdx - Vdxdy = 0, \qquad [3]$$

combined if needful with the equation

$$dz = pdx + qdy,$$

seek to obtain two integrals $u_1 = \alpha$, $v_1 = \beta$. Proceeding in the same way with the second equation of [2], seek two other integrals $u_2 = \alpha_1$, $v_2 = \beta_1$; then the first integrals of the proposed equation will be

$$u_1 = f_1 v_1, \qquad u_2 = f_2 v_2, \tag{4}$$

where f_1 and f_2 denote arbitrary functions.

To deduce the final integral, we must either integrate one of these, or, determining from the two p and q in terms of x, y, and z, substitute those values in the equation

$$dz = pdx + qdy,$$

which will then become integrable. Its solution will give the final integral sought.

If the values of m_1 and m_2 are equal, only one first integral will be obtained, and the final solution must be sought by its integration.

When it is not possible so to combine the auxiliary equations as to obtain two auxiliary integrals $u = \alpha$, $v = \beta$, no first integral of the proposed equation exists, and this method of solution fails.

EXAMPLES.

(1) $\sin x \cos y \,.\, dx - \cos x \sin y \,.\, dy = 0.$ *Ans.* $\cos y = c \cos x.$

(2) $(x+y)^2 \dfrac{dy}{dx} = a^2.$ *Ans.* $y - a \tan^{-1}\dfrac{x+y}{a} = c.$

(3) $\dfrac{d\theta}{d\phi} = \sin(\phi - \theta).$ *Ans.* $\operatorname{ctn}\left[\dfrac{\pi}{4} - \dfrac{\phi - \theta}{2}\right] = \phi + c.$

(4) $x\dfrac{dy}{dx} - y + x\sqrt{x^2 - y^2} = 0.$ *Ans.* $\sin^{-1}\dfrac{y}{x} = c - x.$

(5) $(y - x)(1 + x^2)^{\frac{1}{2}} \dfrac{dy}{dx} = (1 + y^2)^{\frac{3}{2}}.$

Ans. $\operatorname{ctn}\left[\dfrac{\pi}{4} - \dfrac{1}{2}(\tan^{-1}y - \tan^{-1}x)\right] = \tan^{-1}y + c.$

(6) $\left(x\dfrac{dy}{dx} - y\right)^2 = a\left[1 + \left(\dfrac{dy}{dx}\right)^2\right](x^2 + y^2)^{\frac{3}{2}}.$

Ans. $2a(x^2 + y^2) = (x^2 + y^2)^{\frac{1}{2}} - x \cos c + y \sin c.$

(7) $[2\sqrt{(xy)} - x]\, dy + y dx = 0.$ *Ans.* $y = ce^{-\sqrt{\frac{x}{y}}}.$

(8) $(x - y^2) + 2xy\dfrac{dy}{dx} = 0.$ *Ans.* $xe^{\frac{y^2}{x}} = c.$

(9) $(2x - y + 1)dx + (2y - x - 1)\, dy = 0.$

Ans. $x^2 - xy + y^2 + x - y = c.$

(10) $\dfrac{dy}{dx} + y \cos x = \dfrac{\sin 2x}{2}.$ *Ans.* $y = \sin x - 1 + ce^{-\sin x}.$

(11) $(1 - x^2)\dfrac{dy}{dx} - xy = axy^2.$ *Ans.* $y = [c\sqrt{(1 - x^2)} - a]^{-1}.$

(12) $xy(1 + xy^2)\dfrac{dy}{dx} = 1.$ *Ans.* $\dfrac{1}{x} = 2 - y^2 + ce^{\frac{-y^2}{2}}.$

(13) $y(x^2 + y^2 + a^2)\dfrac{dy}{dx} + x(x^2 + y^2 - a^2) = 0.$

Ans. $(x^2 + y^2)^2 - 2a^2(x^2 - y^2) = c.$

(14) $xdx + ydy + \dfrac{xdy - ydx}{x^2 + y^2} = 0.$ *Ans.* $\dfrac{x^2 + y^2}{2} + \tan^{-1}\dfrac{y}{x} = c.$

(15) $\left(\dfrac{dy}{dx}\right)^2 - \dfrac{a^2}{x^2} = 0.$ *Ans.* $(y - a\log x - c)(y + a\log x - c) = 0.$

(16) $\left(\dfrac{dy}{dx}\right)^2 + 2y\operatorname{ctn} x\dfrac{dy}{dx} = y^2.$

Ans. $\left(y\sin^2\dfrac{x}{2} - c\right)\left(y\cos^2\dfrac{x}{2} - c\right) = 0.$

(17) $\dfrac{dy}{dx}\left(\dfrac{dy}{dx} + y\right) = x(x + y).$

Ans. $(2y - x^2 - c)[\log(x + y - 1) + x - c] = 0.$

(18) $\left(\dfrac{dy}{dx}\right)^3 - (x^2 + xy + y^2)\left(\dfrac{dy}{dx}\right)^2 + (x^3y + x^2y^2 + xy^3)\dfrac{dy}{dx} - x^3y^3 = 0.$

Ans. $\left(y - \dfrac{x^3}{3} - c\right)\left(x + \dfrac{1}{y} - c\right)\left(\log y - \dfrac{x^2}{2} - c\right) = 0.$

(19) $\left(1 - y^2 - \dfrac{y^4}{x^2}\right)\left(\dfrac{dy}{dx}\right)^2 - \dfrac{2y}{x}\dfrac{dy}{dx} + \dfrac{y^2}{x^2} = 0.$

Ans. $\left(y + \log\dfrac{x + \sqrt{x^2 + y^2}}{y} - c\right)\left(y - \log\dfrac{x + \sqrt{x^2 + y^2}}{y} - c\right) = 0.$

(20) $y = x\dfrac{dy}{dx} + \dfrac{dy}{dx} - \left(\dfrac{dy}{dx}\right)^2.$ *Ans.* $y = cx + c - c^2.$

Singular solution, $y = \dfrac{(x + 1)^2}{4}.$

(21) $y = y\left(\dfrac{dy}{dx}\right)^2 + 2x\dfrac{dy}{dx}.$ *Ans.* $y^2 = 2cx + c^2.$

(22) $\left[1 - \left(\dfrac{dy}{dx}\right)^2\right]xy = (x^2 - y^2 - a^2)\dfrac{dy}{dx}.$

Ans. $y^2 - cx^2 + \dfrac{a^2c}{1 + c} = 0.$

(23) $y = 2x\dfrac{dy}{dx} + y^2\left(\dfrac{dy}{dx}\right)^3.$ *Ans.* $y^2 = 2cx + c^3.$

(24) $x^3\left(\dfrac{dy}{dx}\right)^2 + x^2y\dfrac{dy}{dx} + a^3 = 0.$ *Ans.* $c^2 + cxy + a^3x = 0.$

(25) $f\left[x\left(\frac{dy}{dx}\right)^2\right]+2x\frac{dy}{dx}-y=0.$
Ans. $(b+y)^2=4ax,\ f(a)+b=0.$

(26) $x^2-\frac{xy}{\frac{dy}{dx}}=f\left[y^2-xy\frac{dy}{dx}\right].$ *Ans.* $\frac{ax^2}{y^2-a}=b,\ f(a)+b=0.$

(27) $\frac{d^4y}{dx^4}-4\frac{d^3y}{dx^3}+6\frac{d^2y}{dx^2}-4\frac{dy}{dx}+y=0.$
Ans. $y=(c_0+c_1x+c_2x^2+c_3x^3)\,e^x.$

(28) $\frac{d^2y}{dx^2}-2k\frac{dy}{dx}+k^2y=e^x.$ *Ans.* $y=(c_1+c_2x)\,e^{kx}+\frac{e^x}{(k-1)^2}.$

(29) $\frac{d^4y}{dx^4}+2\frac{d^2y}{dx^2}+y=e^x.$
Ans. $y=\frac{e^x}{4}+(A+Bx)\cos x+(C+Dx)\sin x.$

(30) $\frac{d^3y}{dx^3}-2\frac{dy}{dx}+4y=e^x\cos x.$
Ans. $y=Ae^{-2x}+e^x\left[\left(B-\frac{x}{20}\right)\cos x+\left(C+\frac{3x}{20}\right)\sin x\right].$

(31) $\frac{d^4y}{dx^4}-2\frac{d^3y}{dx^3}+\frac{d^2y}{dx^2}=x^3.$
Ans. $y=(A+Bx)\,e^x+(C+Dx)+12x^2+3x^3+\frac{x^4}{2}+\frac{x^5}{20}.$

(32) $\frac{d^2y}{dx^2}-4\frac{dy}{dx}+4y=x^2.$
Ans. $y=(A+Bx)\,e^{2x}+\frac{1}{8}(2x^2+4x+3).$

(33) $\frac{d^2y}{dx^2}+y=\cos x.$ *Ans.* $y=A\cos x+B\sin x+\frac{x}{2}\sin x.$

(34) $\frac{d^2y}{dx^2}+4y=x\sin^2x.$
Ans. $y=\left(A-\frac{x}{32}\right)\cos 2x+\left(B-\frac{x^2}{16}\right)\sin 2x+\frac{x}{8}.$

(35) $x^2\frac{d^2y}{dx^2}-x\frac{dy}{dx}+2y=x\log x.$
Ans. $y=Ax\cos(\log x)+Bx\sin(\log x)+x\log x.$

(36) $x^3\frac{d^3y}{dx^3} - x^2\frac{d^2y}{dx^2} + 2x\frac{dy}{dx} - 2y = x^3 + 3x.$

Ans. $y = x(A + B\log x) + Cx^2 + \frac{x^3}{4} - 3x\left(1 + \frac{(\log x)^2}{2}\right).$

(37) $\frac{d^2y}{dx^2} + \frac{2}{x}\frac{dy}{dx} - n^2y = 0.$ *Ans.* $y = \frac{1}{x}(Ae^{nx} + Be^{-nx}).$

(38) $\frac{d^2y}{dx^2} + \tan x\frac{dy}{dx} + \cos^2 x \, . \, y = 0.$

Ans. $y = A\cos(\sin x) + B\sin(\sin x).$

(39) $(1 - x^2)^2\frac{d^2y}{dx^2} + y = 0.$

Ans. $y = \sqrt{1 - x^2}\left(A + B\log\frac{1 + x}{1 - x}\right).$

(40) $(1 + x^2)\frac{d^2y}{dx^2} - 2x\frac{dy}{dx} + 2y = 0.$

Ans. $y = Bx + A(1 - x^2).$

(41) $\frac{d^2y}{dx^2} - \frac{x}{x - 1}\frac{dy}{dx} + \frac{1}{x - 1}y = x - 1.$

Ans. $y = Ae^x + Bx - (1 + x^2).$

(42) $x^2\frac{d^2y}{dx^2} - 2x(1 + x)\frac{dy}{dx} + 2(1 + x)y = x^3.$

Ans $y = Axe^x + Bx - \frac{x^2}{2}.$

(43) $\sin^2 x\frac{d^2y}{dx^2} - 2y = 0.$ *Ans.* $y = A\operatorname{ctn} x + B(1 - x\operatorname{ctn} x).$

(44) $\frac{d^2y}{dx^2} + \frac{1}{x^2\log x}y = e^x\left(\frac{2}{x} + \log x\right).$

Ans. $y = A\log x + e^x\log x + B\left(\log x\int\frac{dx}{\log x} - x\right).$

(45) $\frac{d^2y}{dx^2} - 2\left(n - \frac{a}{x}\right)\frac{dy}{dx} + \left(n^2 - 2\frac{na}{x}\right)y = e^{nx}.$

Ans. $y = e^{nx}\left(A + \frac{B}{x^{2a-1}} + \frac{x^2}{2(2a - 1)}\right).$

(46) $\frac{d^2y}{dx^2} - \frac{2}{x}\frac{dy}{dx} + \left(a^2 + \frac{2}{x^2}\right)y = 0.$

Ans. $y = x(A\cos ax + B\sin ax).$

(47) $\frac{d^2y}{dx^2} - 2bx\frac{dy}{dx} + b^2x^2y = 0.$

Ans. $y = e^{\frac{bx^2}{2}}(A\cos x\sqrt{b} + B\sin x\sqrt{b}).$

(48) $\frac{d^2y}{dx^2} - 4x\frac{dy}{dx} + (4x^2 - 3)y = e^{x^2}.$

Ans. $y = e^{x^2}(Ae^x + Be^{-x} - 1).$

(49) $(1 - x^2)\frac{d^2y}{dx^2} - 4x\frac{dy}{dx} - (1 + x^2)y = x.$

Ans. $y = \frac{1}{1 - x^2}(x + A\cos x + B\sin x).$

(50) $\frac{d^2y}{dx^2} - \frac{1}{\sqrt{x}}\frac{dy}{dx} + \frac{x + \sqrt{x} - 8}{4x^2}y = 0.$

Ans. $y = e^{\sqrt{x}}\left(Ax^2 + \frac{B}{x}\right).$

(51) $\frac{d^2y}{dx^2} + 2n\operatorname{ctn} nx\frac{dy}{dx} + (m^2 - n^2)y = 0.$

Ans. $y = (A\cos mx + B\sin mx)\csc nx.$

(52) $(x^2 - 1)\frac{d^2x}{dx^2} + x\frac{dy}{dx} - c^2y = 0.$

Ans. $y = A(x + \sqrt{x^2 - 1})^c + B(x - \sqrt{x^2 - 1})^c.$

(53) $\frac{d^2y}{dx^2} + \frac{2}{x}\frac{dy}{dx} + \frac{a^2}{x^4}y = 0.$ *Ans.* $y = A\sin\frac{a}{x} + B\cos\frac{a}{x}.$

(54) $\frac{d^2y}{dx^2} - \frac{3x + 1}{x^2 - 1}\frac{dy}{dx} + y\left[\frac{6(x + 1)}{(x - 1)(3x + 5)}\right]^2 = 0.$

Ans. $y = [A + B\log((x - 1)^3(3x + 5))]\sqrt{(x - 1)^3(3x + 5)}.$

(55) $(1 - x^2)\frac{d^2y}{dx^2} - x\frac{dy}{dx} - c^2y = 0.$

Ans. $y = Ae^{c\sin^{-1}x} + Be^{-c\sin^{-1}x}.$

(56) $(1 + ax^2)\frac{d^2y}{dx^2} + ax\frac{dy}{dx} - n^2y = 0.$

Ans. $y = A(\sqrt{1 + ax^2} + x\sqrt{a})^{\frac{n}{\sqrt{a}}} + B(\sqrt{1 + ax^2} + x\sqrt{a})^{-\frac{n}{\sqrt{a}}}.$

(57) $(x - 1)(x - 2)\frac{d^2y}{dx^2} - (2x - 3)\frac{dy}{dx} + 2y = 0.$

Ans. $y = c(x - 2)^2 + c'(x - 2)[(x - 2)\log(x - 2) - 1].$

(58) $(3 - x)\dfrac{d^2 y}{dx^2} - (9 - 4x)\dfrac{dy}{dx} + (6 - 3x)\, y = 0.$

Ans. $y = ce^x + c'e^{3x}\left(\dfrac{183}{8} - \dfrac{75}{4}x + \dfrac{21}{4}x^2 - \dfrac{1}{2}x^3\right).$

(59) $(a^2 - x^2)\dfrac{d^2 y}{dx^2} - 8x\dfrac{dy}{dx} - 12y = 0.$

Ans. $y = \dfrac{c}{(a - x)^3} + c'\dfrac{(a^2 + 3x^2)}{(a^2 - x^2)^3}.$

(60) $\dfrac{d^2 y}{dx^2} + \dfrac{2}{x}\dfrac{dy}{dx} + \left(n^2 - \dfrac{2}{x^2}\right) y = 0.$

Ans. $y = \dfrac{1}{x^2}\left[A(\sin nx - nx \cos nx) + B(\cos nx + nx \sin nx)\right].$

(61) $\dfrac{d^2 y}{dx^2} + \dfrac{1}{x}\dfrac{dy}{dx} = 0.$ *Ans.* $y = c \log x + c'.$

(62) $\left(2x^3\dfrac{dy}{dx} + x^2 y\right)\dfrac{d^2 y}{dx^2} + 4x^2\left(\dfrac{dy}{dx}\right)^2 + 2xy\dfrac{dy}{dx} = 0.$

Ans. $c^2 + cxy = c'x.$

(63) $\left(x^2 + 2y^2\dfrac{dy}{dx}\right)\dfrac{d^2 y}{dx^2} + 2y\left(\dfrac{dy}{dx}\right)^3 + 3x\dfrac{dy}{dx} + y = 0,$

Find a first integral.

Ans. $x^2\dfrac{dy}{dx} + y^2\left(\dfrac{dy}{dx}\right)^2 + xy = c.$

(64) $x^2\dfrac{d^3 y}{dx^3} + x\dfrac{d^2 y}{dx^2} + (2xy - 1)\dfrac{dy}{dx} + y^2 - 0.$

Find a first integral.

Ans. $x^2\dfrac{d^2 y}{dx^2} - x\dfrac{dy}{dx} + xy^2 = c.$

(65) $\dfrac{dx}{x - a} + \dfrac{dy}{y - b} + \dfrac{dz}{z - c} = 0.$ *Ans.* $(x - a)(y - b)(z - c) = c.$

(66) $(y + z)\, dx + dy + dz = 0.$ *Ans.* $e^x (y + z) = c.$

(67) $\dfrac{dx}{dt} + 4x + \dfrac{y}{4} = 0, \quad \dfrac{dy}{dt} + 3y - x = 0.$

Ans. $x = ce^{-\frac{7t}{2}} - \dfrac{y}{2}, \quad y = (ct + c_1)\, e^{-\frac{7t}{2}}.$

(68) $\frac{d^2x}{dt^2} + m^2x = 0, \quad \frac{d^2y}{dt^2} - m^2x = 0.$

Ans. $x = A\sin mt + B\cos mt, \quad x + y = Ct + D.$

(69) $D_x z = \frac{y}{x+z}$. *Ans.* $e^{-\frac{z}{y}}(x+y+z) = \phi y.$

(70) $xzD_x z + yzD_y z = xy.$ *Ans.* $z^2 = xy + \phi\left(\frac{x}{y}\right)$.

(71) $D_x z \,.\, D_y z = 1.$ *Ans.* $z = ax + \frac{y}{a} + b.$

(72) $x^2D_x^2 z + 2xyD_xD_y z + y^2D_y^2 z = 0.$ *Ans.* $z = x\phi\left(\frac{y}{x}\right) + \psi\left(\frac{y}{x}\right)$.

(73) $(D_y z)^2 D_x^2 z - 2D_x z D_y z D_x D_y z + (D_x z)^2 D_y^2 z = 0.$

Ans. $y = x\phi z + \psi z.$

(74) $D_x z \,.\, D_x D_y z - D_y z \,.\, D_x^2 z = 0.$ *Ans.* $x = \phi y + \psi z.$

APPENDIX.

CHAPTER V.

INTEGRATION.

74. We are now able to extend materially our list of *formulas for direct integration* (Art. 55), one of which may be obtained from each of the derivative formulas in our last chapter. The following set contains the most important of these: —

$D_x \log x = \frac{1}{x}$ gives $\int_x \frac{1}{x} = \log x.$

$D_x a^x = a^x \log a$ " $\int_x a^x \log a = a^x.$

$D_x e^x = e^x$ " $\int_x e^x = e^x.$

$D_x \sin x = \cos x$ " $\int_x \cos x = \sin x.$

$D_x \cos x = -\sin x$ " $\int_x (-\sin x) = \cos x.$

$D_x \log \sin x = \operatorname{ctn} x$ " $\int_x \operatorname{ctn} x = \log \sin x.$

$D_x \log \cos x = -\tan x$ " $\int_x (-\tan x) = \log \cos x.$

$D_x \sin^{-1} x = \frac{1}{\sqrt{(1-x^2)}}$ " $\int_x \frac{1}{\sqrt{(1-x^2)}} = \sin^{-1} x.$

$D_x \tan^{-1} x = \frac{1}{1+x^2}$ " $\int_x \frac{1}{1+x^2} = \tan^{-1} x.$

$D_x \operatorname{vers}^{-1} x = \frac{1}{\sqrt{(2x-x^2)}}$ " $\int_x \frac{1}{\sqrt{(2x-x^2)}} = \operatorname{vers}^{-1} x.$

The second, fifth, and seventh in the second group can be written in the more convenient forms,

$$\int_x a^x = \frac{a^x}{\log a};$$

$$\int_x \sin x = -\cos x;$$

$$\int_x \tan x = -\log\cos x.$$

75. When the expression to be integrated does not come under any of the forms in the preceding list, *it can often be prepared for integration by a suitable change of variable*, the new variable, of course, being a function of the old. This method is called *integration by substitution*, and is based upon a formula easily deduced from $D_x(Fy) = D_y Fy \,.\, D_x y$;

which gives immediately

$$Fy = \int_x (D_y Fy . D_x y).$$

Let $u = D_y Fy$,

then $Fy = \int_y u$,

and we have $\int_y u = \int_x (u D_x y)$;

or, interchanging x and y,

$$\int_x u = \int_y (u D_y x). \qquad [1]$$

For example, required $\int_x (a + bx)^n$.

Let $z = a + bx$,

and then $\int_x (a + bx)^n = \int_x z^n = \int_z (z^n \,.\, D_z x)$, by [1];

but $x = \frac{z}{b} - \frac{a}{b}$,

$$D_z x = \frac{1}{b};$$

hence $\int_x (a + bx)^n = \frac{1}{b} \int_z z^n = \frac{1}{b} \frac{z^{n+1}}{n+1}.$

Substituting for z its value, we have

$$\int_x (a+bx)^n = \frac{1}{b}\frac{(a+bx)^{n+1}}{n+1}.$$

EXAMPLE.

Find $\int_x \frac{1}{a+bx}$. *Ans.* $\frac{1}{b}\log(a+bx)$.

76. If fx represents a function that can be integrated, $f(a+bx)$ can always be integrated; for, if

$$z = a + bx,$$

then $$D_z x = \frac{1}{b}$$

and $$\int_x f(a+bx) = \int_x fz = \int_z fz D_z x = \frac{1}{b}\int_z fz.$$

EXAMPLES.

Find

(1) $\int_x \sin ax$. *Ans.* $-\frac{1}{a}\cos ax$.

(2) $\int_x \cos ax$. *Ans.* $\frac{1}{a}\sin ax$.

(3) $\int_x \tan ax$.

(4) $\int_x \operatorname{ctn} ax$.

77. *Required* $\int_x \frac{1}{\sqrt{(a^2-x^2)}}$.

$$\int_x \frac{1}{\sqrt{(a^2-x^2)}} = \frac{1}{a}\int_x \frac{1}{\sqrt{\left[1-\left(\frac{x}{a}\right)^2\right]}}.$$

Let $$z = \frac{x}{a},$$

then $$x = az,$$

$$D_z x = a,$$

$$\frac{1}{a}\int_x \frac{1}{\sqrt{\left[1-\left(\frac{x}{a}\right)^2\right]}} = \frac{1}{a}\int_x \frac{1}{\sqrt{(1-z^2)}} = \frac{1}{a}\int_z \frac{1}{\sqrt{(1-z^2)}}\, D_z x$$

$$= \int_z \frac{1}{\sqrt{(1-z^2)}} = \sin^{-1} z = \sin^{-1}\frac{x}{a}.$$

EXAMPLES.

Find

(1) $\int_x \frac{1}{a^2+x^2}$. *Ans.* $\frac{1}{a}\tan^{-1}\frac{x}{a}$.

(2) $\int_x \frac{1}{\sqrt{(2ax-x^2)}}$. *Ans.* $\text{vers}^{-1}\frac{x}{a}$.

78. *Required* $\int_x \frac{1}{\sqrt{(x^2+a^2)}}$.

Let $$z = x + \sqrt{(x^2+a^2)};$$

then $$z - x = \sqrt{(x^2+a^2)},$$

$$z^2 - 2zx + x^2 = x^2 + a^2,$$

$$2zx = z^2 - a^2,$$

$$x = \frac{z^2-a^2}{2z},$$

$$\sqrt{(x^2+a^2)} = z - x = z - \frac{z^2-a^2}{2z} = \frac{z^2+a^2}{2z},$$

$$D_z x = \frac{z^2+a^2}{2z^2}.$$

$$\int_x \frac{1}{\sqrt{(x^2+a^2)}} = \int_x \frac{2z}{z^2+a^2} = \int_z \frac{2z}{z^2+a^2}\, D_z x$$

$$= \int_z \frac{2z}{z^2+a^2}\,\frac{z^2+a^2}{2z^2} = \int_z \frac{1}{z} = \log z = \log(x+\sqrt{x^2+a^2}).$$

EXAMPLE.

Find $\int_x \frac{1}{\sqrt{(x^2-a^2)}}$. *Ans.* $\log(x+\sqrt{x^2-a^2})$.

79. *When the expression to be integrated can be factored*, the required integral can often be obtained by the use of a formula

deduced from $$D_x(uv) = uD_xv + vD_xu,$$

which gives $$uv = \int_x uD_xv + \int_x vD_xu$$

or $$\int_x uD_xv = uv - \int_x vD_xu. \qquad [1]$$

This method is called *integrating by parts*.

(*a*) For example, required $\int_x \log x$.

$\log x$ can be regarded as the product of $\log x$ by 1.

Call $$\log x = u \text{ and } 1 = D_xv,$$

then $$D_xu = \frac{1}{x},$$

$$v = x;$$

and we have

$$\int_x \log x = \int_x 1 \log x = \int_x uD_xv = uv - \int_x vD_xu$$

$$= x\log x - \int_x \frac{x}{x} = x\log x - x.$$

EXAMPLE.

Find $\int_x x\log x$.

Suggestion: Let $\log x = u$ and $x = D_xv$.

Ans. $\frac{1}{2}x^2\left(\log x - \frac{1}{2}\right)$

80. *Required* $\int_x \sin^2 x$.

Let $$u = \sin x \text{ and } D_xv = \sin x,$$

then $$D_xu = \cos x,$$

$$v = -\cos x,$$

$$\int_x \sin^2 x = -\sin x\cos x + \int_x \cos^2 x;$$

but $$\cos^2 x = 1 - \sin^2 x,$$

so $$\int_x \cos^2 x = \int_x 1 - \int_x \sin^2 x = x - \int_x \sin^2 x$$

and $$\int_x \sin^2 x = x - \sin x \cos x - \int_x \sin^2 x.$$

$$2\int_x \sin^2 x = x - \sin x \cos x.$$

$$\int_x \sin^2 x = \tfrac{1}{2}(x - \sin x \cos x).$$

EXAMPLES.

(1) Find $\int_x \cos^2 x$. *Ans.* $\frac{1}{2}(x + \sin x \cos x)$.

(2) $\int_x \sin x \cos x$. *Ans.* $\frac{\sin^2 x}{2}$.

81. *Very often both methods described above are required in the same integration.*

(*a*) *Required* $\int_x \sin^{-1} x$.

Let $$\sin^{-1} x = y,$$

then $$x = \sin y;$$

$$D_y x = \cos y,$$

$$\int_x \sin^{-1} x = \int_x y = \int_y y \cos y.$$

Let $$u = y \text{ and } D_y v = \cos y;$$

then $$D_y u = 1,$$

$$v = \sin y,$$

and

$$\int_y y \cos y = y \sin y - \int_y \sin y = y \sin y + \cos y = x \sin^{-1} x + \sqrt{(1 - x^2)}.$$

Any inverse or anti-function can be integrated by this method if the direct function is integrable.

(*b*) Thus, $$\int_x f^{-1} x = \int_x y = \int_y y D_y f y = y f y - \int_y f y$$

where $$y = f^{-1} x.$$

EXAMPLES.

(1) Find $\int_x \cos^{-1} x$. *Ans.* $x\cos^{-1}x - \sqrt{(1-x^2)}$.

(2) $\int_x \tan^{-1} x$. *Ans.* $x\tan^{-1}x - \frac{1}{2}\log(1+x^2)$.

(3) $\int_x \operatorname{vers}^{-1} x$. *Ans.* $(x-1)\operatorname{vers}^{-1}x + \sqrt{(2x-x^2)}$.

82. Sometimes an *algebraic transformation, either alone or in combination with the preceding methods, is useful.*

(*a*) *Required* $\int_x \frac{1}{x^2 - a^2}$.

$$\frac{1}{x^2-a^2} = \frac{1}{2a}\left(\frac{1}{x-a} - \frac{1}{x+a}\right),$$

and, by Art. 75 (Ex.),

$$\int_x \frac{1}{x^2-a^2} = \frac{1}{2a}[\log(x-a) - \log(x+a)] = \frac{1}{2a}\log\frac{x-a}{x+a}.$$

(*b*) *Required* $\int_x \sqrt{\left(\frac{1+x}{1-x}\right)}$.

$$\sqrt{\left(\frac{1+x}{1-x}\right)} = \frac{1+x}{\sqrt{(1-x^2)}} = \frac{1}{\sqrt{(1-x^2)}} + \frac{x}{\sqrt{(1-x^2)}},$$

$$\int_x \frac{1}{\sqrt{(1-x^2)}} = \sin^{-1}x.$$

$\int_x \frac{x}{\sqrt{(1-x^2)}}$ can be readily obtained by *substituting* $y = (1-x^2)$, and is $-\sqrt{(1-x^2)}$;

hence $$\int_x \sqrt{\left(\frac{1+x}{1-x}\right)} = \sin^{-1}x - \sqrt{(1-x^2)}.$$

(*c*) *Required* $\int_x \sqrt{(a^2-x^2)}$.

$$\sqrt{(a^2-x^2)} = \frac{a^2-x^2}{\sqrt{(a^2-x^2)}} = \frac{a^2}{\sqrt{(a^2-x^2)}} - \frac{x^2}{\sqrt{(a^2-x^2)}},$$

and $$\int_x \sqrt{(a^2-x^2)} = a^2 \int_x \frac{1}{\sqrt{(a^2-x^2)}} - \int_x \frac{x^2}{\sqrt{(a^2-x^2)}},$$

whence $$\int_x \sqrt{(a^2-x^2)} = a^2 \sin^{-1}\frac{x}{a} - \int_x \frac{x^2}{\sqrt{(a^2-x^2)}}, \text{ by Art. 77;}$$

but $$\int_x \sqrt{(a^2-x^2)} = x\sqrt{(a^2-x^2)} + \int_x \frac{x^2}{\sqrt{(a^2-x^2)}},$$

by integration by parts, if we let

$$u = \sqrt{(a^2-x^2)} \text{ and } D_x v = 1.$$

Adding our two equations, we have

$$2\int_x \sqrt{(a^2-x^2)} = x\sqrt{(a^2-x^2)} + a^2 \sin^{-1}\frac{x}{a};$$

and $$\therefore \int_x \sqrt{(a^2-x^2)} = \frac{1}{2}\left(x\sqrt{a^2-x^2} + a^2\sin^{-1}\frac{x}{a}\right).$$

EXAMPLES.

Find

(1) $\int_x \sqrt{(x^2+a^2)}$.

Ans. $\frac{1}{2}[x\sqrt{(x^2+a^2)} + a^2\log(x+\sqrt{x^2+a^2})]$.

(2) $\int_x \sqrt{(x^2-a^2)}$.

Ans. $\frac{1}{2}[x\sqrt{(x^2-a^2)} - a^2\log(x+\sqrt{x^2-a^2})]$.

Applications.

83. *To find the area of a segment of a circle.*

Let the equation of the circle be

$$x^2+y^2=a^2,$$

and let the required segment be cut off by the double ordinates through (x_0,y_0) and (x,y). Then the required area

$$A = 2\int_x y + C.$$

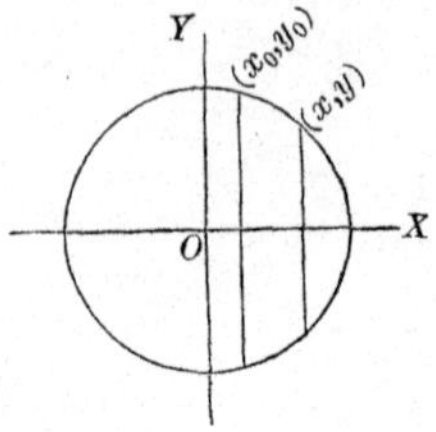

From the equation of the circle,

$$y = \sqrt{(a^2 - x^2)},$$

hence $$A = 2 \int_x \sqrt{(a^2 - x^2)} + C;$$

and therefore, by Art. 82 (c),

$$A = x\sqrt{(a^2 - x^2)} + a^2 \sin^{-1}\frac{x}{a} + C.$$

As the area is measured from the ordinate y_0 to the ordinate y,

$$A = 0 \text{ when } x = x_0;$$

therefore $$0 = x_0\sqrt{(a^2 - x_0^2)} + a^2 \sin^{-1}\frac{x_0}{a} + C,$$

$$C = -x_0\sqrt{(a^2 - x_0^2)} - a^2 \sin^{-1}\frac{x_0}{a},$$

and we have

$$A = x\sqrt{(a^2 - x^2)} + a^2 \sin^{-1}\frac{x}{a} - x_0\sqrt{(a^2 - x_0^2)} - a^2 \sin^{-1}\frac{x_0}{a}$$

If $x_0 = 0$, and *the segment begins with the axis of* Y,

$$A = x\sqrt{(a^2 - x^2)} + a^2 \sin^{-1}\frac{x}{a}.$$

If, at the same time, $x = a$, *the segment becomes a semicircle*, and

$$A = a\sqrt{(a^2 - a^2)} + a^2 \sin^{-1}\frac{a}{a} = \frac{\pi a^2}{2}.$$

The area of the whole circle is πa^2.

EXAMPLES.

(1) Show that, in the case of an ellipse,

$$\frac{x^2}{a^2}+\frac{y^2}{b^2}=1,$$

the area of a segment beginning with any ordinate y_0 is

$$A=\frac{b}{a}\left[x\sqrt{(a^2-x^2)}+a^2\sin^{-1}\frac{x}{a}-x_0\sqrt{(a^2-x_0^2)}-a^2\sin^{-1}\frac{x_0}{a}\right].$$

That if the segment begins with the minor axis,

$$A=\frac{b}{a}\left[x\sqrt{(a^2-x^2)}+a^2\sin^{-1}\frac{x}{a}\right].$$

That the area of the whole ellipse is πab.

(2) The area of a segment of the hyperbola

$$\frac{x^2}{a^2}-\frac{y^2}{b^2}=1$$

is

$$A=\frac{b}{a}[x\sqrt{(x^2-a^2)}-a^2\log(x+\sqrt{x^2-a^2})$$
$$-x_0\sqrt{(x_0^2-a^2)}+a^2\log(x_0+\sqrt{x_0^2-a^2})].$$

If $x_0=a$, and the segment begins at the vertex,

$$A=\frac{b}{a}[x\sqrt{(x^2-a^2)}-a^2\log(x+\sqrt{x^2-a^2})+a^2\log a].$$

84. *To find the length of any arc of a circle*, the coördinates of its extremities being (x_0,y_0) and (x,y).

By Art. 52, $$s=\int_x\sqrt{[1+(D_xy)^2]}.$$

From the equation of the circle,

$$x^2+y^2=a^2,$$

we have
$$2x + 2yD_x y = 0,$$

$$D_x y = -\frac{x}{y},$$

$$1 + (D_x y)^2 = \frac{x^2 + y^2}{y^2} = \frac{a^2}{y^2},$$

$$s = \int_x \frac{a}{y} = a\int_x \frac{1}{\sqrt{(a^2 - x^2)}} = a \sin^{-1}\frac{x}{a} + C. \quad \text{(Art. 77.)}$$

When
$$x = x_0, \qquad s = 0;$$

hence
$$0 = a \sin^{-1}\frac{x_0}{a} + C,$$

$$C = -a \sin^{-1}\frac{x_0}{a},$$

and
$$s = a\left(\sin^{-1}\frac{x}{a} - \sin^{-1}\frac{x_0}{a}\right).$$

If $x_0 = 0$, and *the arc is measured from the highest point of the circle,*
$$s = a \sin^{-1}\frac{x}{a}.$$

If the arc is a quadrant, $x = a$,
$$s = a \sin^{-1}(1) = \frac{\pi a}{2},$$

and the whole circumference $= 2\pi a$.

85. *To find the length of an arc of the parabola* $y^2 = 2mx$.

We have
$$2yD_x y = 2m;$$

$$D_x y = \frac{m}{y};$$

$$\sqrt{[1 + (D_x y)^2]} = \sqrt{\left(\frac{m^2 + y^2}{y^2}\right)} = \frac{1}{y}\sqrt{(m^2 + y^2)};$$

$$s = \int_x \left[\frac{1}{y}\sqrt{(m^2+y^2)}\right] = \int_y \left[\frac{1}{y}\sqrt{(m^2+y^2)}\ D_y x\right];$$

$$D_y x = \frac{1}{D_x y} = \frac{y}{m}, \qquad \text{by Art. 73;}$$

$$s = \frac{1}{m}\int_y \sqrt{m^2+y^2} = \frac{1}{2m}[y\sqrt{m^2+y^2} + m^2\log(y+\sqrt{m^2+y^2})] + C,$$

by Art. 82, Ex. 1.

If the arc is measured from the vertex,

$$s = 0 \text{ when } y = 0;$$

$$0 = \frac{1}{2m}(m^2 \log m) + C,$$

$$C = -\frac{1}{2} m \log m,$$

and
$$s = \frac{1}{2}\left[\frac{y\sqrt{(m^2+y^2)}}{m} + m \log \frac{y+\sqrt{(m^2+y^2)}}{m}\right].$$

EXAMPLE.

Find the length of the arc of the curve $x^3 = 27y^2$ included between the origin and the point whose abscissa is 15.

Ans. 19.

A

SHORT TABLE OF INTEGRALS.

COMPILED BY

B. O. PEIRCE,

HOLLIS PROFESSOR OF MATHEMATICS AND NATURAL PHILOSOPHY
IN HARVARD UNIVERSITY.

BOSTON, U.S.A.:
PUBLISHED BY GINN & COMPANY.
1892.

The compiler will be grateful to any person who may send notice of errors in these formulas to

B. O. PEIRCE,

Harvard College, Cambridge.

I. FUNDAMENTAL FORMS.

---

1. $\int a\,dx = ax.$

2. $\int af(x)\,dx = a\int f(x)\,dx.$

3. $\int \frac{dx}{x} = \log x.$

4. $\int x^m\,dx = \frac{x^{m+1}}{m+1}$, when m is different from -1.

5. $\int e^x\,dx = e^x.$

6. $\int a^x \log a\,dx = a^x.$

7. $\int \frac{dx}{1+x^2} = \tan^{-1} x.$

8. $\int \frac{dx}{\sqrt{1-x^2}} = \sin^{-1} x.$

9. $\int \frac{dx}{x\sqrt{x^2-1}} = \sec^{-1} x.$

10. $\int \frac{dx}{\sqrt{2x-x^2}} = \operatorname{versin}^{-1} x.$

11. $\int \cos x\,dx = \sin x.$

12. $\int \sin x\,dx = -\cos x.$

13. $\int \operatorname{ctn} x\, dx = \log \sin x.$

14. $\int \tan x\, dx = -\log \cos x.$

15. $\int \tan x \sec x\, dx = \sec x.$

16. $\int \sec^2 x\, dx = \tan x.$

17. $\int \csc^2 x\, dx = -\operatorname{ctn} x.$

In the following formulas, u, v, w, and y represent any functions of x:

18. $\int (u + v + w + \text{etc.})\, dx = \int u dx + \int v dx + \int w dx + \text{etc.}$

19a. $\int u dv = uv - \int v du.$

19b. $\int u \frac{dv}{dx} dx = uv - \int v \frac{du}{dx} dx.$

20. $\int f(y)\, dx = \int \frac{f(y)\, dy}{\frac{dy}{dx}}.$

II. RATIONAL ALGEBRAIC FUNCTIONS.

A.— Expressions Involving $(a+bx)$.

The substitution of y or z for x, where $y = xz = a + bx$, gives

21. $\int (a+bx)^m \, dx = \frac{1}{b}\int y^m \, dy.$

22. $\int x(a+bx)^m \, dx = \frac{1}{b^2}\int y^m (y-a) dy.$

23. $\int x^n (a+bx)^m dx = \frac{1}{b^{n+1}}\int y^m (y-a)^n \, dy.$

24. $\int \frac{x^n \, dx}{(a+bx)^m} = \frac{1}{b^{n+1}}\int \frac{(y-a)^n \, dy}{y^m}.$

25. $\int \frac{dx}{x^n (a+bx)^m} = -\frac{1}{a^{m+n-1}}\int \frac{(z-b)^{m+n-2} \, dz}{z^m}.$

Whence

26. $\int \frac{dx}{a+bx} = \frac{1}{b}\log(a+bx).$

27. $\int \frac{dx}{(a+bx)^2} = -\frac{1}{b(a+bx)}.$

28. $\int \frac{dx}{(a+bx)^3} = -\frac{1}{2b(a+bx)^2}.$

29. $\int \frac{x \, dx}{a+bx} = \frac{1}{b^2}[a + bx - a\log(a+bx)].$

30. $\int \frac{x \, dx}{(a+bx)^2} = \frac{1}{b^2}\left[\log(a+bx) + \frac{a}{a+bx}\right].$

31. $\displaystyle\int\frac{x\,dx}{(a+bx)^3}=\frac{1}{b^2}\left[-\frac{1}{a+bx}+\frac{a}{2(a+bx)^2}\right].$

32. $\displaystyle\int\frac{x^2\,dx}{a+bx}=\frac{1}{b^3}\left[\tfrac{1}{2}(a+bx)^2-2a(a+bx)+a^2\log(a+bx)\right].$

33. $\displaystyle\int\frac{x^2\,dx}{(a+bx)^2}=\frac{1}{b^3}\left[a+bx-2a\log(a+bx)-\frac{a^2}{a+bx}\right].$

34. $\displaystyle\int\frac{dx}{x(a+bx)}=-\frac{1}{a}\log\frac{a+bx}{x}.$

35. $\displaystyle\int\frac{dx}{x(a+bx)^2}=\frac{1}{a(a+bx)}-\frac{1}{a^2}\log\frac{a+bx}{x}.$

36. $\displaystyle\int\frac{dx}{x^2(a+bx)}=-\frac{1}{ax}+\frac{b}{a^2}\log\frac{a+bx}{x}.$

B. — Expressions Involving $(a+bx^n)$.

37. $\displaystyle\int\frac{dx}{c^2+x^2}=\frac{1}{c}\tan^{-1}\frac{x}{c}.$

38. $\displaystyle\int\frac{dx}{c^2-x^2}=\frac{1}{2c}\log\frac{c+x}{c-x}.$

39. $\displaystyle\int\frac{dx}{a+bx^2}=\frac{1}{\sqrt{ab}}\tan^{-1}x\sqrt{\frac{b}{a}},$ if $a>0,\ b>0.$

40. $\displaystyle\int\frac{dx}{a+bx^2}=\frac{1}{2\sqrt{-ab}}\log\frac{\sqrt{a}+x\sqrt{-b}}{\sqrt{a}-x\sqrt{-b}},$ if $a>0,\ b<0.$

41. $\displaystyle\int\frac{dx}{(a+bx^2)^2}=\frac{x}{2a(a+bx^2)}+\frac{1}{2a}\int\frac{dx}{a+bx^2}.$

42. $\displaystyle\int\frac{dx}{(a+bx^2)^{m+1}}=\frac{1}{2ma}\,\frac{x}{(a+bx^2)^m}+\frac{2m-1}{2ma}\int\frac{dx}{(a+bx^2)^m}.$

43. $\displaystyle\int\frac{x\,dx}{a+bx^2}=\frac{1}{2b}\log\left(x^2+\frac{a}{b}\right).$

44. $$\int \frac{x\,dx}{(a+bx^2)^{m+1}} = \frac{1}{2}\int \frac{dz}{(a+bz)^{m+1}}, \text{ where } z = x^2.$$

45. $$\int \frac{dx}{x(a+bx^2)} = \frac{1}{2a}\log\frac{x^2}{a+bx^2}.$$

46. $$\int \frac{x^2\,dx}{a+bx^2} = \frac{x}{b} - \frac{a}{b}\int \frac{dx}{a+bx^2}.$$

47. $$\int \frac{dx}{x^2(a+bx^2)} = -\frac{1}{ax} - \frac{b}{a}\int \frac{dx}{a+bx^2}.$$

48. $$\int \frac{x^2\,dx}{(a+bx^2)^{m+1}} = \frac{-x}{2mb(a+bx^2)^m} + \frac{1}{2mb}\int \frac{dx}{(a+bx^2)^m}.$$

49. $$\int \frac{dx}{x^2(a+bx^2)^{m+1}} = \frac{1}{a}\int \frac{dx}{x^2(a+bx^2)^m} - \frac{b}{a}\int \frac{dx}{(a+bx^2)^{m+1}}.$$

50. $$\int \frac{dx}{a+bx^3} = \frac{k}{3a}\left[\tfrac{1}{2}\log\left(\frac{(k+x)^2}{k^2-kx+x^2}\right) + \sqrt{3}\tan^{-1}\frac{2x-k}{k\sqrt{3}}\right],$$

where $bk^3 = a$.

51. $$\int \frac{x\,dx}{a+bx^3} = \frac{1}{3bk}\left[\tfrac{1}{2}\log\left(\frac{k^2-kx+x^2}{(k+x)^2}\right) + \sqrt{3}\tan^{-1}\frac{2x-k}{k\sqrt{3}}\right],$$

where $bk^3 = a$.

52. $$\int \frac{dx}{x(a+bx^n)} = \frac{1}{an}\log\frac{x^n}{a+bx^n}.$$

53. $$\int \frac{dx}{(a+bx^n)^{m+1}} = \frac{1}{a}\int \frac{dx}{(a+bx^n)^m} - \frac{b}{a}\int \frac{x^n\,dx}{(a+bx^n)^{m+1}}.$$

54. $$\int \frac{x^m\,dx}{(a+bx^n)^{p+1}} = \frac{1}{b}\int \frac{x^{m-n}}{(a+bx^n)^p} - \frac{a}{b}\int \frac{x^{m-n}\,dx}{(a+bx^n)^{p+1}}.$$

55. $$\int \frac{dx}{x^m(a+bx^n)^{p+1}} = \frac{1}{a}\int \frac{dx}{x^m(a+bx^n)^p} - \frac{b}{a}\int \frac{dx}{x^{m-n}(a+bx^n)^{p+1}}.$$

56. $$\int x^{m-1}(a+bx^n)^p\,dx = \begin{cases} \dfrac{1}{b(m+np)}\left[x^{m-n}(a+bx^n)^{p+1}-(m-n)a\displaystyle\int x^{m-n-1}(a+bx^n)^p\,dx\right]. \\[2ex] \dfrac{1}{m-np}\left[x^m(a+bx^n)^p+npa\displaystyle\int x^{m-1}(a+bx^n)^{p-1}dx\right]. \\[2ex] \dfrac{1}{ma}\left[x^m(a+bx^n)^{p+1}-(m+np+n)b\displaystyle\int x^{m+n-1}(a+bx^n)^p\,dx\right]. \\[2ex] \dfrac{1}{an(p+1)}\left[-x^m(a+bx^n)^{p+1}+(m+np+n)\displaystyle\int x^{m-1}(a+bx^n)^{p+1}dx\right]. \end{cases}$$

C. — Expressions Involving $(a+bx+cx^2)$.

Let $X=a+bx+cx^2$ and $q=4ac-b^2$, then

57. $\displaystyle\int\frac{dx}{X}=\frac{2}{\sqrt{q}}\tan^{-1}\frac{2cx+b}{\sqrt{q}}$, when $q>0$.

58. $\displaystyle\int\frac{dx}{X}=\frac{1}{\sqrt{-q}}\log\frac{2cx+b-\sqrt{-q}}{2cx+b+\sqrt{-q}}$, when $q<0$.

59. $\displaystyle\int\frac{dx}{X^2}=\frac{2cx+b}{qX}+\frac{2c}{q}\int\frac{dx}{X}.$

60. $\displaystyle\int\frac{dx}{X^3}=\frac{2cx+b}{q}\left(\frac{1}{2X^2}+\frac{3c}{qX}\right)+\frac{6c^2}{q^2}\int\frac{dx}{X}.$

61. $\displaystyle\int\frac{dx}{X^{n+1}}=\frac{2cx+b}{nqX^n}+\frac{2(2n-1)c}{qn}\int\frac{dx}{X^n}.$

62. $\displaystyle\int\frac{x\,dx}{X}=\frac{1}{2c}\log X-\frac{b}{2c}\int\frac{dx}{X}.$

63. $$\int \frac{x\,dx}{X^2} = -\frac{bx + 2a}{qX} - \frac{b}{q}\int \frac{dx}{X}.$$

64. $$\int \frac{x\,dx}{X^{n+1}} = -\frac{2a + bx}{nqX^n} - \frac{b(2n-1)}{nq}\int \frac{dx}{X^n}.$$

65. $$\int \frac{x^2}{X}dx = \frac{x}{c} - \frac{b}{2c^2}\log X + \frac{b^2 - 2ac}{2c^2}\int \frac{dx}{X}.$$

66. $$\int \frac{x^2}{X^2}dx = \frac{(b^2 - 2ac)x + ab}{cqX} + \frac{2a}{q}\int \frac{dx}{X}.$$

67. $$\int \frac{x^m\,dx}{X^{n+1}} = -\frac{x^{m-1}}{(2n - m + 1)cX^n} - \frac{n - m + 1}{2n - m + 1}\cdot\frac{b}{c}\int \frac{x^{m-1}dx}{X^{n+1}}$$
$$+ \frac{m-1}{2n - m + 1}\cdot\frac{a}{c}\int \frac{x^{m-2}dx}{X^{n+1}}.$$

68. $$\int \frac{dx}{xX} = \frac{1}{2a}\log\frac{x^2}{X} - \frac{b}{2a}\int \frac{dx}{X}.$$

69. $$\int \frac{dx}{x^2X} = \frac{b}{2a^2}\log\frac{X}{x^2} - \frac{1}{ax} + \left(\frac{b^2}{2a^2} - \frac{c}{a}\right)\int \frac{dx}{X}.$$

70. $$\int \frac{dx}{x^mX^{n+1}} = -\frac{1}{(m-1)ax^{m-1}X^n} - \frac{n + m - 1}{m-1}\cdot\frac{b}{a}\int \frac{dx}{x^{m-1}X^{n+1}}$$
$$- \frac{2n + m - 1}{m-1}\cdot\frac{c}{a}\int \frac{dx}{x^{m-2}X^{n+1}}.$$

D. — Rational Fractions.

Every proper fraction can be represented by the general form:

$$\frac{f(x)}{F(x)} = \frac{g_1x^{n-1} + g_2x^{n-2} + g_3x^{n-3} + \cdots + g_n}{x^n + k_1x^{n-1} + k_2x^{n-2} + \cdots + k_n}.$$

a, b, c, etc., are the roots of the equation $F(x) = 0$, so that

$$F(x) = (x - a)^p\,(x - b)^q\,(x - c)^r \cdots,$$

then $$\frac{f(x)}{F(x)} = \frac{A_1}{(x-a)^p} + \frac{A_2}{(x-a)^{p-1}} + \frac{A_3}{(x-a)^{p-2}} + \cdots + \frac{A_p}{x-a}$$
$$+ \frac{B_1}{(x-b)^q} + \frac{B_2}{(x-b)^{q-1}} + \frac{B_3}{(x-b)^{q-2}} + \cdots + \frac{B_q}{x-b}$$
$$+ \frac{C_1}{(x-c)^r} + \frac{C_2}{(x-c)^{r-1}} + \frac{C_3}{(x-c)^{r-2}} + \cdots + \frac{C_r}{x-c}$$
$$+ \cdots \quad \cdots \quad \cdots \quad \cdots \quad \cdots \quad \cdots$$

Where the numerators of the separate fractions may be determined by the equations

$$A_m = \frac{\phi_1^{[m-1]}(a)}{(m-1)\,!}, \quad B_m = \frac{\phi_2^{[m-1]}(b)}{(m-1)\,!}, \text{ etc., etc.}$$

$$\phi_1(x) = \frac{f(x)(x-a)^p}{F(x)}, \quad \phi_2(x) = \frac{f(x)(x-b)^q}{F(x)}, \text{ etc., etc.}$$

If a, b, c, etc., are single roots, then $p = q = r = \cdots = 1$, and

$$\frac{f(x)}{F(x)} = \frac{A}{x-a} + \frac{B}{x-b} + \frac{C}{x-c} \cdots$$

where $$A = \frac{f(a)}{F'(a)}, \quad B = \frac{f(b)}{F'(b)}, \text{ etc.}$$

The simpler fractions, into which the original fraction is thus divided, may be integrated by means of the following formulas:

71. $$\int \frac{h\,dx}{(mx+n)^l} = \int \frac{h\,d(mx+n)}{m\,(mx+n)^l} = \frac{h}{m\,(1-l)(mx+n)^{l-1}}.$$

72. $$\int \frac{h\,dx}{mx+n} = \frac{h}{m} \log(mx+n).$$

If any of the roots of the equation $f(x) = 0$ are imaginary, the parts of the integral which arise from conjugate roots can be combined together and the integral brought into a real form. The following formula, in which $i = \sqrt{-1}$, is often useful in combining logarithms of conjugate complex quantities:

73. $$\log(x \pm yi) = \tfrac{1}{2} \log(x^2 + y^2) \pm i \tan^{-1}\frac{y}{x}.$$

III. IRRATIONAL ALGEBRAIC FUNCTIONS.

A.—Expressions Involving $\sqrt{a+bx}$.

The substitution of a new variable of integration, $y=\sqrt{a+bx}$, gives

74. $$\int\sqrt{a+bx}\,dx=\frac{2}{3b}\sqrt{(a+bx)^3}.$$

75. $$\int x\sqrt{a+bx}\,dx=-\frac{2(2a-3bx)\sqrt{(a+bx)^3}}{15b^2}.$$

76. $$\int x^2\sqrt{a+bx}\,dx=\frac{2(8a^2-12abx+15b^2x^2)\sqrt{(a+bx)^3}}{105b^3}.$$

77. $$\int\frac{\sqrt{a+bx}}{x}dx=2\sqrt{a+bx}+a\int\frac{dx}{x\sqrt{a+bx}}.$$

78. $$\int\frac{dx}{\sqrt{a+bx}}=\frac{2\sqrt{a+bx}}{b}.$$

79. $$\int\frac{x\,dx}{\sqrt{a+bx}}=-\frac{2(2a-bx)}{3b^2}\sqrt{a+bx}.$$

80. $$\int\frac{x^2\,dx}{\sqrt{a+bx}}=\frac{2(8a^2-4abx+3b^2x^2)}{15b^3}\sqrt{a+bx}.$$

81. $$\int\frac{dx}{x\sqrt{a+bx}}=\frac{1}{\sqrt{a}}\log\left(\frac{\sqrt{a+bx}-\sqrt{a}}{\sqrt{a+bx}+\sqrt{a}}\right),\ \text{for } a>0.$$

82. $$\int\frac{dx}{x\sqrt{a+bx}}=\frac{2}{\sqrt{-a}}\tan^{-1}\sqrt{\frac{a+bx}{-a}},\ \text{for } a<0.$$

83. $$\int\frac{dx}{x^2\sqrt{a+bx}}=-\frac{\sqrt{a+bx}}{ax}-\frac{b}{2a}\int\frac{dx}{x\sqrt{a+bx}}.$$

84. $\int (a+bx)^{\pm\frac{n}{2}}\,dx = \frac{2}{b}\int y^{1\pm n}\,dy = \frac{2\,(a+bx)^{\frac{2\pm n}{2}}}{b\,(2\pm n)}.$

85. $\int x\,(a+bx)^{\pm\frac{n}{2}}\,dx = \frac{2}{b^2}\left[\frac{(a+bx)^{\frac{4\pm n}{2}}}{4\pm n} - \frac{a\,(a+bx)^{\frac{2\pm n}{2}}}{2\pm n}\right].$

86. $\int \frac{x^m\,dx}{\sqrt{a+bx}} = \frac{2\,x^m\sqrt{a+bx}}{(2m+1)b} - \frac{2\,ma}{(2m+1)b}\int\frac{x^{m-1}\,dx}{\sqrt{a+bx}}.$

87. $\int \frac{dx}{x^n\sqrt{a+bx}} = -\frac{\sqrt{a+bx}}{(n-1)\,ax^{n-1}} - \frac{(2n-3)\,b}{(2n-2)\,a}\int\frac{dx}{x^{n-1}\sqrt{a+bx}}.$

88. $\int \frac{(a+bx)^{\frac{n}{2}}\,dx}{x} = b\int (a+bx)^{\frac{n-2}{2}}\,dx + a\int\frac{(a+bx)^{\frac{n-2}{2}}}{x}\,dx.$

89. $\int \frac{dx}{x\,(a+bx)^{\frac{m}{2}}} = \frac{1}{a}\int\frac{dx}{x\,(a+bx)^{\frac{m-2}{2}}} - \frac{b}{a}\int\frac{dx}{(a+bx)^{\frac{m}{2}}}.$

B. — Expressions* Involving $\sqrt{x^2 \pm a^2}$ and $\sqrt{a^2 - x^2}$.

90. $\int \sqrt{x^2 \pm a^2}\,dx = \tfrac{1}{2}\left[x\sqrt{x^2\pm a^2} \pm a^2\log\left(x+\sqrt{x^2\pm a^2}\right)\right].$

91. $\int \sqrt{a^2 - x^2}\,dx = \tfrac{1}{2}\left(x\sqrt{a^2-x^2} + a^2\sin^{-1}\frac{x}{a}\right).$

92. $\int \frac{dx}{\sqrt{x^2\pm a^2}} = \log\left(x+\sqrt{x^2\pm a^2}\right).$

93. $\int \frac{dx}{\sqrt{a^2-x^2}} = \sin^{-1}\frac{x}{a}.$

94. $\int \frac{dx}{x\sqrt{x^2-a^2}} = \frac{1}{a}\cos^{-1}\frac{a}{x}.$

95. $\int \frac{dx}{x\sqrt{a^2\pm x^2}} = -\frac{1}{a}\log\left(\frac{a+\sqrt{a^2\pm x^2}}{x}\right).$

96. $\int \frac{\sqrt{a^2\pm x^2}}{x}\,dx = \sqrt{a^2\pm x^2} - a\log\frac{a+\sqrt{a^2\pm x^2}}{x}.$

* These equations are all special cases of more general equations given in the next section.

97. $\int \frac{\sqrt{x^2 - a^2}}{x}\,dx = \sqrt{x^2 - a^2} - a\cos^{-1}\frac{a}{x}.$

98. $\int \frac{x\,dx}{\sqrt{a^2 \pm x^2}} = \pm\sqrt{a^2 \pm x^2}.$

99. $\int \frac{x\,dx}{\sqrt{x^2 - a^2}} = \sqrt{x^2 - a^2}.$

100. $\int x\sqrt{x^2 \pm a^2}\,dx = \frac{1}{3}\sqrt{(x^2 \pm a^2)^3}.$

101. $\int x\sqrt{a^2 - x^2}\,dx = -\frac{1}{3}\sqrt{(a^2 - x^2)^3}.$

102. $\int \sqrt{(x^2 \pm a^2)^3}\,dx$

$$= \frac{1}{4}\left[x\sqrt{(x^2 \pm a^2)^3} \pm \frac{3a^2x}{2}\sqrt{x^2 \pm a^2} + \frac{3a^4}{2}\log\left(x + \sqrt{x^2 \pm a^2}\right)\right].$$

103. $\int \sqrt{(a^2 - x^2)^3}\,dx$

$$= \frac{1}{4}\left[x\sqrt{(a^2 - x^2)^3} + \frac{3a^2x}{2}\sqrt{a^2 - x^2} + \frac{3a^4}{2}\sin^{-1}\frac{x}{a}\right].$$

104. $\int \frac{dx}{\sqrt{(x^2 \pm a^2)^3}} = \frac{\pm x}{a^2\sqrt{x^2 \pm a^2}}.$

105. $\int \frac{dx}{\sqrt{(a^2 - x^2)^3}} = \frac{x}{a^2\sqrt{a^2 - x^2}}.$

106. $\int \frac{x\,dx}{\sqrt{(x^2 \pm a^2)^3}} = \frac{-1}{\sqrt{x^2 \pm a^2}}.$

107. $\int \frac{x\,dx}{\sqrt{(a^2 - x^2)^3}} = \frac{1}{\sqrt{a^2 - x^2}}.$

108. $\int x\sqrt{(x^2 \pm a^2)^3}\,dx = \frac{1}{5}\sqrt{(x^2 \pm a^2)^5}$

109. $\int x\sqrt{(a^2 - x^2)^3}\,dx = -\frac{1}{5}\sqrt{(a^2 - x^2)^5}$

110. $\int x^2\sqrt{x^2 \pm a^2}\,dx$

$$= \frac{x}{4}\sqrt{(x^2 \pm a^2)^3} \mp \frac{a^2}{8}\left(x\sqrt{x^2 \pm a^2} \pm a^2 \log\left(x + \sqrt{x^2 \pm a^2}\right)\right).$$

111. $\int x^2\sqrt{a^2 - x^2}\,dx$

$$= -\frac{x}{4}\sqrt{(a^2 - x^2)^3} + \frac{a^2}{8}\left(x\sqrt{a^2 - x^2} + a^2 \sin^{-1}\frac{x}{a}\right).$$

112. $\int \frac{x^2\,dx}{\sqrt{x^2 \pm a^2}} = \frac{x}{2}\sqrt{x^2 \pm a^2} \mp \frac{a^2}{2}\log\left(x + \sqrt{x^2 \pm a^2}\right).$

113. $\int \frac{x^2\,dx}{\sqrt{a^2 - x^2}} = -\frac{x}{2}\sqrt{a^2 - x^2} + \frac{a^2}{2}\sin^{-1}\frac{x}{a}.$

114. $\int \frac{dx}{x^2\sqrt{x^2 \pm a^2}} = \pm\frac{\sqrt{x^2 \pm a^2}}{a^2 x}.$

115. $\int \frac{dx}{x^2\sqrt{a^2 - x^2}} = -\frac{\sqrt{a^2 - x^2}}{a^2 x}.$

116. $\int \frac{\sqrt{x^2 \pm a^2}\,dx}{x^2} = -\frac{\sqrt{x^2 \pm a^2}}{x} + \log\left(x + \sqrt{x^2 \pm a^2}\right).$

117. $\int \frac{\sqrt{a^2 - x^2}}{x^2}\,dx = -\frac{\sqrt{a^2 - x^2}}{x} - \sin^{-1}\frac{x}{a}.$

118. $\int \frac{x^2\,dx}{\sqrt{(x^2 \pm a^2)^3}} = \frac{-x}{\sqrt{x^2 \pm a^2}} + \log\left(x + \sqrt{x^2 \pm a^2}\right).$

119. $\int \frac{x^2\,dx}{\sqrt{(a^2 - x^2)^3}} = \frac{x}{\sqrt{a^2 - x^2}} - \sin^{-1}\frac{x}{a}.$

C. — Expressions Involving $\sqrt{a + bx + cx^2}$.

Let $X = a + bx + cx^2$, $q = 4ac - b^2$, and $k = \frac{4c}{q}$. In order to rationalize the function $f(x, \sqrt{a + bx + cx^2})$ we may put $\sqrt{a + bx + cx^2} = \sqrt{\pm c}\sqrt{A + Bx \pm x^2}$, according as c is positive or negative, and then substitute for x a new variable z, such that

$$z = \sqrt{A + Bx + x^2} - x, \text{ if } c > 0.$$

$$z = \frac{\sqrt{A + Bx - x^2} - \sqrt{A}}{x}, \text{ if } c < 0 \text{ and } \frac{a}{-c} > 0.$$

$$z = \sqrt{\frac{x - \beta}{\alpha - x}}, \text{ where } \alpha \text{ and } \beta \text{ are the roots of the equation}$$

$$A + Bx - x^2 = 0, \text{ if } c < 0 \text{ and } \frac{a}{-c} < 0.$$

By rationalization, or by the aid of reduction formulas, may be obtained the values of the following integrals:

120. $\int \frac{dx}{\sqrt{X}} = \frac{1}{\sqrt{c}} \log\left(\sqrt{X} + x\sqrt{c} + \frac{b}{2\sqrt{c}}\right)$, if $c > 0$.

121. $\int \frac{dx}{\sqrt{X}} = \frac{1}{\sqrt{-c}} \sin^{-1}\left(\frac{-2cx - b}{\sqrt{b^2 - 4ac}}\right)$, if $c < 0$.

122. $\int \frac{dx}{X\sqrt{X}} = \frac{2(2cx + b)}{q\sqrt{X}}.$

123. $\int \frac{dx}{X^2\sqrt{X}} = \frac{2(2cx + b)}{3q\sqrt{X}}\left(\frac{1}{X} + 2k\right).$

124. $\int \frac{dx}{X^n\sqrt{X}} = \frac{2(2cx + b\sqrt{X}}{(2n - 1)qX^n} + \frac{2k(n-1)}{2n - 1}\int \frac{dx}{X^{n-1}\sqrt{X}}.$

125. $\int \sqrt{X}dx = \frac{(2cx + b)\sqrt{X}}{4c} + \frac{1}{2k}\int \frac{dx}{\sqrt{X}}.$

126. $\int X\sqrt{X}dx = \frac{(2cx + b)\sqrt{X}}{8c}\left(X + \frac{3}{2k}\right) + \frac{3}{8k^2}\int \frac{dx}{\sqrt{X}}.$

127. $\int X^2\sqrt{X}dx = \frac{(2cx + b)\sqrt{X}}{12c}\left(X^2 + \frac{5X}{4k} + \frac{15}{8k^2}\right) + \frac{5}{16k^3}\int \frac{dx}{\sqrt{X}}.$

128. $\int X^n\sqrt{X}dx = \frac{(2cx + b)X^n\sqrt{X}}{4(n + 1)c} + \frac{2n + 1}{2(n + 1)k}\int \frac{X^n dx}{\sqrt{X}}.$

129. $\int \frac{x\,dx}{\sqrt{X}} = \frac{\sqrt{X}}{c} - \frac{b}{2c}\int \frac{dx}{\sqrt{X}}.$

130. $$\int \frac{x\,dx}{X\sqrt{X}} = -\frac{2(bx+2a)}{q\sqrt{X}}.$$

131. $$\int \frac{x\,dx}{X^n\sqrt{X}} = -\frac{\sqrt{X}}{(2n-1)cX^n} - \frac{b}{2c}\int \frac{dx}{X^n\sqrt{X}}.$$

132. $$\int \frac{x^2\,dx}{\sqrt{X}} = \left(\frac{x}{2c} - \frac{3b}{4c^2}\right)\sqrt{X} + \frac{3b^2-4ac}{8c^2}\int \frac{dx}{\sqrt{X}}.$$

133. $$\int \frac{x^2\,dx}{X\sqrt{X}} = \frac{(2b^2-4ac)x+2ab}{cq\sqrt{X}} + \frac{1}{c}\int \frac{dx}{\sqrt{X}}.$$

134. $$\int \frac{x^2\,dx}{X^n\sqrt{X}} = \frac{(2b^2-4ac)x+2ab}{(2n-1)cqX^{n-1}\sqrt{X}} + \frac{4ac+(2n-3)b^2}{(2n-1)cq}\int \frac{dx}{X^{n-1}\sqrt{X}}.$$

135. $$\int \frac{x^3\,dx}{\sqrt{X}} = \left(\frac{x^2}{3c} - \frac{5bx}{12c^2} + \frac{5b^2}{8c^3} - \frac{2a}{3c^2}\right)\sqrt{X} + \left(\frac{3ab}{4c^2} - \frac{5b^3}{16c^3}\right)\int \frac{dx}{\sqrt{X}}.$$

136. $$\int x\sqrt{X}\,dx = \frac{X\sqrt{X}}{3c} - \frac{b}{2c}\int \sqrt{X}\,dx.$$

137. $$\int xX\sqrt{X}\,dx = \frac{X^2\sqrt{X}}{5c} - \frac{b}{2c}\int X\sqrt{X}\,dx.$$

138. $$\int \frac{xX^n\,dx}{\sqrt{X}} = \frac{X^n\sqrt{X}}{(2n+1)c} - \frac{b}{2c}\int \frac{X^n\,dx}{\sqrt{X}}.$$

139. $$\int x^2\sqrt{X}\,dx = \left(x - \frac{5b}{6c}\right)\frac{X\sqrt{X}}{4c} + \frac{5b^2-4ac}{16c^2}\int \sqrt{X}\,dx.$$

140. $$\int \frac{x^2X^n\,dx}{\sqrt{X}} = \frac{xX^n\sqrt{X}}{2(n+1)c} - \frac{(2n+3)b}{4(n+1)c}\int \frac{xX^n\,dx}{\sqrt{X}} - \frac{a}{2(n+1)c}\int \frac{X^n\,dx}{\sqrt{X}}.$$

141. $$\int x^3\sqrt{X}\,dx = \left(x^2 - \frac{7bx}{8c} + \frac{35b^2}{48c^2} - \frac{2a}{3c}\right)\frac{X\sqrt{X}}{5c} + \left(\frac{3ab}{8c^2} - \frac{7b^3}{32c^3}\right)\int \sqrt{X}\,dx.$$

142. $\int \frac{dx}{x\sqrt{X}} = -\frac{1}{\sqrt{a}} \log\left(\frac{\sqrt{X}+\sqrt{a}}{x} + \frac{b}{2\sqrt{a}}\right)$, if $a > 0$.

143. $\int \frac{dx}{x\sqrt{X}} = \frac{1}{\sqrt{-a}} \sin^{-1}\left(\frac{bx+2a}{x\sqrt{b^2-4ac}}\right)$, if $a < 0$.

144. $\int \frac{dx}{x\sqrt{X}} = -\frac{2\sqrt{X}}{bx}$, if $a = 0$.

145. $\int \frac{dx}{xX^n\sqrt{X}} = \frac{\sqrt{X}}{(2n-1)aX^n} + \frac{1}{a}\int \frac{dx}{xX^{n-1}\sqrt{X}} - \frac{b}{2a}\int \frac{dx}{X^n\sqrt{X}}$.

146. $\int \frac{dx}{x^2\sqrt{X}} = -\frac{\sqrt{X}}{ax} - \frac{b}{2a}\int \frac{dx}{x\sqrt{X}}$.

147. $\int \frac{\sqrt{X}\,dx}{x} = \sqrt{X} + \frac{b}{2}\int \frac{dx}{\sqrt{X}} + a\int \frac{dx}{x\sqrt{X}}$.

148. $\int \frac{X^n\,dx}{x\sqrt{X}} = \frac{X^n}{(2n-1)\sqrt{X}} + a\int \frac{X^{n-1}\,dx}{x\sqrt{X}} + \frac{b}{2}\int \frac{X^{n-1}\,dx}{\sqrt{X}}$.

149. $\int \frac{\sqrt{X}\,dx}{x^2} = -\frac{\sqrt{X}}{x} + \frac{b}{2}\int \frac{dx}{x\sqrt{X}} + c\int \frac{dx}{\sqrt{X}}$.

150. $\int \frac{x^m\,dx}{X^n\sqrt{X}} = \frac{1}{c}\int \frac{x^{m-2}\,dx}{X^{n-1}\sqrt{X}} - \frac{b}{c}\int \frac{x^{m-1}\,dx}{X^n\sqrt{X}} - \frac{a}{c}\int \frac{x^{m-2}\,dx}{X^n\sqrt{X}}$.

151. $\int \frac{x^m X^n\,dx}{\sqrt{X}} = \frac{x^{m-1}X^n\sqrt{X}}{(2n+m)c} - \frac{(2n+2m-1)b}{2c(2n+m)}\int \frac{x^{m-1}X^n\,dx}{\sqrt{X}}$

$$- \frac{(m-1)a}{(2n+m)c}\int \frac{x^{m-2}X^n\,dx}{\sqrt{X}}.$$

152. $\int \frac{dx}{x^m X^n\sqrt{X}} = -\frac{\sqrt{X}}{(m-1)ax^{m-1}X^n}$

$$-\frac{(2n+2m-3)b}{2a(m-1)}\int \frac{dx}{x^{m-1}X^n\sqrt{X}} - \frac{(2n+m-2)c}{(m-1)a}\int \frac{dx}{x^{m-2}X^n\sqrt{X}}.$$

153. $\int \frac{X^n\,dx}{x^m\sqrt{X}} = -\frac{X^{n-1}\sqrt{X}}{(m-1)\,x^{m-1}} + \frac{(2n-1)\,b}{2\,(m-1)}\int \frac{X^{n-1}\,dx}{x^{m-1}\sqrt{X}}$
$+\frac{(2n-1)\,c}{m-1}\int \frac{X^{n-1}\,dx}{x^{m-2}\sqrt{X}}.$

D. — Miscellaneous Expressions.

154. $\int \sqrt{2\,ax - x^2}\,dx = \frac{x-a}{2}\sqrt{2\,ax - x^2} + \frac{a^2}{2}\sin^{-1}\frac{x-a}{a}.$

155. $\int \frac{dx}{\sqrt{2\,ax - x^2}} = \text{versin}^{-1}\frac{x}{a}.$

156. $\int \frac{dx}{(x+1)\sqrt{x^2-1}} = +\sqrt{\frac{x-1}{x+1}}.$

157. $\int \frac{dx}{(x-1)\sqrt{x^2-1}} = -\sqrt{\frac{x+1}{x-1}}.$

158. $\int \sqrt{\frac{1+x}{1-x}}\,dx = \sin^{-1}x - \sqrt{1-x^2}.$

159. $\int \sqrt{\frac{x+a}{x+b}}\,dx = \sqrt{(x+a)\,(x+b)}$
$+\,(a-b)\log\left(\sqrt{x+a} + \sqrt{x+b}\right).$

160. $\int \frac{dx}{\sqrt{(x-\alpha)\,(\beta - x)}} = 2\sin^{-1}\sqrt{\frac{x-\alpha}{\beta-\alpha}}.$

161. $\int \frac{dx}{\sqrt{(a+bx)\,(\alpha - \beta x)}} = \frac{2}{\sqrt{b\beta}}\sin^{-1}\sqrt{\frac{\beta\,(a+bx)}{a\beta + b\alpha}}.$

162. $\int \sqrt[3]{a+bx}\,dx = \frac{3}{4b}\sqrt[3]{(a+bx)^4}.$

163. $\int \frac{dx}{\sqrt[3]{a+bx}} = \frac{3}{2b}\sqrt[3]{(a+bx)^2}.$

164. $$\int \frac{x\,dx}{\sqrt[3]{a+bx}} = -\frac{3(3a-2bx)}{10b^2}\sqrt[3]{(a+bx)^2}.$$

165. $$\int \frac{dx}{x\sqrt{x^n-a^2}} = \frac{2}{an}\sec^{-1}\left(\frac{x^{\frac{n}{2}}}{a}\right).$$

166. $$\int \frac{dx}{x\sqrt{x^n+a^2}} = \frac{1}{an}\log\frac{\sqrt{a^2+x^n}-a}{\sqrt{a^2+x^n}+a}.$$

IV. TRANSCENDENTAL FUNCTIONS.

167. $\int \sin x \, dx = -\cos x.$

168. $\int \sin^2 x \, dx = -\frac{1}{2}\cos x \sin x + \frac{1}{2}x.$

169. $\int \sin^3 x \, dx = -\frac{1}{3}\cos x (\sin^2 x + 2).$

170. $\int \sin^n x \, dx = -\frac{\sin^{n-1} x \cos x}{n} + \frac{n-1}{n}\int \sin^{n-2} x \, dx.$

171. $\int \cos x \, dx = \sin x.$

172. $\int \cos^2 x \, dx = \frac{1}{2}\sin x \cos x + \frac{1}{2}x.$

173. $\int \cos^3 x \, dx = \frac{1}{3}\sin x (\cos^2 x + 2).$

174. $\int \cos^n x \, dx = \frac{1}{n}\cos^{n-1} x \sin x + \frac{n-1}{n}\int \cos^{n-2} x \, dx.$

175. $\int \sin x \cos x \, dx = \frac{1}{2}\sin^2 x.$

176. $\int \sin^2 x \cos^2 x \, dx = -\frac{1}{8}(\frac{1}{4}\sin 4x - x).$

177. $\int \sin x \cos^m x \, dx = -\frac{\cos^{m+1} x}{m+1}.$

178. $\int \sin^m x \cos x \, dx = \frac{\sin^{m+1} x}{m+1}.$

179. $$\int \cos^m x \sin^n x \, dx = \frac{\cos^{m-1} x \sin^{n+1} x}{m+n} + \frac{m-1}{m+n}\int \cos^{m-2} x \sin^n x \, dx.$$

180. $$\int \cos^m x \sin^n x \, dx = -\frac{\sin^{n-1} x \cos^{m+1} x}{m+n} + \frac{n-1}{m+n}\int \cos^m x \sin^{n-2} x \, dx$$

181. $\int \frac{\cos^m x\,dx}{\sin^n x} = -\frac{\cos^{m+1} x}{(n-1)\sin^{n-1} x} - \frac{m-n+2}{n-1}\int \frac{\cos^m x\,dx}{\sin^{n-2} x}.$

182. $\int \frac{\cos^m x\,dx}{\sin^n x} = \frac{\cos^{m-1} x}{(m-n)\sin^{n-1} x} + \frac{m-1}{m-n}\int \frac{\cos^{m-2} x\,dx}{\sin^n x}.$

183. $\int \frac{\sin^m x\,dx}{\cos^n x} = -\int \frac{\cos^m\left(\frac{\pi}{2}-x\right)d\left(\frac{\pi}{2}-x\right)}{\sin^n\left(\frac{\pi}{2}-x\right)}.$

184. $\int \frac{dx}{\sin^m x \cos^n x}$

$$= \frac{1}{n-1}\cdot\frac{1}{\sin^{m-1} x \,.\, \cos^{n-1} x} + \frac{m+n-2}{n-1}\int \frac{dx}{\sin^m x \,.\, \cos^{n-2} x}$$

$$= -\frac{1}{m-1}\cdot\frac{1}{\sin^{m-1} x \,.\, \cos^{n-1} x} + \frac{m+n-2}{m-1}\int \frac{dx}{\sin^{m-2} x \,.\, \cos^n x}.$$

185. $\int \frac{dx}{\sin^m x} = -\frac{1}{m-1}\cdot\frac{\cos x}{\sin^{m-1} x} + \frac{m-2}{m-1}\int \frac{dx}{\sin^{m-2} x}.$

186. $\int \frac{dx}{\cos^n x} = \frac{1}{n-1}\cdot\frac{\sin x}{\cos^{n-1} x} + \frac{n-2}{n-1}\int \frac{dx}{\cos^{n-2} x}.$

187. $\int \tan x\,dx = -\log\cos x.$

188. $\int \tan^2 x\,dx = \tan x - x.$

189. $\int \tan^n x\,dx = \frac{\tan^{n-1} x}{n-1} - \int \tan^{n-2} x\,dx.$

190. $\int \operatorname{ctn} x\,dx = \log\sin x.$

191. $\int \operatorname{ctn}^2 x\,dx = -\operatorname{ctn} x - x.$

192. $\int \operatorname{ctn}^n x\,dx = -\frac{\operatorname{ctn}^{n-1} x}{n-1} - \int \operatorname{ctn}^{n-2} x\,dx.$

193. $\int \sec x\,dx = \log\tan\left(\frac{\pi}{4}+\frac{x}{2}\right).$

194. $\int \sec^2 x\,dx = \tan x.$

195. $\int \sec^n x\,dx = \int \frac{dx}{\cos^n x}.$

196. $\int \csc x\,dx = \log \tan \tfrac{1}{2} x.$

197. $\int \csc^2 x\,dx = -\operatorname{ctn} x.$

198. $\int \csc^n x\,dx = \int \frac{dx}{\sin^n x}.$

199. $\int \frac{dx}{a + b\cos x} = \frac{-1}{\sqrt{a^2 - b^2}} \cdot \sin^{-1}\left[\frac{b + a\cos x}{a + b\cos x}\right],$

or $\frac{1}{\sqrt{b^2 - a^2}} \log\left[\frac{b + a\cos x + \sqrt{b^2 - a^2} \,.\, \sin x}{a + b\cos x}\right].$

200. $\int \frac{dx}{a + b\cos x + c\sin x}$

$$= \frac{-1}{\sqrt{a^2 - b^2 - c^2}} \cdot \sin^{-1}\left[\frac{b^2 + c^2 + a(b\cos x + c\sin x)}{\sqrt{b^2 + c^2}\,(a + b\cos x + c\sin x)}\right]$$

$$= \frac{1}{\sqrt{b^2 + c^2 - a^2}} \cdot \log$$

$$\left[\frac{b^2 + c^2 + a(b\cos x + c\sin x) + \sqrt{b^2 + c^2 - a^2}\,(b\sin x - c\cos x)}{\sqrt{b^2 + c^2}\,(a + b\cos x + c\sin x)}\right].$$

201. $\int x\sin x\,dx = \sin x - x\cos x.$

202. $\int x^2 \sin x\,dx = 2x\sin x - (x^2 - 2)\cos x.$

203. $\int x^3 \sin x\,dx = (3x^2 - 6)\sin x - (x^3 - 6x)\cos x.$

204. $\int x^m \sin x\,dx = -x^m \cos x + m\int x^{m-1}\cos x\,dx.$

205. $\int x\cos x\,dx = \cos x + x\sin x.$

206. $\int x^2 \cos x\,dx = 2x\cos x + (x^2 - 2)\sin x.$

207. $\int x^3 \cos x\,dx = (3x^2 - 6)\cos x + (x^3 - 6x)\sin x.$

208. $\int x^m \cos x\,dx = x^m \sin x - m\int x^{m-1}\sin x\,dx.$

209. $\int \frac{\sin x}{x^m}dx = -\frac{1}{m-1}\cdot\frac{\sin x}{x^{m-1}} + \frac{1}{m-1}\int\frac{\cos x}{x^{m-1}}dx.$

210. $\int \frac{\cos x}{x^m}dx = -\frac{1}{m-1}\cdot\frac{\cos x}{x^{m-1}} - \frac{1}{m-1}\int\frac{\sin x}{x^{m-1}}dx.$

211. $\int \frac{\sin x}{x}dx = x - \frac{x^3}{3\cdot 3!} + \frac{x^5}{5\cdot 5!} - \frac{x^7}{7\cdot 7!} + \frac{x^9}{9\cdot 9!}\cdots.$

212. $\int \frac{\cos x}{x}dx = \log x - \frac{x^2}{2\cdot 2!} + \frac{x^4}{4\cdot 4!} - \frac{x^6}{6\cdot 6!} + \frac{x^8}{8\cdot 8!}\cdots.$

213. $\int \sin mx \sin nx\,dx = \frac{\sin(m-n)x}{2(m-n)} - \frac{\sin(m+n)x}{2(m+n)}.$

214. $\int \cos mx \cos nx\,dx = \frac{\sin(m-n)x}{2(m-n)} + \frac{\sin(m+n)x}{2(m+n)}.$

215. $\int \sin^{-1}x\,dx = x\sin^{-1}x + \sqrt{1-x^2}.$

216. $\int \cos^{-1}x\,dx = x\cos^{-1}x - \sqrt{1-x^2}.$

217. $\int \tan^{-1}x\,dx = x\tan^{-1}x - \frac{1}{2}\log(1+x^2).$

218. $\int \text{ctn}^{-1}x\,dx = x\,\text{ctn}^{-1}x + \frac{1}{2}\log(1+x^2).$

219. $\int \text{versin}^{-1}x\,dx = (x-1)\,\text{versin}^{-1}x + \sqrt{2x-x^2}.$

220. $\int (\sin^{-1}x)^2\,dx = x(\sin^{-1}x)^2 - 2x + 2\sqrt{1-x^2}\sin^{-1}x.$

221. $\int x.\sin^{-1}x\,dx = \frac{1}{4}[(2x^2-1)\sin^{-1}x + x\sqrt{1-x^2}].$

222. $\int x^n \sin^{-1}x\,dx = \frac{x^{n+1}\sin^{-1}x}{n+1} - \frac{1}{n+1}\int\frac{x^{n+1}dx}{\sqrt{1-x^2}}.$

223. $\int x^n \cos^{-1}x\,dx = \frac{x^{n+1}\cos^{-1}x}{n+1} + \frac{1}{n+1}\int\frac{x^{n+1}dx}{\sqrt{1-x^2}}.$

224. $\int x^n \tan^{-1}x\,dx = \frac{x^{n+1}\tan^{-1}x}{n+1} - \frac{1}{n+1}\int\frac{x^{n+1}dx}{1+x^2}.$

225. $\int \log x\, dx = x \log x - x.$

226. $\int \frac{(\log x)^n}{x} dx = \frac{1}{n+1} (\log x)^{n+1}.$

227. $\int \frac{dx}{x \log x} = \log . \log x.$

228. $\int \frac{dx}{x (\log x)^n} = -\frac{1}{(n-1)(\log x)^{n-1}}.$

229. $\int x^m \log x\, dx = x^{m+1} \left[\frac{\log x}{m+1} - \frac{1}{(m+1)^2} \right].$

230. $\int e^{ax} dx = \frac{e^{ax}}{a}.$

231. $\int x e^{ax} dx = \frac{e^{ax}}{a^2} (ax - 1).$

232. $\int x^m e^{ax} dx = \frac{x^m e^{ax}}{a} - \frac{m}{a} \int x^{m-1} e^{ax} dx.$

233. $\int \frac{e^{ax}}{x^m} dx = -\frac{1}{m-1} \frac{e^{ax}}{x^{m-1}} + \frac{a}{m-1} \int \frac{e^{ax}}{x^{m-1}} dx.$

234. $\int e^{ax} \log x\, dx = \frac{e^{ax} \log x}{a} - \frac{1}{a} \int \frac{e^{ax}}{x} dx.$

235. $\int e^{ax} \sin x\, dx = \frac{e^{ax} (a \sin x - \cos x)}{a^2 + 1}.$

236. $\int e^{ax} \cos x\, dx = \frac{e^{ax} (a \cos x + \sin x)}{a^2 + 1}.$

DEFINITE INTEGRALS.

237. $\int_0^\infty \frac{a\,dx}{a^2+x^2}=\frac{\pi}{2}$, if $a>0$; 0, if $a=0$; $-\frac{\pi}{2}$, if $a<0$.

238. $\int_0^\infty x^{n-1}e^{-x}\,dx=\int_0^1\left[\log\frac{1}{x}\right]^{n-1}dx=\Gamma(n)$.

$\Gamma(n+1)=n\cdot\Gamma(n)$. $\Gamma(2)=\Gamma(1)=1$.

$\Gamma(n+1)=n!$, if n is an integer $\Gamma(\frac{1}{2})=\sqrt{\pi}$.

239. $\int_0^1 x^{m-1}(1-x)^{n-1}dx=\int_0^\infty \frac{x^{m+1}\,dx}{(1+x)^{m+n}}=\frac{\Gamma(m)\,\Gamma(n)}{\Gamma(m+n)}$.

240. $\int_0^{\frac{\pi}{2}}\sin^n x\,dx=\int_0^{\frac{\pi}{2}}\cos^n x\,dx$

$=\frac{1.3.5\ldots(n-1)}{2.4.6\ldots(n)}\cdot\frac{\pi}{2}$, if n is an even integer.

$=\frac{2.4.6\ldots(n-1)}{1.3.5.7\ldots n}$, if n is an odd integer.

$=\frac{1}{2}\sqrt{\pi}\frac{\Gamma\left(\frac{n+1}{2}\right)}{\Gamma\left(\frac{n}{2}+1\right)}$, for any value of n.

241. $\int_0^\infty \frac{\sin mx\,dx}{x}=\frac{\pi}{2}$, if $m>0$; 0, if $m=0$; $-\frac{\pi}{2}$, if $m<0$.

242. $\int_0^\infty \frac{\sin x\,.\cos mx\,dx}{x}=0$, if $m<-1$ or $m>1$;

$\frac{\pi}{4}$, if $m=-1$ or $m=1$; $\frac{\pi}{2}$, if $-1<m<1$.

243. $\int_0^\infty \frac{\sin^2 x\,dx}{x^2}=\frac{\pi}{2}$.

244. $\int_0^\infty \cos(x^2)\,dx=\int_0^\infty \sin(x^2)\,dx=\frac{1}{2}\sqrt{\frac{\pi}{2}}$.

245. $\int_0^\infty \frac{\cos mx\,dx}{1+x^2} = \frac{\pi}{2} \cdot e^{-m}.$

246. $\int_0^\infty \frac{\cos x\,dx}{\sqrt{x}} = \int_0^\infty \frac{\sin x\,dx}{\sqrt{x}} = \sqrt{\frac{\pi}{2}}.$

247. $\int_0^{\frac{\pi}{2}} \frac{dx}{\sqrt{1-k^2\sin^2 x}}$

$$= \frac{\pi}{2}\left[1 + (\tfrac{1}{2})^2 k^2 + \left(\frac{1.3}{2.4}\right)^2 k^4 + \left(\frac{1.3.5}{2.4.6}\right)^2 k^6 + \dots\right], \text{ if } k^2 < 1.$$

$$= K.$$

248. $\int_0^{\frac{\pi}{2}} \sqrt{1-k^2\sin^2 x}\,.\,dx$

$$= \frac{\pi}{2}\left[1 - (\tfrac{1}{2})^2 k^2 - \left(\frac{1.3}{2.4}\right)^2 \frac{k^4}{3} - \left(\frac{1.3.5}{2.4.6}\right)^2 \frac{k^6}{5} - \dots\right], \text{ if } k^2 < 1.$$

$$= E.$$

249. $\int_0^\infty e^{-a^2x^2}\,dx = \frac{1}{2a}\sqrt{\pi} \cdot = \frac{1}{2a}\Gamma(\tfrac{1}{2}).$

250. $\int_0^\infty x^n e^{-ax}\,dx = \frac{\Gamma(n+1)}{a^{n+1}} = \frac{n\,!}{a^{n+1}}.$

251. $\int_0^\infty x^{2n} e^{-ax^2}\,dx = \frac{1.3.5\dots(2n-1)}{2^{n+1}a^n}\sqrt{\frac{\pi}{a}}.$

252. $\int_0^\infty e^{-x^2-\frac{a^2}{x^2}}\,dx = \frac{e^{-2a}\sqrt{\pi}}{2}.$

253. $\int_0^\infty e^{-ax}\cos mx\,dx = \frac{a}{a^2+m^2}$, if $a > 0.$

254. $\int_0^\infty e^{-ax}\sin mx\,dx = \frac{m}{a^2+m^2}$, if $a > 0.$

255. $\int_0^\infty e^{-a^2x^2}\cos bx\,dx = \frac{\sqrt{\pi}\,.\,e^{-\frac{b^2}{4a^2}}}{2a}.$

256. $\int_0^1 \frac{\log x}{1-x}\,dx = -\frac{\pi^2}{6}.$

257. $\int_0^1 \frac{\log x}{1+x}\,dx = -\frac{\pi^2}{12}.$

258. $\int_0^1 \frac{\log x}{1-x^2}dx = -\frac{\pi^2}{8}.$

259. $\int_0^1 \log\left(\frac{1+x}{1-x}\right) \cdot \frac{dx}{x} = \frac{\pi^2}{4}.$

260. $\int_0^\infty \log\left(\frac{e^x+1}{e^x-1}\right) dx = \frac{\pi^2}{4}.$

261. $\int^1 \frac{dx}{\sqrt{\log\left(\frac{1}{x}\right)}} = \sqrt{\pi}.$

262. $\int_0^1 x^m \log\left(\frac{1}{x}\right)^n dx = \frac{\Gamma(n+1)}{(m+1)^{n+1}}.$

263. $\int_0^{\frac{\pi}{2}} \log \sin x\, dx = \int_0^{\frac{\pi}{2}} \log \cos x\, dx = -\frac{\pi}{2} \cdot \log 2.$

264. $\int_0^\pi x \,.\log \sin x\, dx = -\frac{\pi^2}{2} \log 2.$

AUXILIARY FORMULAS.

The following formulas are sometimes useful in the reduction of integrals:

265. $\log u = \log cu + \text{a constant}.$

266. $\log(-u) = \log u + \text{a constant}.$

267. $$\sin^{-1}u = \begin{cases} -\sin^{-1}\sqrt{1-u^2} & + \text{a constant}. \\ -\frac{1}{2}\sin^{-1}(2u^2-1) & + \text{a constant}. \\ \frac{1}{2}\sin^{-1}2u\sqrt{1-u^2} & + \text{a constant}. \end{cases}$$

268. $$\tan^{-1}u = \begin{cases} -\tan^{-1}\frac{1}{u} & + \text{a constant}. \\ \tan^{-1}\frac{u+c}{1-cu} & + \text{a constant}. \end{cases}$$

269. $\log(x \pm yi) = \frac{1}{2}\log(x^2+y^2) \pm i\tan^{-1}\frac{y}{x}.$

270. $\sin^{-1}u = \cos^{-1}\sqrt{1-u^2} = \tan^{-1}\frac{u}{\sqrt{1-u^2}} = \csc^{-1}\frac{1}{u}.$

271. $\cos^{-1}u = \sin^{-1}\sqrt{1-u^2} = \tan^{-1}\sqrt{\frac{1}{u^2}-1} = \sec^{-1}\frac{1}{u}.$

272. $\tan^{-1}x \pm \tan^{-1}y = \tan^{-1}\left(\frac{x \pm y}{1 \mp xy}\right).$

273. $\sin^{-1}x \pm \sin^{-1}y = \sin^{-1}(x\sqrt{1-y^2} \pm y\sqrt{1-x^2}).$

274. $\cos^{-1}x \pm \cos^{-1}y = \cos^{-1}(xy \mp \sqrt{(1-x^2)(1-y^2)}).$

275. $\sin x = \frac{e^{xi} - e^{-xi}}{2i}.$

276. $\cos x = \frac{e^{xi} + e^{-xi}}{2}.$

277. $\sin xi = \frac{1}{2}i(e^x - e^{-x}) = i\sinh x.$

278. $\cos xi = \frac{1}{2}(e^x + e^{-x}) = \cosh x.$

279. $\log_e x = (2.3025851)\log_{10}x.$

The Natural Logarithms of Numbers between 1.0 and 9.9.

N.	0	1	2	3	4	5	6	7	8	9
1.	0.000	0.095	0.182	0.262	0.336	0.405	0.470	0.531	0.588	0.642
2.	0.693	0.742	0.788	0.833	0.875	0.916	0.956	0.993	1.030	1.065
3.	1.099	1.131	1.163	1.194	1.224	1.253	1.281	1.308	1.335	1.361
4.	1.386	1.411	1.435	1.459	1.482	1.504	1.526	1.548	1.569	1.589
5.	1.609	1.629	1.649	1.668	1.686	1.705	1.723	1.740	1.758	1.775
6.	1.792	1.808	1.825	1.841	1.856	1.872	1.887	1.902	1.917	1.932
7.	1.946	1.960	1.974	1.988	2.001	2.015	2.028	2.041	2.054	2.067
8.	2.079	2.092	2.104	2.116	2.128	2.140	2.152	2.163	2.175	2.186
9.	2.197	2.208	2.219	2.230	2.241	2.251	2.262	2.272	2.282	2.293

The Natural Logarithms of Whole Numbers from 10 to 109.

N.	0	1	2	3	4	5	6	7	8	9
1	2.303	2.398	2.485	2.565	2.639	2.708	2.773	2.833	2.890	2.944
2	2.996	3.045	3.091	3.135	3.178	3.219	3.258	3.296	3.332	3.367
3	3.401	3.434	3.466	3.497	3.526	3.555	3.584	3.611	3.638	3.664
4	3.689	3.714	3.738	3.761	3.784	3.807	3.829	3.850	3.871	3.892
5	3.912	3.932	3.951	3.970	3.989	4.007	4.025	4.043	4.060	4.078
6	4.094	4.111	4.127	4.143	4.159	4.174	4.190	4.205	4.220	4.234
7	4.248	4.263	4.277	4.290	4.304	4.317	4.331	4.344	4.357	4.369
8	4.382	4.394	4.407	4.419	4.431	4.443	4.454	4.466	4.477	4.489
9	4.500	4.511	4.522	4.533	4.543	4.554	4.564	4.575	4.585	4.595
10	4.605	4.615	4.625	4.635	4.644	4.654	4.663	4.673	4.682	4.691

The Values in Circular Measure of Angles which are given in Degrees and Minutes.

1′	0.0003	9′	0.0026	3°	0.0524	20°	0.3491	100°	1.7453
2′	0.0006	10′	0.0029	4°	0.0698	30°	0.5236	110°	1.9199
3′	0.0009	20′	0.0058	5°	0.0873	40°	0.6981	120°	2.0944
4′	0.0012	30′	0.0087	6°	0.1047	50°	0.8727	130°	2.2689
5′	0.0015	40′	0.0116	7°	0.1222	60°	1.0472	140°	2.4435
6′	0.0017	50′	0.0145	8°	0.1396	70°	1.2217	150°	2.6180
7′	0.0020	1°	0.0175	9°	0.1571	80°	1.3963	160°	2.7925
8′	0.0023	2°	0.0349	10°	0.1745	90°	1.5708	170°	2.9671

NATURAL TRIGONOMETRIC FUNCTIONS.

Angle.	Sin.	Csc.	Tan.	Ctn.	Sec.	Cos.	
0°	0.000	∞	0.000	∞	1.000	1.000	90°
1	0.017	57.30	0.017	57.29	1.000	1.000	89
2	0.035	28.65	0.035	28.64	1.001	0.999	88
3	0.052	19.11	0.052	19.08	1.001	0.999	87
4	0.070	14.34	0.070	14.30	1.002	0.998	86
5°	0.087	11.47	0.087	11.43	1.004	0.996	85°
6	0.105	9.567	0.105	9.514	1.006	0.995	84
7	0.122	8.206	0.123	8.144	1.008	0.993	83
8	0.139	7.185	0.141	7.115	1.010	0.990	82
9	0.156	6.392	0.158	6.314	1.012	0.988	81
10°	0.174	5.759	0.176	5.671	1.015	0.985	80°
11	0.191	5.241	0.194	5.145	1.019	0.982	79
12	0.208	4.810	0.213	4.705	1.022	0.978	78
13	0.225	4.445	0.231	4.331	1.026	0.974	77
14	0.242	4.134	0.249	4.011	1.031	0.970	76
15°	0.259	3.864	0.268	3.732	1.035	0.966	75°
16	0.276	3.628	0.287	3.487	1.040	0.961	74
17	0.292	3.420	0.306	3.271	1.046	0.956	73
18	0.309	3.236	0.325	3.078	1.051	0.951	72
19	0.326	3.072	0.344	2.904	1.058	0.946	71
20°	0.342	2.924	0.364	2.747	1.064	0.940	70°
21	0.358	2.790	0.384	2.605	1.071	0.934	69
22	0.375	2.669	0.404	2.475	1.079	0.927	68
23	0.391	2.559	0.424	2.356	1.086	0.921	67
24	0.407	2.459	0.445	2.246	1.095	0.914	66
25°	0.423	2.366	0.466	2.145	1.103	0.906	65°
26	0.438	2.281	0.488	2.050	1.113	0.899	64
27	0.454	2.203	0.510	1.963	1.122	0.891	63
28	0.469	2.130	0.532	1.881	1.133	0.883	62
29	0.485	2.063	0.554	1.804	1.143	0.875	61
30°	0.500	2.000	0.577	1.732	1.155	0.866	60°
31	0.515	1.942	0.601	1.664	1.167	0.857	59
32	0.530	1.887	0.625	1.600	1.179	0.848	58
33	0.545	1.836	0.649	1.540	1.192	0.839	57
34	0.559	1.788	0.675	1.483	1.206	0.829	56
35°	0.574	1.743	0.700	1.428	1.221	0.819	55°
36	0.588	1.701	0.727	1.376	1.236	0.809	54
37	0.602	1.662	0.754	1.327	1.252	0.799	53
38	0.616	1.624	0.781	1.280	1.269	0.788	52
39	0.629	1.589	0.810	1.235	1.287	0.777	51
40°	0.643	1.556	0.839	1.192	1.305	0.766	50°
41	0.656	1.524	0.869	1.150	1.325	0.755	49
42	0.669	1.494	0.900	1.111	1.346	0.743	48
43	0.682	1.466	0.933	1.072	1.367	0.731	47
44	0.695	1.440	0.966	1.036	1.390	0.719	46
45°	0.707	1.414	1.000	1.000	1.414	0.707	45°
	Cos.	Sec.	Ctn.	Tan.	Csc.	Sin.	Angle.

Values of the Complete Elliptic Integrals, K and E, for Different Values of the Modulus, k.

$\sin^{-1}k$	K	E	$\sin^{-1}k$	K	E	$\sin^{-1}k$	K	E
0°	1.5708	1.5708	30°	1.6858	1.4675	60°	2.1565	1.2111
1°	1.5709	1.5707	31°	1.6941	1.4608	61°	2.1842	1.2015
2°	1.5713	1.5703	32°	1.7028	1.4539	62°	2.2132	1.1920
3°	1.5719	1.5697	33°	1.7119	1.4469	63°	2.2435	1.1826
4°	1.5727	1.5689	34°	1.7214	1.4397	64°	2.2754	1.1732
5°	1.5738	1.5678	35°	1.7312	1.4223	65°	2.3088	1.1638
6°	1.5711	1.5665	36°	1.7415	1.4248	66°	2.3439	1.1545
7°	1.5767	1.5649	37°	1.7522	1.4171	67°	2.3809	1.1453
8°	1.5785	1.5632	38°	1.7633	1.4092	68°	2.4198	1.1362
9°	1.5805	1.5611	39°	1.7748	1.4013	69°	2.4610	1.1272
10°	1.5828	1.5589	40°	1.7868	1.3931	70°	2.5046	1.1184
11°	1.5854	1.5564	41°	1.7992	1.3849	71°	2.5507	1.1096
12°	1.5882	1.5537	42°	1.8122	1.3765	72°	2.5998	1.1011
13°	1.5913	1.5507	43°	1.8256	1.3680	73°	2.6521	1.0927
14°	1.5946	1.5476	44°	1.8396	1.3594	74°	2.7081	1.0844
15°	1.5981	1.5442	45°	1.8541	1.3506	75°	2.7681	1.0764
16°	1.6020	1.5405	46°	1.8691	1.3418	76°	2.8327	1.0686
17°	1.6061	1.5367	47°	1.8848	1.3329	77°	2.9026	1.0611
18°	1.6105	1.5326	48°	1.9011	1.3238	78°	2.9786	1.0538
19°	1.6151	1.5283	49°	1.9180	1.3147	79°	3.0617	1.0468
20°	1.6200	1.5238	50°	1.9356	1.3055	80°	3.1534	1.0401
21°	1.6252	1.5191	51°	1.9539	1.2963	81°	3.2553	1.0338
22°	1.6307	1.5141	52°	1.9729	1.2870	82°	3.3699	1.0278
23°	1.6365	1.5090	53°	1.9927	1.2776	83°	3.5004	1.0223
24°	1.6426	1.5037	54°	2.0133	1.2681	84°	3.6519	1.0172
25°	1.6490	1.4981	55°	2.0347	1.2587	85°	3.8317	1.0127
26°	1.6557	1.4924	56°	2.0571	1.2492	86°	4.0528	1.0086
27°	1.6627	1.4864	57°	2.0804	1.2397	87°	4.3387	1.0053
28°	1.6701	1.4803	58°	2.1047	1.2301	88°	4.7427	1.0026
29°	1.6777	1.4740	59°	2.1300	1.2206	89°	5.4349	1.0008

The Common Logarithms of $\Gamma(n)$ for Values of n between 1 and 2.

n	$\log_{10}\Gamma(n)$	n	$\log_{10}\Gamma(n)$	n	$\log_{10}\Gamma(n)$	n	$\log_{10}\Gamma(n)$	n	$\log_{10}\Gamma(n)$
1.01	$\bar{1}.9975$	1.21	$\bar{1}.9617$	1.41	$\bar{1}.9478$	1.61	$\bar{1}.9517$	1.81	$\bar{1}.9704$
1.02	$\bar{1}.9951$	1.22	$\bar{1}.9605$	1.42	$\bar{1}.9476$	1.62	$\bar{1}.9523$	1.82	$\bar{1}.9717$
1.03	$\bar{1}.9928$	1.23	$\bar{1}.9594$	1.43	$\bar{1}.9475$	1.63	$\bar{1}.9529$	1.83	$\bar{1}.9730$
1.04	$\bar{1}.9905$	1.24	$\bar{1}.9583$	1.44	$\bar{1}.9473$	1.64	$\bar{1}.9536$	1.84	$\bar{1}.9743$
1.05	$\bar{1}.9883$	1.25	$\bar{1}.9573$	1.45	$\bar{1}.9473$	1.65	$\bar{1}.9543$	1.85	$\bar{1}.9757$
1.06	$\bar{1}.9862$	1.26	$\bar{1}.9564$	1.46	$\bar{1}.9472$	1.66	$\bar{1}.9550$	1.86	$\bar{1}.9771$
1.07	$\bar{1}.9841$	1.27	$\bar{1}.9554$	1.47	$\bar{1}.9473$	1.67	$\bar{1}.9558$	1.87	$\bar{1}.9786$
1.08	$\bar{1}.9821$	1.28	$\bar{1}.9546$	1.48	$\bar{1}.9473$	1.68	$\bar{1}.9566$	1.88	$\bar{1}.9800$
1.09	$\bar{1}.9802$	1.29	$\bar{1}.9538$	1.49	$\bar{1}.9474$	1.69	$\bar{1}.9575$	1.89	$\bar{1}.9815$
1.10	$\bar{1}.9783$	1.30	$\bar{1}.9530$	1.50	$\bar{1}.9475$	1.70	$\bar{1}.9584$	1.90	$\bar{1}.9831$
1.11	$\bar{1}.9765$	1.31	$\bar{1}.9523$	1.51	$\bar{1}.9477$	1.71	$\bar{1}.9593$	1.91	$\bar{1}.9846$
1.12	$\bar{1}.9748$	1.32	$\bar{1}.9516$	1.52	$\bar{1}.9479$	1.72	$\bar{1}.9603$	1.92	$\bar{1}.9862$
1.13	$\bar{1}.9731$	1.33	$\bar{1}.9510$	1.53	$\bar{1}.9482$	1.73	$\bar{1}.9613$	1.93	$\bar{1}.9878$
1.14	$\bar{1}.9715$	1.34	$\bar{1}.9505$	1.54	$\bar{1}.9485$	1.74	$\bar{1}.9623$	1.94	$\bar{1}.9895$
1.15	$\bar{1}.9699$	1.35	$\bar{1}.9500$	1.55	$\bar{1}.9488$	1.75	$\bar{1}.9633$	1.95	$\bar{1}.9912$
1.16	$\bar{1}.9684$	1.36	$\bar{1}.9495$	1.56	$\bar{1}.9492$	1.76	$\bar{1}.9644$	1.96	$\bar{1}.9929$
1.17	$\bar{1}.9669$	1.37	$\bar{1}.9491$	1.57	$\bar{1}.9496$	1.77	$\bar{1}.9656$	1.97	$\bar{1}.9946$
1.18	$\bar{1}.9655$	1.38	$\bar{1}.9487$	1.58	$\bar{1}.9501$	1.78	$\bar{1}.9667$	1.98	$\bar{1}.9964$
1.19	$\bar{1}.9642$	1.39	$\bar{1}.9483$	1.59	$\bar{1}.9506$	1.79	$\bar{1}.9679$	1.99	$\bar{1}.9982$
1.20	$\bar{1}.9629$	1.40	$\bar{1}.9481$	1.60	$\bar{1}.9511$	1.80	$\bar{1}.9691$	2.00	0.0000

Printed in Dunstable, United Kingdom

91220101R00231